AF598677

3289380900

WITHDRAWN
from
STIRLING UNIVERSITY LIBRARY

Applied Fish Pharmacology

AQUACULTURE

Volume 3

1. **Fish Nutrition in Aquaculture**
   Sena S. De Silva and Trevor A. Anderson
   ISBN 0-412-55030-X

2. **Environmental Management for Aquaculture**
   Alex Midlen and Theresa Redding
   ISBN 0-412-59500-1

XC
4.2
TRE

Applied Fish Pharmacology

K. M. Treves-Brown
MA, Vet MB (Cantab),
Master of Arts,
Bachelor of Veterinary Medicine,
Cambridge University
MRCVS, Member of the Royal College of Veterinary Surgeons, London

KLUWER ACADEMIC PUBLISHERS
DORDRECHT / BOSTON / LONDON

63767
4/00

Library of Congress Cataloging-in-Publication Data

ISBN 0-412-62180-0

Published by Kluwer Academic Publishers,
P.O. Box 17, 3300 AA Dordrecht, The Netherlands.

Sold and distributed in North, Central and South America
by Kluwer Academic Publishers,
101 Philip Drive, Norwell, MA 02061, U.S.A.

In all other countries, sold and distributed
by Kluwer Academic Publishers,
P.O. Box 322, 3300 AH Dordrecht, The Netherlands.

Printed on acid-free paper

All Rights Reserved
© 2000 Kluwer Academic Publishers
No part of the material protected by this copyright notice may be reproduced or
utilized in any form or by any means, electronic or mechanical,
including photocopying, recording or by any information storage and
retrieval system, without written permission from the copyright owner.

Printed in the Netherlands.

SERIES EDITOR'S PREFACE

Fish culture is a growing agribusiness. Considering the value of that business, the frequency and often devastating nature of fish diseases, it is perhaps surprising that a comprehensive text explaining the pharmacology of fish medicines has not been available until now.

In the space of little over 300 pages, some 22 Chapters have been organised into four parts: General Considerations (3 Chapters); Antibacterial Drugs (8 Chapters); Other Chemotherapeutic Agents (4 Chapters) and Pharmacodynamic Agents (7 Chapters).

Keith Treves-Brown is to be congratulated on producing a comprehensive yet readable text on a topic few have ventured to touch on to date. An explanation of the pharmacology of fish medicines is indeed long overdue and I feel sure that Keith Treves-Brown's book will be a source used by Fish Health Managers for many years to come. Indeed, I fully expect it to become a part of the essential reference collection on a great many fish farms. Veterinarians specialising in fish diseases will find it invaluable. Students studying fish diseases, whether as part of a general aquaculture course or as a specialist subject, will also wish to have access to it. Post-graduates will find it a constant source of much useful information, as I am sure will managers working in 'Regulatory Agencies'.

While Keith Treves-Brown's experience and examples are naturally European, the topic of 'Applied Fish Pharmacology' has a universal appeal and this book should be of great value throughout the world of fish culture. As such it is a very welcome addition to the 'Aquaculture Book Series'.

Michael G. Poxton

Editor, Aquaculture Series

CONTENTS

PART TWO ANTIBACTERIAL DRUGS

ACKNOWLEDGEMENTS

Advice on chapters on which they have particular expertise has been freely given to me by Dr. David Alderman of The Centre for Environment, Fisheries and Aquaculture Science (CEFAS) and Mr. Edward Branson, MRCVS. All the other chapters have been reviewed by my son, Dr. Bernard Treves-Brown, a chemist; and he has produced the chemical formulae which are the subject of several of my figures. My wife, Ruth, has produced computer print-outs of over 20 tables. I am grateful to them all for the time and attention they have given. Mrs. Sue Walker, librarian at CEFAS, has been unfailingly patient and helpful, not only during my many visits to her library but also on the innumerable occasions when I have telephoned and asked her to drop everything and look up an odd point in a journal for me.

PREFACE

In writing a book with any technical content at all one has to decide what degree of prior knowledge the reader will have. This book is intended to be of worldwide application, and it is hoped that it will prove of interest to numerous people in fish farm management and also some involved with other aspects of aquaculture such as the supply of ornamental fish. This means that the hoped-for readership will have wide variations in prior knowledge both qualitatively and quantitatively. As a veterinarian myself I have chosen as the person to whom to address the book a newly qualified veterinarian. I am assuming that he or she will be very well versed in the medicine of homoiotherms, but know very little about poikilothermic vertebrates despite their great, and increasing, importance as sources of protein for mankind. This means I am assuming considerable understanding of the different types of medicines used by veterinarians, but occasionally pointing out some aspects of fish behaviour and physiology, particularly osmo-regulation, which non-veterinarian readers may regard as elementary. Above all one must keep emphasizing the fundamental importance of temperature to the rate of biochemical, including pharmacological, processes.

This book is about what is known, not what is legal. Readers who are aware of my previous employment in the Veterinary Medicines Directorate (V.M.D.), the UK government regulatory body, may be surprised at my advocating medicinal applications which are illegal in UK. Three points need to be made: first, I am not advocating but stating what has been reported; secondly, even though something is illegal in one or a few countries it is probably legal in many others, and thirdly, law is continually changing. In the last context it must be remarked that the European Union legislators have acknowledged that current medicines law is unsatisfactory in its application to fish; and significant changes have been made to the procedures for determining maximum residue limits in fish tissues during the time taken to write this book.

Another group who may be surprised at my writing this book are the active fish pharmacologists. I am not one; in fact I am happy to admit that I have never done any laboratory or field research in my life. I have nevertheless spent most of my career studying veterinary pharmacological research reports and deciding what they prove in terms of justifiable claims and necessary warnings and precautions for veterinary medicinal products. I have done this in pharmaceutical industry, writing many labels and directions-for-use leaflets which have been widely used and translated into many languages; I have done the same thing in V.M.D., and I have done the same thing again in writing this book.

Many fish pharmacologists may notice their results quoted selectively and sometimes without acknowledgement. To address the second point first, the fact is that a majority of readers would not study research papers in the scientific literature even if the references were given. Those who read references are research workers who wish either to use the same techniques or to dispute the conclusions; this book is not primarily intended for them.

The first word of the title is 'Applied' because I have aimed to cover those pharmacological data, and only those data, which can be used in the prevention and treatment of fish disease or the enhancement of fish productivity. Where I have listed research reports under 'Further Reading' I believe the reader can obtain additional information relevant to the aims of the book.

In keeping with those aims I have tended to use units which modern scientific purists will regard as old-fashioned. My experience leads me to believe that to be easily read is more important than to be scientifically fashionable, and where dose rates are concerned clarity is paramount. To the veterinarian in practice and to the fish farmer I believe 'mg/kg/day' is preferable to '$mg.kg^{-1}.d^{-1}$', and 'ppm' obviates problems of whether to use 'mg/kg' or 'µg/g' (or should it be '$10^{-6}g.g^{-1}$'?). Even scientists (both authors and reviewers) sometimes get confused by the modern notation: one paper listed in this book gives the units of the concentration of azamethiphos in water correctly as $mg.L^{-1}$ on three occasions and wrongly as '$mg^{-1}.L$' on over 40 occasions!

Another foreseeable criticism of this book is that it is patchy in its coverage of the subject - some aspects are covered extensively and some not at all. This is regrettably true; it reflects the available data. Two aspects of the research reports I have read have been disappointing to one with the particular interests I hold. I have read at times of elegant and pains-taking research which does not have any obvious application. How many studies have been made of elimination half-lives of residues of drugs in serum or plasma? We do not drink fish blood; we eat the musculature! How often do we read computer calculations to many significant figures of transference rates of drugs to deep or superficial compartments of a fish? The equations may give remarkable correlations with observed plasma concentrations but what the clinician needs to know is the identity of the tissues or organs represented by those compartments. A small volume of distribution is relevant, albeit negative, information for the clinician; a large volume of distribution says nothing - is the drug widely distributed at low concentration or is it selectively concentrated in one organ?

The second disappointment is the narrow range of fish species and drugs on which research has been concentrated. The former is perhaps inevitable; well-equipped and well-staffed laboratories tend to be in places where salmonids or channel catfish are the predominant farmed species, although one must not overlook the important work the Japanese have done with the ayu and yellowtail. Nevertheless it is extraordinary how much has been published about oxytetracycline, a rather unsuitable antibiotic for in-feed administration to fish, and how little has been published about amoxycillin which seems to me to be far superior. There is much more literature about oxolinic acid than about flumequine, but it is the latter which now has an MRL assigned in the EU*.

I hope this book will help veterinarians to prescribe wisely and fish farmers to administer drugs safely and effectively. Although I have written that the book is not primarily intended for research pharmacologists I hope it will help them to concentrate their attentions where they are most needed.

*An MRL has recently been assigned to oxolinic acid.

PART ONE

GENERAL CONSIDERATIONS

1. METHODS OF DRUG ADMINISTRATION

1.1 Water Medication

The commonest method by which drugs have traditionally been administered to fish is by medication of the water which the fish inhabit. It has the outstanding advantage of simplicity, which is why it is normally recommended for the low molecular weight medicinal products supplied to amateur aquarists. The only expertise required of the user is knowledge of the volume of water into which the product is to be mixed, a correct measurement of the corresponding quantity of concentrated medicinal product and care in ensuring that it is well distributed in the water.

Drugs are added to water for two distinct purposes. The first and most obvious one is so that the drug will be absorbed by, and so medicate, the fish; the second is to kill free-living and hence transmissible stages of parasites. Absorption by fish may occur across the epithelia of the gills and possibly the olfactory tract. Water passes across fish skin but the extent to which this occurs is limited by the mucus and scales. Ions probably pass across to only a very limited extent; but recent work with radio-iodine-labelled bovine serum albumin suggests that fish skin is permeable to water-soluble non-polar molecules (Ototake *et al.* 1996). Fish absorb water and some drugs through the gastro-intestinal mucosa, but the extent to which a drug given in the water is absorbed depends on not only the chemistry of the drug but also the amount of medicated water reaching the lumen of the intestine. Fish in freshwater do not drink; they absorb an excess of water through osmosis and have to excrete it to maintain homeostasis. In contrast marine fish are continually losing water through osmosis; they have to drink continually and excrete electrolytes.

1.1.1 MERITS AND DEMERITS

In addition to its simplicity water medication has the advantage that it is adaptable to mass medication of large numbers of fish. Furthermore unlike mass in-feed medication it does not depend on the fish feeding, so it can be applied to ailing fish and also to herbivorous fish which are not fed but live on plant life in the water.

Nevertheless absorption varies between drugs. As a broad generalization it may be said that lipophilic compounds will diffuse across gill membranes but ions with a molecular weight greater than about 100 will not. Absorption also varies between fish species; for example juvenile channel catfish (*Ictalurus punctatus*) exposed to various antimicrobial drugs were found not to absorb chloramphenicol at all; erythromycin and gentamycin were poorly absorbed, but kanamycin, nifurpirinol and oxytetracycline were absorbed in proportion to their concentrations in the water. Although channel catfish readily absorb oxytetracycline, common carp (*Cyprinus carpio*) do not to absorb therapeutic concentrations of it, nor of ampicillin or nitrofurazone, despite being kept in high concentrations of them for 24 hours (Blasiola *et al.*, 1980). In contrast rainbow trout (*Oncorhynchus mykiss*) actively absorb malachite green and after 60 minutes exposure homogenized whole fish contain a higher concentration of the drug than the water (Clifton-Hadley and Alderman, 1987). Thus it is important to ascertain that any drug proposed for administration in the water will in fact be absorbed by the species concerned. (See Section 4.2.5 with reference to antibacterial drugs).

Even where absorption is known to occur the technique does have some important disadvantages. In particular, in most cases less than 5% of the administered dose will be absorbed by a fish, and in consequence:

1. The technique is wasteful.
2. It is expensive. At least twenty times the dose required by the fish must be provided. Conversely, only inexpensive drugs can be afforded.
3. It is environmentally undesirable. The unabsorbed moiety of the dose will enter the environment either as the unchanged drug or its degradation products.

As recommended for use by aquarists water medication also results in overdosage, and not infrequently the unnecessary medication of fish. Once water has been medicated fish will absorb the drug very quickly. T_{max} (plasma) will not be more than 30 minutes in any species and in small tropical ornamentals may be less than 60 seconds. From that time onwards the fish will be swimming in medicine which it does not need, as will all other fish in the same tank or pond whether they need medication or not. This will continue until a complete water change.

1.1.2 IMMERSION OR DIPPING

These disadvantages may be mitigated, although the first two will not be eliminated, by ‘dipping’. This term is used to mean the preparation of a small volume of medicated water in a separate container from that holding the fish. The fish, usually held in a net, are immersed in it for a short period of time and then returned to their normal environment. It is advisable that the water to be medicated should be taken from the tank or pond containing the fish so that chemical and temperature stresses are minimized. Also, depending on the number and size of the fish and the chemical nature of the drug, it may be necessary to aerate the medicated water artificially, for example with an airstone. Even with these precautions dipping has the disadvantage compared to direct water medication that the fish are exposed to the stresses of chasing, handling and netting. Furthermore the additional manpower required will be significant to commercial enterprises.

Dipping has a particular advantage in the use of certain antibacterial drugs in aquaria where filters are used to effect bacterial oxidation of ammonia to nitrites and nitrates. Direct medication of the aquarium water might well sterilize and inactivate the filter. Similar filters are used for some ponds and raceways where the water is recirculated. Although dipping may not be a feasible alternative method of drug administration in such installations, the question of filter inactivation remains. In experiments on aquaria with recirculating water systems and containing channel catfish it was found that therapeutic concentrations of methylene blue or erythromycin resulted in a rapid rise in ammonia due to inactivation of the filter. However therapeutic concentrations of the following had no effect of the action of the filter:

chloramphenicol	oxytetracycline
sulphamerazine	nifurpirinol
formalin	malachite green
formalin and malachite green in combination	
copper sulphate	potassium permanganate
sodium chloride	

1.1.3 HYPEROSMOTIC INFILTRATION

Hyperosmotic infiltration (HI) is a development of immersion designed to accelerate the absorption of macromolecules or even of particles such as antigenic bacteria. The procedure as originally devised consisted of two separate immersions. The first was in a pharmacologically inert solution, hypertonic to fish plasma; 10% urea and 5.23% sodium chloride, both being 1650 mOsm/l, have been used. This immersion was for 3 minutes and was followed immediately by the solution to be absorbed. More recently trials have been conducted with a 'one-step' procedure of a single immersion in a combination of the 'carrier' and drug solutes.

The early studies, conducted on the absorption of bovine serum albumin (BSA) by rainbow trout at 20°C, showed plasma levels to be low immediately after the second immersion. T_{max} (plasma) was about 1 hour. The apparent absence of BSA from other possible entry points led to the conclusion that it entered through the lateral line and thence into the lymphatic system. It was hypothesized that the first step caused both egress of water from and ingress of carrier solute into the lateral line, and that the hypertonic lateral line contents facilitated the ingress of drug solution in the second step. The hypothesis carried the implication that proteins could be absorbed through the lateral line. More recent work has involved slaughter of fish and dividing the body into sections immediately after the second step. This has shown no absorption of human gamma-globulin or bacteria by the body (*i.e.* without the head); ingress of these materials was found only at the gill lamellae, and the more recent theory of hyperosmotic infiltration is that it increases the permeability of the gill epithelia to these large particles.

Recent work (Ototake *et al.*, 1996) on radio-labelled BSA has confirmed the low plasma levels immediately after HI, higher levels being found at 2 hours than at 1 hour. However they dispute the finding that ingress did not occur other than through the lateral line. 10 minutes after HI they found levels in trunk skin as high as in lateral line skin, and these had fallen to less than half within an hour after HI. The gills showed lower levels than the trunk skin 10 minutes after HI and these had fallen significantly within 30 minutes. Comparative results following direct immersion (*i.e.* without the HI procedure) showed lower absorption by all routes except the stomach, and again the trunk skin showed higher levels than any other tissue examined throughout the 2 hour observation period. The researchers acknowledge that the findings are relevant to soluble antigens but have no application to large particles such as killed bacteria.

The following procedural points have been noted:

1. For a one-step solution to show significantly better results than conventional immersion for rainbow trout, its concentration must exceed 1200 mOsm/L.
2. In a one-step procedure with rainbow trout uptake increases with increasing pH up to 9.
3. In a two-step procedure with rainbow trout uptake does not increase with pH, but for optimum results the two solutions must have the same pH.
4. In a two-step procedure with rainbow trout 30 seconds for the second step is sufficient; longer exposures gave no better results.
5. In a two-step procedure with channel catfish a short (*e.g.* 30 seconds) second step worked best with a short (*e.g.* 60 seconds) first step, whereas a 3 minute second step required a 3 minute first step.
6. In a one-step procedure with channel catfish increasing exposure time increased uptake. The limit was not determined but there is not likely to be any practical advantage in an exposure of more than 5 minutes at most.

Hyperosmotic infiltration has been shown to give better results than conventional immersion for killed bacterial vaccines although it is an uneconomic procedure for larger fish. For it to be economic for large numbers of fish a one-step procedure is necessary, but points 1 and 2 above indicate that it would be stressful to the fish; this in itself would depress the immune response.

1.1.4 FLUSHING

Where fish are kept in running water which is not recirculated, for example in a raceway, immersion can be achieved by flushing, or, as the process is sometimes called, a California flush. This means shutting off the flow, medicating the water and, after an appropriate interval, restarting the flow and hence removing the medicated water. The effect is a rapid rise in drug concentration in the water and a slow fall. Flushing is more wasteful, and hence more environmentally polluting, than dipping; and it may be difficult to obtain a homogeneous distribution of the medication in the water. However there are husbandry conditions where it is the only possible means of administering drugs. Where a farm's water supply is actually or potentially contaminated, for example where it is sited downstream from another fish farm, flushing with a disinfectant may be used continuously or on a routine intermittent regimen as a prophylactic measure. An adaptation of this procedure is commonly used in hatcheries for the control of fungal (*Saprolegnia*) infection.

1.1.5 BATH TREATMENT

Where large fish or large numbers, for example in excess of a thousand in one cage, are to be treated, dipping as described above is impracticable; the bathing technique has to be resorted to. Bathing differs from dipping in that the fish are kept in the water in which they are living. It also differs from medicating the water in an aquarium in that, quite apart from differences in scale, the exposure of the fish to medicated water is for a limited time, never more than 60 minutes.

In bathing the bottom of the net cage is raised, typically to 2 metres, thus limiting the volume of water to be medicated. This reduces the weight of drug required and hence reduces both the cost and the degree of environmental contamination. Tarpaulins are placed around the cage, separating the contained from the surrounding water, and the drug is added, preferably at several points in the bath to ensure rapid mixing. It is normally essential to sparge oxygen into the water to prevent the fish crowding together. At the end of the exposure period the tarpaulins are lowered allowing the inflow of unmedicated water. Oxygen should continue to be provided for a time.

Bath treatment is wasteful and environmentally contaminating, and additionally is labour-intensive. Furthermore, experience has shown that the volume of water contained within a tarpaulin cannot be estimated with any degree of accuracy. This means that the calculated requirement of concentrated drug may be seriously in error. Where the final concentration is critical - too little being insufficiently effective; too much being toxic to the target fish species - a 'bankside' chemical test is sometimes necessary. This means a titration which can be conducted very rapidly, at the cage site, on a sample of medicated water. Too little concentrate is used initially; the test is made, and on the result of it the volume of the bath water and the weight of additional concentrate required can be read from tables. Efficacy is usually a function of exposure period as well as concentration; so where a bankside test is not available it may be possible to work with an estimated concentration at the lower end of the effective range and a long exposure time. This approach naturally increases both the labour and oxygen costs. Whatever method is used to determine the required weight of concentrate, the fish must be kept under observation throughout the exposure time. If there are signs of toxicity due to the volume of water having been less than estimated, the tarpaulins must be dropped immediately.

Where, as is usually the case, bathing is conducted in tidal waters, the tarpaulin should be set at slack tide. This facilitates the operation and makes it more efficient. It also means that the tarpaulin will be removed as the tide begins to run; this will help both to flush the medicated water out of the cage minimizing its toxicity to the fish, and to promote its dilution and dispersal thus minimizing its environmental impact.

1.1.6 SUBMERGED BAGS OR BASKETS

Control of bacteria and the transmissible stages of parasites can be achieved by hanging bags or baskets of simple disinfecting agents in the water. Typical compounds used are bleaching powder in baskets for bacteria and copper sulphate or ferrous sulphate in bags for protozoa. There is no control over concentration but the fish keep away voluntarily if it is too high at the basket site. The technique is used in pond culture of fish such as Chinese perch and silver carp which do not eat artificial pelleted food, and is regarded as "indispensable for culturing fish" in P.R. China.

1.1.7 MEDICINAL FORMULATION

The excipients (see Appendix 3) present in normal medicinal formulations perform no useful function in water, and they will add to the inevitable environmental contamination when the medicated water is discarded. Wherever possible pure drugs should be used for water medication.

Absorption of drugs by fish from water can sometimes be enhanced by the addition of low concentrations (*e.g.* 0.01 %) of surfactants. Different surfactants appear to enhance the absorption of different drugs. For example in a comparative trial of Aerosil OT®, an anionic surfactant, with Tween 80® (non-ionic), it was found that the former gave the better enhancement of absorption of erythromycin phosphate but the latter was better for sarafloxacin. The difference was attributed to the different solubilities of the two drugs, erythromycin salts being very soluble but sarafloxacin having a solubility only of the order of 0.1 %.

1.1.8 WATER TREATMENT

There are few if any contagious diseases of fish, and the transmissible stages of parasites and microbial infections reach new host fish through the water. In aquaria it is possible to prevent the spread of the diseases by killing the transmissible stages of the infectious agents in the water. In this procedure the drug is applied to all the water in the aquarium. It is therefore not applicable to antibacterial drugs as these would inactivate the filter. It is used for some antiprotozoal drugs to prevent the spread of certain ectoparasitisms. The parasites in question have free-living stages sensitive to concentrations of drug which are safe for the fish, while the parasitic stages are insensitive to such concentrations.

In UK law an animal medicine is defined as a substance or article "administered to animals" for a medicinal purpose. Although the prevention of the spread of a parasitism is unquestionably a medicinal purpose, when the drug is used to kill *free-living* organisms it is not being administered to animals. Material labelled for use exclusively in this way is not therefore subject to UK medicines legislation even though it may get into the fish.

1.2 In-feed medication

1.2.1 MERITS AND DEMERITS

In-feed medication or the provision of medicated feed is a much less wasteful method of administration than water medication. It is adopted whenever possible on fish farms. For aquaria it is less popular because small quantities of food are used and accurate measurement of the very small quantities of drug required is difficult. There is an important limit to its applicability which is that the fish to be treated must be feeding. Thus it has no place in the medication of eggs or sac-fry; and it is of limited applicability to fry because with the exception of salmonids fish larvae have to be fed live organisms. It is also useless for mature adult anadromous fish such as salmonids returning to freshwater to breed, as these too do not feed. (In fact Pacific salmon returning to freshwater have degenerating intestines).

Surpassing any of these exclusions is the fact that in almost all diseases affected fish cease eating. This means that disease therapy cannot normally be in-feed. In-feed medication is standard practice for a large number of diseases but it is actually prophylactic not therapeutic; it is medicating healthy in-contact fish to prevent them contracting the disease. Affected fish are allowed to die unless they are particularly valuable, such as some ornamentals or farm broodstock. In such cases therapy must perforce be by other routes such as immersion or injection.

As in other classes of vertebrates so in fish, the bigger and more vigorous individuals eat more and are more competitive in obtaining food than smaller or debilitated ones. This is true even if the fish are not diseased and the feed not medicated, and is the reason why regular grading of fish is commonly regarded as an essential husbandry practice. Incipient disease makes a fish less competitive and this exacerbates the problem described in the previous paragraph.

Finally the efficacy of any oral medication depends on the drug concerned not being digested before absorption. This is also true in other classes of vertebrates but in fish there are important species differences especially for proteinaceous drugs such as gonadotropins and gonadotrophin-releasing hormones, and some vaccines. Suzuki *et al.* (1988) showed that sGTH (see Section 17.1.4b) was absorbed to a pharmacologically active extent by goldfish. This was attributed to the absence of a stomach in this species.

1.2.2 BIOAVAILABILITY

The point that in-feed antimicrobial medication is prophylactic rather than therapeutic has profound implications on what may be considered desirable pharmacokinetic profiles. High concentrations of drug are desirable not in the organs and tissues affected by the disease but at the normal portals of entry of the pathogenic micro-organism. If entry is normally through the gut wall then the ideal drug will be bioavailable in the gut lumen; any absorbed from the gut by feeding (and therefore healthy) fish will be a source of undesirable tissue residues but of little prophylactic value. In this context it is important to consider what the portals of entry of potentially pathogenic micro-organisms might be. They are in the water so the gills are always a possible portal. The gut is a portal only if the water gets there: as already mentioned, freshwater fish do not drink; marine fish drink continually. It may be argued that freshwater fish require drugs absorbed into the system to protect them against entrants at the gills; marine fish primarily require non-absorbed drugs to prevent ingress of pathogens from the gut.

Despite this, the term “bioavailability” is normally used by fish pharmacologists to mean the fraction of an orally administered dose which is absorbed into the circulation. It is conventionally represented by the letter F and is determined by comparing blood levels following single oral and intravenous doses. Since absorption from the gut is never instantaneous the blood concentrations measured must have a time element; the “area under the curve” (AUC) is used. Furthermore the calculation must take into account any difference between the doses given by the two routes. The formula is:

$$F = \frac{AUC_{oral} \text{ x } dose_{i/v}}{AUC_{i/v} \text{ x } dose_{oral}}$$

In calculations of F for some drugs C_{max} is considered easier to determine experimentally than AUC, and so $C_{max\ oral}$ and $C_{max\ i/v}$ are used in the formula in place of the corresponding AUC figures. It can be shown mathematically that the resultant F is the same using either parameter. This formula, and indeed the whole concept of bioavailability, can be confounded by "first pass metabolism". This is metabolism of part of the absorbed drug either in the gut mucosa or, after its transport in the hepatic portal vein, in the liver - in either case before it has reached the general circulation. Several published determinations of F for various drug/species/temperature combinations have been criticized for failure to take first pass metabolism into account. From an academic pharmacological standpoint this criticism may be valid, but unless one is treating a pure enteritis or hepatitis what matters is the drug concentration reaching the general circulation.

1.2.3 MEDICATION RATES

Dosages are normally expressed in mg/kg b.w./day where kg b.w. means kg of bodyweight. Feed medication is on the basis of g/tonne of feed, and the required rate will depend on the fishes' feed intake. Furthermore to facilitate accurate weighing and to ensure homogeneous mixing it is normal to use a premix rather than 100% drug.

Given that:

1. The fish to be medicated are feeding at x% of bodyweight per day;
2. The required dose rate is y mg/kg b.w./day;
3. The premix contains z% active ingredient;

Then:

$$\text{1 tonne of feed should contain } \frac{10y}{xz} \text{ kg premix.}$$

1.2.4 PELLETED MEDICATED FEED

The ideal way to medicate feed is to add the medicinal product to the feed mix prior to pelleting. Since farms are rarely equipped with pelleting plant, this has to be done at a feed mill where there will also be sophisticated mixing equipment which will ensure a homogeneous distribution of the drug through the feed. However pelleting involves high temperatures and hence pellets can only be medicated with compounds stable to heat. Furthermore since the plant has to be cleaned down after the production of medicated feed, the process is economic only for large batches.

1.2.5 SURFACE-COATING PELLETED FEED

This process is suitable to the medication of small batches of feed and can be used for drugs which are heat-labile. It is therefore the normal means of medicating feed on fish farms.

Surface coating of feed pellets with a medicinal product involves mixing the pellets and drug and the use of an adhesive or "binding agent", which is usually gelatin or an edible oil such as sunflower oil or cod liver oil. On fish farms the mixing is normally done in a concrete mixer. The pellets are loaded first, followed by either the powdered medicinal product or the adhesive depending on the physical characteristics of the product. Only when the first two ingredients are thoroughly mixed is the third ingredient added.

(a) Binding agents

Problems of surface-coating as a method of drug administration include leaching (see Section 1.2.7), and the palatabilities of both the drug and the binding agent. The possibility must be borne in mind that the palatabilities of a particular drug and a particular binding agent may enhance each other or be antagonistic, and these effects may differ according to fish species.

Comparison of the suitabilities of gelatin, a fish oil, a vegetable oil and water as binding agents for amoxycillin showed water, as might be expected, to be very little different from the control with food medicated but with no binding agent. Gelatin was clearly unpalatable to rainbow trout but the fish oil and the vegetable oil resulted in virtually all the medicated food being consumed. Amoxycillin residues in the fish after 10 days medication reflected the weights of the medicated diet consumed. Since the consumption of food with fish oil binder was some 60-65% higher than of food with gelatin, the binding agent clearly has a significant effect on the rate at which a feed should be medicated to achieve the required dose of a drug. In practice vegetable and fish oils are normally used, and these would appear to be the ideals at least where amoxycillin in rainbow trout is concerned.

In 1996 a new binding agent, MEDI-TAK® Oil was put on the market. This is a mixture of refined, stabilized, polyunsaturated fatty acids and is claimed to stimulate the appetite of inappetant fish and overcome problems of drug palatability.

(b) The problem of homogeneity

A further disadvantage of surface-coating is that it is not easy to achieve homogeneity. In consequence even if a pen of fish given surface-coated feed all eat the same quantity they will not receive the same dose of drug. It may be noted that, precisely because of the problem of achieving homogeneity, UK law allows the mixing of medicinal products into mammal and poultry feeds only on registered premises. There are two categories of premises, each with its own code of practice which must be adhered to in order to maintain registration. Category A registration allows the incorporation of medicinal products into feed at less than 2 kg/tonne; this requires advanced mixing techniques and the category is primarily intended for feed mills. Category B registration allows mixing at 2 kg/tonne or more and is for premises such as farms which have less elaborate equipment. Fish farmers are exempt from this registration requirement because frequently the only formulations available to them are pure drugs or highly concentrated premixes. However there is no reason why they could not mix at 2-3 kg/tonne if premixes of the appropriate concentration were made available to them; and a report to the government in 1993 recommended that the exemption from registration should be withdrawn.

1.2.6 SPRAY-MEDICATION OF PELLETED FEED

Sex hormones are important examples of a class of drugs which are, for practical purposes, insoluble in water and which are used in very small doses. A standard method of administration in pelleted feed is to dissolve them in an alcohol, usually ethanol or isopropanol, spray the solution over the feed and allow the alcohol to evaporate. This is not strictly surface medication as the drugs will be absorbed into the lipids in the pellets.

1.2.7 LEACHING

Leaching of drug into the water occurs with all forms of in-feed medication but is a particular problem with surface-coated feed. The extent of leaching varies according to the solubility of the active ingredient in water and the time for which the feed is in the water. This latter parameter is in inverse proportion to the voracity of the fish and many people consider that for healthy salmonids the maximum period over which leaching can take place is about 10 seconds. The period for slower feeding species such as carp and for ailing salmonids may be quite significant. A further factor affecting the rate of leaching from medicated pellets is the size of the pellets and hence the ratio of surface area to weight. The smaller the pellets the faster will be the leaching.

A study has been made (Duis *et al.*, 1995a) of the use of certain alginates instead of edible oils in the surface-coating of drugs onto feed as a means of reducing leaching. These compounds are soluble alkali metal salts of alginic acid which are commonly used in the manufacture of human as well as animal foods. They form gelatinous precipitates with divalent cations such as Ca^{++}. A typical use was to mix the drug into the pellets first; then add 2% sodium alginate solution at 150 ml per kilo, and after this has been well mixed to add calcium chloride powder at 30g per kilo. Comparisons were made of leaching rates of commonly used antibacterial drugs surface-coated using oil as adhesive and alginate as an outer coating. Leaching of co-trimazine, oxytetracycline hydrochloride and amoxycillin were high but were significantly less with alginate than with oil; for oxolinic acid losses were low with oil and nil with alginate. An important finding was that the *in vitro* antibacterial activity of amoxycillin was significantly reduced by calcium chloride. Various possible reasons for the effect were suggested and whether it would occur *in vivo* was not determined.

1.2.8 MICRO-ENCAPSULATION OF DRUGS

Although there are no micro-encapsulated drugs for fish on the market at present, this technique has been investigated for two different purposes - to achieve slow release and to pass through the stomach undigested. One micro-capsule structure which has been well researched has a calcium alginate core into which the drug is mixed and a chitosan-alginate shell. The latter determines both the physical characteristics of the micro-capsule and the rate of drug release. It has been found that the chitosan-alginate should have a molecular weight no greater than 200 kDal. Higher molecular weight material produces brittle micro-capsule shells.

Whether a micro-capsule is retained in the stomach or not is largely a question of its size. The larger a food particle is the longer it will be retained in the stomach. Small micro-capsules will therefore pass rapidly to the small intestine having lost very little drug *en route*.

1.2.9 *ARTEMIA* ENRICHMENT

Many farmed marine species, notably sea-bass and sea-bream, are fed, particularly as juveniles, on rotifers or on brine-shrimp (*Artemia* spp.) nauplii. These are in fact of relatively low nutritional value and for cultured cod it is essential to 'enrich' them with unsaturated fatty acids. Even for fish species where enrichment is not essential the unsaturated fatty acids can be used as a vehicle for drugs and the technique made into a convenient method of medicating this type of feed. As the drug is inside the food rather than surface-coated the technique is useful for vaccines and other unpalatable drugs.

To enrich *Artemia* cysts they are decapsulated and placed in seawater which has been pre-sterilized by UV light and is well oxygenated. Nauplii hatch in about 24 hours; they are separated from unhatched cysts, rinsed and placed in seawater into which 0.6% of the enrichment diet has been emulsified. After 24 hours they are ready for use. Where medication is to be used the enrichment diet should consist of 60% unsaturated fatty acids and 40% pure drug.

A reported analysis of *Artemia* medicated in this way with one of three antibacterial agents showed that they contained 0.67% oxolinic acid, 0.18% sarafloxacin or 0.21% trimethoprim/sulphamethoxazole. When fed prophylactically to turbot (*Scophthalmus maximus*) larvae for 10 days the medicated diets conferred significant protection against a vibriosis challenge (Duis *et al.*, 1995b).

Artemia enrichment has the disadvantage of other forms of in-feed medication that only healthy fish can be treated. In addition it is extremely wasteful: in the case mentioned above where analyses were made the results showed only about 0.2% of the drugs used in the enrichment diet had been absorbed into the *Artemia* nauplii. Apart from the cost of the process there is the environmental safety problem of disposal of the unused 99.8% of the drug.

1.2.10 ENVIRONMENTAL IMPACT

In-feed medication is mainly used on farms so the weight of food used is large, and there is always some wastage of it whether medicated or not. Uneaten food accumulates on the floor of a freshwater pond or on the seabed under a marine netpen. Such accumulations will also contain medicated faeces if the eaten food is medicated either with a drug which is not absorbed or with one, such as oxytetracycline, which is absorbed but is eliminated unmetabolized. (The word "accumulations" is used because the more obvious word "sediments" is ambiguous: it can mean the natural material of the upper layers of the seabed).

Techniques have been developed for the recovery of uneaten food. While these may have a significant effect on feed costs and the degree of environmental contamination by feed, they will rarely mitigate environmental contamination by drugs. Drugs which are not extensively absorbed will eventually reach the same environmental compartment irrespective of whether they are in uneaten medicated feed or in faeces. Although the *proportion* of drug used which is wasted in in-feed medication is much lower than in water medication, the *absolute* weight of drug used is the more significant index of environmental contamination.

1.3 Gavage

Gavage is a form of oral administration extensively used in experimental work because the dose can be known accurately. It is rarely used in routine fish management as it is labour intensive and stressful to the fish. It involves the use of a stomach tube of calibre appropriate to the individual fish. This is attached to a hypodermic syringe containing the drug in solution or suspension and the drug is pumped into the fish stomach. It is usually necessary to anaesthetize, or at least sedate, the fish beforehand.

Gavage is a useful technique where a few fish such as brood fish or valuable ornamentals have to be dosed orally, especially if there is no suitable formulation of the drug available for in-feed medication. For example it can be used for drugs available only as tablets; these can be crushed, suspended in water and given by gavage.

1.4 Injection

1.4.1 INDICATIONS

Venepuncture and even cardiac puncture are occasionally used for experimental purposes; but routine injections are avoided where possible, because they are either labour-intensive or capital-intensive or both and are always stressful to the fish. Experimental procedures aside, the indications for injection are:

1. Vaccination.
2. Treatment of limited numbers of valuable fish with drugs which cannot be given by other routes of administration. Examples are hormone treatment of brood fish, and therapy of inherently expensive types of fish, such as koi, which are ailing and therefore not feeding.
3. The 'Hormone and Heat Stress' test, a diagnostic procedure conducted on a sample of salmonid smolts to check for 'stress inducible furunculosis' (SIB), in effect the carrier state which is the epidemiological reservoir of this common bacterial infection. The hormone used is methyl-prednisolone which is active only by injection; and the sample of smolts tested has to be large, typically of the order of 150, in order to ensure detection of infection in small numbers of them.

Injections cannot be made into fish of less than 10-15 g liveweight; this lower limit varies with the viscosity of the material to be injected. The concentration of the inoculum is normally adjusted to give dose volumes of the order of 0.1-0.2 ml for fish near the lower limit of liveweight.

1.4.2 MANUAL INJECTION

A prerequisite for injection is that the fish should be anaesthetized; without this precaution injury is likely to be caused to the fish and possibly also to the operator. The normal procedure is to anaesthetize a batch of fish by immersion. They are then taken one by one from the anaesthetic-medicated water, given an injection and returned to their normal water. The needle is always directed forward between the scales; it should never pierce a scale.

(a) Intramuscular injection

Intramuscular injections are given into the epaxial musculature, normally approximately mid-way between the mid-dorsal line and the lateral line. Sites below the dorsal fin and in the tail musculature have their advocates. At the former the muscle mass is deep: at the latter the scales are smaller.

(b) Intraperitoneal injection

Intraperitoneal injections are made into the mid-ventral line just cranial to the vent. This is a widely used route of injection but it can sometimes cause peritoneal adhesions. These may be of no consequence in most fish but severe adhesions in brood fish may obstruct ovulation (*i.e.* the release of eggs into the peritoneal cavity) or spawning (*i.e.* the voiding of eggs or milt).

In carp extensive peritoneal adhesions are normal. In consequence intraperitoneal injection should be avoided; the dose is likely to be made into a viscus or the lumen of the gut.

(c) Injection into the dorso-median sinus

There are now available on the market some injectable vaccines containing irritant oil adjuvants; these may cause sterile abscesses if injected intramuscularly or adhesions if injected intraperitoneally. A route of administration which is being recommended for salmonids is into the dorso-median sinus (DMS). The point of insertion of the needle is in the mid-dorsal line in the angle at the caudal margin of the cranial dorsal fin; it is directed cranio-ventrally at an angle of 45° to the mid-dorsal line.

This route is unsuitable for most fish other than salmonids because of their different anatomy.

1.4.3 AUTOMATIC INJECTORS

Multiple dose syringes such as are available for injecting drugs into mammals rarely deliver sufficiently accurate doses of the volumes used in fish. However automatic injectors of the type used for vaccinating poultry are suitable for fish. As with the use of simple hypodermic syringes, the fish need to be anaesthetized first.

1.4.4 MACHINE INJECTION

Machines are available for the rapid injection of large numbers of fish. In practice this means vaccination by the intraperitoneal route. An example is the Marivax® (Diamond Engineering Company, Saffron Walden, UK). This contains a specially designed automatic injector actuated by compressed air when a fish is slid against a sensitive guarded button. Anaesthesia is not needed with the machine, but even so use of the machine is safer for the operator than manual procedures with anaesthesia.

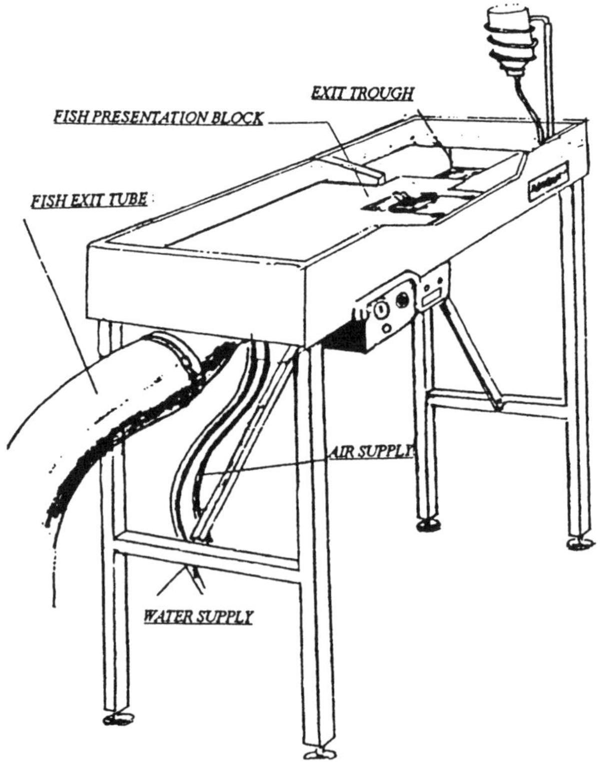

Figure 1.1. The Marivax.

1.4.5 IMPLANTATION

Where there is a need for prolonged medication with a drug which, for either economic or biochemical reasons, can only be administered by injection, it is sometimes formulated as a pellet or capsule for implantation. Administration is effected with the formulated product in a hypodermic needle with just sufficient internal diameter, and ejection is with a wire trocar. As with other injections, implantations may be intramuscular or intraperitoneal. Pellets differ from capsules in that the former consist entirely of drug and pharmaceutical vehicle, both of which will eventually be absorbed; slow-release capsules contain the drug in a permeable outer membrane of a non-absorbable material such as Silastic®.

Because of the nature of their intended use, pellets would have a very long withdrawal period and slow-release capsules are completely unacceptable for fish intended for the table. In practice these implantations are mainly used to administer hormones to induce spawning; and the issue of consumer safety does not arise because mature broodfish are both too valuable and unsuitable for use as food.

1.5 Topical application

Topical application of drugs to fish is rare; where it is done it is usually for the treatment of skin ulcers on valuable ornamental fish. Anaesthesia is an essential preliminary procedure. Debridement and cleaning may be conducted with cotton wool soaked in a suitable antiseptic. Exposed flesh may then be covered with antiseptic or antibacterial ointment. This should preferably be oil-based. It is always desirable, and with water-soluble formulations essential, to cover the medicinal application by a water-proofer, for example Orabase® (Squibb).

Further Reading

Blasiola, G.C., Dempster, R.P. and Morales, P. (1980) Uptake of ampicillin, nitrofurazone and oxytetracycline in carp, *Cyprinus carpio*, with special reference to use of antibacterials in aquaria and notes on heparin as an anticoagulant. *Journal of Aquariculture* **1**, 1-5.

Clifton-Hadley, R.S. and Alderman, D.J. (1987) The effects of malachite green upon proliferative kidney disease. *Journal of Fish Diseases* **10**, 101-107.

Duis, K., Inglis, V., Beveridge, M.C.M. and Hammer, C. (1995a) Leaching of four different antibacterials from oil- and alginate-coated fish-feed pellets. *Aquaculture Research* **26**, 549-556.

Duis, K., Hammer, C., Beveridge, M.C.M., Inglis, V. and Braun, E. (1995b) Delivery of quinolone antibacterials to turbot, *Scophthalmus maximus* (L.), via bioencapsulation: quantification and efficacy trial. *Journal of Fish Diseases* **18**, 229-238.

Michel, C. (1986) Practical value, potential dangers and methods of using antibacterial drugs in fish. *Revue Scientifique et Technique de l'Office International des Epizooties* **5**, 659-675.

Ototake, M., Iwama, G.K. and Nakanishi, T. (1996) The uptake of bovine serum albumin by the skin of bath-immunized rainbow trout *(Oncorhynchus mykiss). Fish and Shellfish Immunology* **6**, 321-333.

Suzuki, Y., Kobayashi, M., Nakamura, O., Aida, K. and Hanyu, I. (1998) Induced ovulation of the goldfish by oral administration of salmon pituitary extract. *Aquaculture* **74**, 379-384.

2. SAFETY OF FISH MEDICINES

2.1 Aspects of safety

Safety of medicines is conventionally subdivided into four aspects, namely:
1. Safety to the target species,
2. Safety to the operator, *i.e.* the person administering the medicine,
3. Safety to the consumer - in the case of food-producing target species,
4. Safety to the environment.

A fifth aspect, safety to the manufacturing personnel, is outside the scope of this book.

2.1.1 DEFINITIONS

In this book the meanings assigned to certain words will follow official British practice and it is important to clarify them:

The HAZARD of a substance is a statement of the nature of its effect. For example the hazard of organo-phosphorus compounds is neuro-toxicity; that of hydrogen peroxide is corrosive oxidation. In the environmental safety context hazard includes a statement of the taxonomic groups of organisms affected.

The RISK is a quantitative statement of the probability of the hazard occurring. While the hazard to operators is a feature of the substance, the risk to them depends, among other things, on the quantity used. The risk may be minimized by wearing appropriate protective clothing; what is appropriate depends on the hazard. In the environmental safety context the risk includes consideration of the quantities and frequency of use of the substance and the likelihood of it entering any particular environmental compartment.

The SAFETY assigned to the use of a substance is a subjective assessment of its acceptability taking into account both hazard and risk. The more serious the hazard the lower the risk must be for the use to be considered safe.

It should be noted that European Union usage of the word, RISK, is rather different from the British usage explained above. Council Directive 67/548/EEC and Commission Directive 93/67/EC, which are concerned with operator and environmental safety and are reviewed in the next chapter, require that, "Assessment of risks should be based on a comparison of the potential adverse effects of a substance with the reasonably foreseeable exposure of man and the environment to that substance." This usage of RISK or RISKS is the inverse of the British usage of SAFETY.

2.2 Safety to the target species

2.2.1 WATER MEDICATION

Water medication is often used with the intention that the drug should be absorbed, but it is also use for the control of external infections and infestations varying from *Flexibacter* and myxobacterial infections of the buccal cavity and fins to protozoan and metazoan parasitisms of the gills or skin, including ichthyophthyriasis and anchor-worm and sea-lice infestations. While any drug may be toxic in gross overdose, the issue of safety to the target species arises especially with chemotherapeutic agents added to water for external use, because in addition to their reaching their intended site of action they will almost certainly be absorbed through the gills to some extent.

Drugs which are intended to be absorbed may cause stress when added to water through altering the pH. An important example is the anaesthetic, benzocaine, which for this reason is often used neutralized with hydrochloric acid.

Fish differ significantly from terrestrial animals in that they have to maintain homeostasis while living in a medium at a different osmotic pressure. Any erosion of any epithelium interfacing with the water will result in osmotic transfer of water (in or out depending on whether the fish is in freshwater or seawater respectively), and the fish will have to expend energy in correcting this. The erosion will also allow the influx of the chemical causing it; and this will occur to a greater extent than, for example, in the localized contact of the skin of a terrestrial animal with a corrosive substance, because diffusion of the substance through the water and movement of water over the erosion will allow continuation of the influx, theoretically until it has reached the same concentration in the fish as in the water.

Fish also differ from terrestrial animals in terms of the organ likely to be affected by a toxic pollutant or a toxic concentration of a drug. In mammals and birds the chemical may enter the body by, or affect, the skin, the respiratory mucosa, the alimentary mucosa etc. In fish there is one organ, the gills, which is permanently in contact with the water containing the chemical. Furthermore the gills are a very sensitive organ, and so water-borne materials which are toxic to, for example, the alimentary or olfactory tracts, will almost certainly also have adverse effects on the gills. The gills are more important than the kidneys as organs of osmo-regulation and so this system is always vulnerable to chemical toxicity. In the case of some chemicals the hazard to the gills is to cause excessive mucus secretion which reduces gaseous exchange and causes respiratory embarrassment; in some other cases, notably quaternary ammonium compounds, the normal mucus layer is removed from the gills leading to loss of osmotic control as described in the previous paragraph.

2.2.2 IN-FEED MEDICATION

Safety of drugs given in feed to fish is rarely a problem. Few of the drugs normally administered to fish in feed are toxic other than in very gross overdose, and grossly overmedicated food will usually be rejected by the fish. Fish have acute chemical senses and will refuse or regurgitate food medicated with many compounds, especially if they are surface-coated.

2.2.3 INJECTIONS

As in mammals and birds so in fish some substances are irritant and can produce sterile abscesses in muscle, or adhesions following intraperitoneal injection. Fish differ from mammals and birds in that substances are absorbed from intramuscular injections rather slowly.

Injection site abscesses are uncommon in fish because injections are usually done in air and air-borne bacteria are rarely pathogenic to fish. However when abscesses are formed and rupture the lesions are a hazard to osmo-regulation.

2.3 Safety to the operator

2.3.1 WATER MEDICATION

Some compounds widely used for sea-lice control present specific hazards to operators. Hydrogen peroxide, at 35% concentration as well as at 50%, is extremely corrosive to skin and tissues; organo-phosphorus compounds have potent acute toxic effects as anti-cholinesterases, and may, in some individuals, have other chronic toxic effects. Such materials may be used by lay fish farm staff and it is essential that they are trained to use the specially designed equipment and wear recommended protective clothing.

Other water medication is mainly on the aquarium scale and hence usually done by lay amateurs including children. For such people it is important to emphasize the need to wear rubber gloves when handling fish medicines, especially when, as is usually the case in the UK, the composition is not disclosed on the label. There are real hazards from breathing formaldehyde or contaminating the fingers with malachite green.

2.3.2 IN-FEED MEDICATION

In countries where the sale of fish medicines is legally regulated, authorization is given for specific formulations. The specifications for premixes (or pure drugs) for in-feed medication will include limits for particle size to ensure that the drug will not be respired or ingested by operators. Where such legal controls are not in place prescribers should be aware of this hazard and ensure that the products made available to operators are safe in this respect. Masks and clothing to give adequate protection are cumbersome and uncomfortable and are unlikely to be used under farm conditions.

2.3.3 INJECTIONS

Unless injections are done with automatic equipment the fish should always be anaesthetized first. This is an operator safety precaution as well as an animal welfare procedure; a fish wriggling in one hand while a syringe with needle attached is held in the other constitute a positive hazard that the operators will inject medication into themselves.

The Fishguard® needle protector is a simple and useful device for protecting operators from injecting themselves. It was developed jointly by Kaycee Veterinary Products Ltd.

and Vetrepharm Ltd. (see Appendix II) as an attachment to Kaycee syringes. It has two loops of copper wire which keep the needle away from the operator's hands and which can be bent to the appropriate shape for any size of fish being inoculated.

2.4 Safety to the consumer

2.4.1 WITHDRAWAL PERIODS

This aspect of safety is concerned with residues, either of the drug used or of its metabolites, in any tissue which may be used for human food. Safety is achieved by applying a "withdrawal period" between the last use of the drug in an animal and the earliest time at which it may be slaughtered for food. (Provisions for withdrawal periods for "products" such as milk and eggs do not apply to fish).

The establishment of a withdrawal period for a medicine is a governmental function. It has to take into account a number of variables, some related to the medicine, some to the fish and, most importantly for fish, one relating to the environment, the temperature. The withdrawal period is such that at the end of it the residue level of the drug or metabolite in question is safe for the consumer. Thus determination of a withdrawal period presupposes the determination of a level, known as the maximum residue limit (MRL) of that drug or metabolite, which is safe for the consumer. As 'safe' in this context is a subjective concept there is a political element in the determination of MRLs, and different countries apply different MRL values for their determinations of withdrawal periods.

2.4.2 MAXIMUM RESIDUE LIMITS (MRLs)

At one time some countries, including notably USA, applied a policy of "zero tolerance", that is, that no residue of any drug was acceptable. This may have sounded comforting to the electorate but it was scientific nonsense. Since nearly all drugs are eliminated from the animal body exponentially with respect to time there is theoretically no time after dosing at which a drug level will reach zero. The level was assumed to be zero if the drug could not be detected chemically; so in practice the MRL was not zero but the limit of chemical detection. Thus the MRLs actually applied were subject to variation - reduction - with developments in chemical techniques.

In all countries where there is statutory control over medicinal residues in foods an attempt is now made to base withdrawal periods on rationally determined MRLs. Such an MRL is calculated from two separate parameters, one, the "acceptable daily intake" or ADI, related to the toxicology of the drug, and the other to the relative concentrations of it found in different edible tissues. The ADI is determined by testing the drug in "lifetime" feeding studies in laboratory animals; these are conventionally conducted for 2 years in species other than mice. The aim is to arrive at the "maximum no effect level", or NOEL, the absence of detectable effect meaning not only clinical signs but also histological and haematological changes. The daily NOEL is divided by a safety factor to give the ADI; the safety factor is normally between 100 and 1000 and depends on the qualitative nature of the effect detectable immediately above the NOEL.

The procedure for determining MRLs from an ADI is very much better established for mammals and poultry than it is for fish. In mammals four edible tissues are recognized: muscle, liver, kidney and fat; in birds the kidney is often ignored. (For injectable products the injection site tissues must also be considered). Where drugs are deposited at markedly different levels in different tissues account can be taken of the different weights of those tissues in a carcass; for example in mammals a higher MRL is acceptable for kidney than for muscle because there is much less kidney available to be eaten.

2.4.3 DETERMINATION OF MRLs FOR FISH

Fish have hitherto received scant attention from the authorities who set MRLs, and where fish MRLs do exist there has been no differentiation between tissues. In the European Union where differential MRLs are set at Community level, member states have sometimes interpreted "other tissues" and "all food producing species" as including fish although in all probability only mammals and possibly birds were originally envisaged. Nevertheless in some cases separate residue data have been required for fish muscle and skin for the purposes of setting withdrawal periods. Trimethoprim is selectively concentrated in skin and at one time in the UK the withdrawal period for potentiated sulfonamides in fish was based on the assumption that some consumers might eat nothing but fish skin for the whole of their lives!

2.4.4 PARTICULARS OF WITHDRAWAL PERIODS FOR FISH

Theoretically every statement of a withdrawal period for fish should include particulars of the tissue, the fish species and the temperature to which it applies. For anadromous and catadromous species the salinity of the water should also be specified. The number of determinations needed to meet this ideal is enormous and quite uneconomic; in practice certain reasonable but unproven assumptions have to be made.

In most farmed fish species there are only two tissues, muscle and skin, which need be considered as edible; and to avoid the need to determine a separate MRL for each edible tissue it is always possible, as in the example in the previous section, to err on the side of caution. A far more serious problem derives from the range of different species of fish which are farmed for food. The ADI is assigned to a drug; but the residue levels of that drug will vary widely between species. Where a drug is to be used in more than one farmed species of mammal or bird separate residue studies are conducted and a separate withdrawal period is determined for each, but there are relatively few species in total. Appendix I is a taxonomic list of commonly farmed fish species; no pharmaceutical company could ever afford to conduct residue studies on all of them for any drug. To predict the pharmacokinetics of a drug in a poorly researched species it just has to be assumed that the more closely related two species are the more similar will be their absorption, distribution, metabolism and excretion (ADME) of a drug.

Temperature is another variable with a profound effect on ADME; the rate of each element of ADME rises with higher temperatures. So for fish no withdrawal period can be considered valid without an accompanying statement of the temperature at which it applies. On the assumption that the rate of elimination of a drug is approximately proportional to the temperature in °C the convention has developed of quoting withdrawal periods in degree-days. The assumption on which the convention is based is now generally regarded as untenable, and the more modern practice is to quote a withdrawal period for a specified temperature range. Nevertheless the convention has become enshrined in EU legislation where, when no withdrawal period has been determined for a drug-species combination, 500 degree-days is applied.

The withdrawal period specified for an orally administered drug should apply to only one product. A withdrawal period will be profoundly affected by bioavailability and determinations of F for oxytetracycline in rainbow trout have varied from 4% to 7%. An important cause of variation in bioavailability in many drugs is particle size, an aspect which appears to be of much less importance in mammals and poultry.

2.4.5 DETERMINATION OF WITHDRAWAL PERIODS FOR FISH

Given the MRL for a drug in any particular species, be it mammal, bird, fish or indeed invertebrate, the method of establishing the withdrawal period is the same. A large number of individuals are dosed at the proposed regimen; samples of them are slaughtered at intervals, and their tissues are assayed for the drug. The aim is to demonstrate a time point at which the levels in all tissues are below the MRL. It has generally been accepted that as few as four cattle may be assayed at any one time but that for poultry and fish ten or more are required. This policy is based on the cost of the individual animals, but an important reason for using more fish than mammals at any one time point is the wider range of residue titres which is normally found between individuals given the same treatment (NicGabhainn *et al.* 1996). Not infrequently all the individuals at one time point may be clear but some at a subsequent time point show residues in excess of the MRL. This variation occurs particularly when the drug is administered in feed; the main cause is variation in medicated feed intake, but a further cause is variability in absorption, especially where the bioavailability of the drug is low.

A practical solution to this problem has been proposed by Salte and Liestøl (1983). Assuming that the decline of any drug in any individual is exponential, when the experimental residue data are plotted with log concentration against time it is possible to calculate a linear regression line. It is also possible to calculate parallel lines on each side of the regression line which are the 90% confidence limits - 5% of individuals being expected to be above the upper line and 5% below the lower line. So at the time point where the upper line meets the MRL only 5% of fish are expected to contain residues in excess of the MRL, and this is proposed as the withdrawal period.

Figure 2.1 is an example plot of the results from ten fish assayed at each of five time points. AB is the calculated regression line; CD and EF are the 90% confidence limits. G is the point at which the upper confidence limit reaches the MRL; at this time the statistical probability is that only 5% of fish will contain residues in excess of the MRL.

The fact that this approach has not been widely adopted is no reflection on its merits. Few if any new drugs are being developed for fish at the present time; and for those already approved no-one wants to incur further expenditure to redetermine withdrawal periods, especially since any such determination will apply to only one temperature.

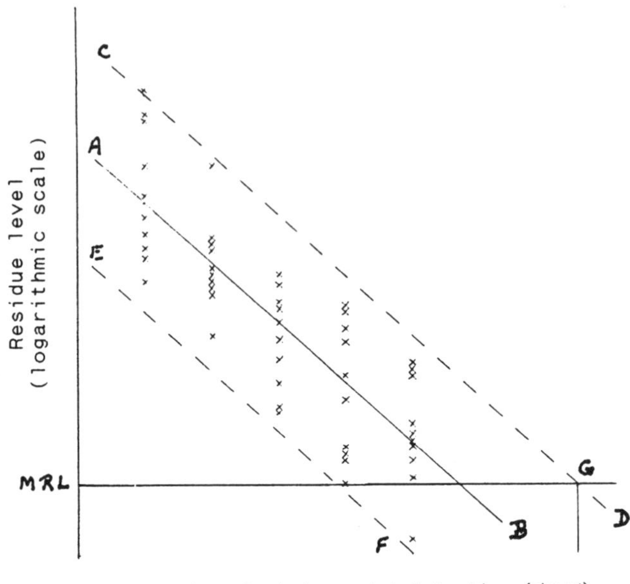

Figure 2.1. An example of Salte and Liestøl, methods for determining withdrawal periods of drugs used in fish.

2.5 Safety to the environment

2.5.1 HAZARDS ATTRIBUTABLE TO MEDICINES

The aquaculture industry in most parts of the world faces continuous criticism and opposition from environmentalists because of the water pollution which is an inevitable concomitant of its activities. Frequently the criticism focuses on the discharge of "chemicals", a term which is used to include everything from antibiotics to agricultural lime. The issue of the environmental safety of fish medicines cannot be ignored, but in fact the most important discharges from fish farms in terms of their effect on the environment are of uneaten food and excreta. The water-soluble fractions of these materials are the cause of algal and plankton blooms and violent swings in dissolved oxygen; the suspended solids clog the gills of wildlife both vertebrate and invertebrate in the water column; and the sedimentary fractions, with their high biological oxygen demand, smother and produce anoxic conditions in the pre-existing sediments below the water column.

Nevertheless medicinal substances used on fish can have an adverse effect on the environment, and an assessment of this is a requirement in all countries where there is legal regulation of the distribution and use of animal medicines. It needs to be recognized that whatever the means by which a medicine is administered to fish it can never be recovered from the environment. Of those medicinal products given orally very few are 100% absorbed; of those absorbed from the gut or given by injection a large proportion is excreted either unchanged or as conjugates which can hydrolyse back to the parent compound. In an environmental impact assessment it is reasonable to work from the entire dose given.

The hazards which can be attributed to medicines have been listed by Redshaw (1995) as:

1. Direct toxicity to non-target organisms;
2. Uptake of contaminants by wild fish and shellfish;
3. Inhibition of microbiological activity in sediments below cages;
4. Induction of antibiotic resistance;
5. Contamination of river water affecting its quality as a potable supply.

Direct toxicity to non-target organisms is a hazard potentially attributable to all classes of chemicals but is perhaps of greatest significance in disinfectants and anti-parasitic drugs. It is for example very difficult to find a compound effective against sea-lice but safe for crustacea since both are arthropods. A specific example of a fish medicine having direct toxicity to non-target species is furazolidone which is extremely toxic to crustacea.

Two points should be noted about the induction of antibiotic resistance. In the first place this is only a hazard if the micro-organisms in which resistance is induced are potential pathogens, either of Man or of other fauna which Man is likely to want to treat with antibiotics, or if there is a significant risk of the resistance being transferable to such potential pathogens. For micro-organisms performing useful scavenging activity on sedimentary organic matter antibiotic resistance may be positively desirable. In the second place the cause of antibiotic resistance is not invariably antibiotic discharge. It has been shown that accumulation of uneaten fish food on the sea-bed may lead to the development of bacteria resistant to one or more of oxytetracycline, oxolinic acid, potentiated sulfonamides and furazolidone despite the total absence of these medicinal substances. In consequence sediments from rivers with no fish farms cannot be used as controls in assessing the environmental effects of antibiotic usage; they can only be used as controls in assessing the effects of fish farming.

2.5.2 ENVIRONMENTAL COMPARTMENTS

The risk assessment of the use of a drug must include consideration of whether it will ever get into the same environmental compartment as the non-target organism to which it presents a hazard. At first view the environmental compartments which might be contaminated by medicines would appear to be the water and the sediment below the water, but this is an over-simplification. The water column needs to be considered as at least three compartments and the sediment as at least two.

Lipophilic compounds are unlikely to be used for water medication, but if they leach from food or faeces or are contained in the lipid fragments of disintegrated pellets then they are likely to form slicks on the water surface. In this compartment the compounds are very exposed to photo-degradation and the slicks are easily broken up by waves, but while they remain they lower the amenity value of the area.

Water-soluble compounds may be excreted in fish urine and are much more liable than lipophilic compounds to leaching from food and especially from faeces. Their positions in the water column will vary with solubility, whether they are incorporated in or surface-coated onto food, particle size of food and faeces, and the rate of descent of the particles.

Solutes and suspended particles in the upper layers of the water column will be more exposed to light than those lower down; the former will also disperse more rapidly because water currents and tides are always stronger at the surface. Waters at the bottom of the column are subject to contamination by less soluble compounds which have remained in the food or faeces as they descended to become sediments; they gradually leach from those sediments over a long period and are not photo-degraded.

Sediments below the water column are of two types: the water bed present whether there is a fish farm or not, and the accumulated detritus (uneaten food and to a lesser extent faeces) from a farm. The former will normally carry an invertebrate fauna whose movements will aerate the sediment; the latter may have only a microbial flora creating a high biological oxygen demand. Medicinal compounds may be expected to diffuse from one to the other only slowly.

Some fish species, notably sea-bass, sea-bream and turbot, are often farmed in salt marsh areas, and these constitute a different environmental compartment from the open sea. There is less water movement and hence depuration through dispersal in water is slower. Concomitant with the higher drug concentrations in the water the areas are frequently colonized by large numbers of molluscs. At least in the case of the most extensively used of all drugs in fish farming, oxytetracycline, shellfish in salt marshes are recognized as being more liable to contamination than those in other areas, even on salmon cages and their anchor lines.

2.5.3 DEPURATION OF ENVIRONMENTAL COMPARTMENTS

The environmental impact assessment of a medicinal product must take into account not only the compartment in which it is found but also the rate at which it degrades or disperses. This may be a multi-factorial process; for example near the bottom of the water column a drug may decompose, be dispersed by water movements and diffuse (leach) out of sediments all at the same time. If the site is below an operating fish cage additional sedimentary deposits may also be arriving.

For any one substance in any one compartment one or more of the following processes may occur:

1. Chemical decomposition - usually hydrolysis, which may be accelerated by sunlight;
2. Microbial degradation;
3. Diffusion;
4. Dispersal - by currents and tides in water; by movements of invertebrates in sediments.

Once an affected sector of the environment has depurated it will be recolonized very rapidly by individual organisms which would otherwise have failed in the struggle for existence. An environmental impact assessment of a medicinal product will therefore include the dimensions of an environmental sector likely to be affected by the normal use of the product, and the duration of the effect in relation to the frequency of use. While biological studies will ascertain such data they are very expensive. It is normal to conduct initial laboratory studies of the physical, chemical and toxicological properties of the product to assess whether there is likely to be a problem.

2.5.4 DEVELOPMENTS IN ENVIRONMENTAL SAFETY PROCEDURES

In the context of the UK Redshaw (1995) has called for:
1. Closer liaison between regulatory bodies;
2. A list of approved chemicals;
3. An archive of the types and quantities of chemicals used;
4. A standardized procedure for environmental risk assessment;
5. Research into the fate and effects of priority chemicals;
6. Development of Environmental Quality Standards (EQSs);
7. Development of analytical methods;
8. Fate modelling.

The pharmaceutical industry would argue that a list of approved chemicals (or at least of approved medicinal products) already exists (see Section 3.2). It too would like to see closer liaison between regulatory bodies, although it is questionable whether the author intended to include bodies regulating medicinal products in item 1 above. An archive of the types and quantities of medicinal products used probably exists only in Norway, where the distribution of drugs to fish farms is conducted by a state monopoly. In a freer economy it is difficult to see how such an archive could be policed.

In the UK development is proceeding, albeit slowly, on EQSs. An Environmental Quality Objective (EQO) is a use of freshwater which needs to be protected; an EQS is a limit concentration of a particular substance outside a mixing zone which must be maintained in order to meet the EQO. The priority list of compounds for the establishment of EQSs consists mainly of non-medicinal substances but it does include formalin, malachite green, oxolinic acid and oxytetracycline.

Further reading

Council Directive 76/464/EEC of 4th May 1976 on pollution caused by certain dangerous substances discharged into the aquatic environment of the Community. *Official Journal of the European Communities* No. L 129/23.

Greenlees, K.J. (1997) Laboratory studies for the approval of aquaculture drugs. *Progressive Fish-culturist* **59**, 141-148.

NicGabhainn, S., Pursell, L., McCormack, F., Barker, G., Alderman, D. and Smith, P. (1996) How many fish are there in an adequate sample? In *Proceedings of EuroResidue III conference, Veldhoven, Netherlands, 6-8 May 1996* (eds. Haagsma, N. and Ruiter, A.), CIP-Gegevens Koninklijke Bibliotheek, The Hague.

Redshaw, C.J. (1995) Ecotoxicological risk assessment of chemicals used in aquaculture: a regulatory viewpoint. *Aquaculture Research* **26**, 629-637.

Salte, R. and Liestøl, K. (1983) Drug withdrawal from farmed fish: depletion of oxytetracycline, sulphadiazine and trimethoprim from muscular tissue of rainbow trout (*Salmo gairdneri*). *Acta Veterinaria Scandinavica* **24**, 418-430.

Vennekens-Capkova, J. (1990) Report No. EUR 12547: *Dangerous substances: Chameleons in water policy.* Office for Official Publications of the European Communities, Luxembourg.

3. THE LAW

3.1 The aims of legislation

3.1.1 KINDS OF LEGISLATION

Legislation covering the medication of fish is designed to ensure the quality of available medicinal substances and their safety and efficacy in use. Safety in this context comprises all four aspects listed in Section 2.1.1 and elaborated in Chapter 2. These legislative aims have resulted in enactments of two distinct kinds: in the first case control is exercised over who may supply and use medicinal substances and how they should be labelled; in the second case control is exercised over the circumstances in which those substances may be used. The provisions needing to be incorporated in enactments of the first kind are not significantly different for fish from those needed for mammals and birds, and indeed, Man; in many countries a single enactment covers medicinal substances for all target species. The second kind tends to be specific to fish medicines: it arises from the problems of environmental safety which are peculiar to fish medication and may control the quantities and frequency of use of certain medicaments and manner of discharge of water from farms where they have been used.

Legislation controlling the supply and use of medicines provides for the creation of a finite list of permitted substances, such substances being said to be registered, to have a New Animal Drug Authorization (NADA) or to have a Market Authorization (MA) etc. MA is the term now used in the European Union (EU) and it will be used in this book except where specific national legislation is considered. The quantity of data necessary to obtain an MA is very extensive and outside the scope of this book. Suffice it to say that generating this data is very expensive, and without an assured monopoly market no company will undertake the investment. This is the reason for the paucity of MA medicines for fish, and users who "shop around" for cheap alternatives to MA products are a positive disincentive to companies investing in development of fish medicines. Two fish products once the subject of NADAs in the USA have been withdrawn because fish farmers used cheaper poultry or agricultural pesticide formulations. At the same time the authorities started to enforce the law so neither active ingredient is now available to fish farmers (Meyer, 1989).

A further pharmaceutical industry requirement before it will invest in generating data for an MA is a stable regulatory environment. Generating data is time-consuming and there needs to be the assurance that the regulatory requirements will not change over the course of the work. This applies to both types of legislation mentioned above.

3.1.2 DEFINITIONS OF A MEDICINE

While the above aims of legislation are fairly universally recognized, the extent of legislative control over the use of medicines varies widely between countries. This variation arises largely from differing legal concepts of what constitutes a medicine. The legal definitions in the UK and USA form a striking example of this.

The original medicines legislation in the UK was The Medicines Act of 1968. This was found to be largely adequate to comply with the requirements of Directive 81/851/EEC. The amending Directive 90/676/EEC necessitated some secondary legislation in the UK but the definition of a medicine has remained from the 1968 act. This is that a "'medicinal product' means any substance ... for use by being administered to one or more human beings or animals for a medicinal purpose" (The Medicines Act, Section 130). Two points should be emphasized: that it is the intended purpose for which it is used which defines a medicine, and that it must be administered to one or more human beings or animals. An adjunct to this latter point which is of particular relevance to fish is that while the point has never been determined in a court of law, it has always been accepted by the UK regulatory body, the Veterinary Medicines Directorate (VMD), that eggs are not animals. Sac-fry presumably are animals even though they do not take externally provided food.

In the case of a number of external protozoal parasitisms of freshwater fish the parasitic stages of the life cycle tend to be resistant to medication but the free-swimming infectious stages are sensitive. So while the infected fish cannot be treated, transmission of infection can be prevented by medication *of the water*. A substance used to treat water in this way is not administered to an animal and so is not legally a medicine in UK. Nevertheless it has an unquestioned medicinal purpose and will be considered in this book.

In USA medicines are defined as in UK by the intended use. However the definition is not limited as to the method of use; "fish" does include eggs, and the legislation covers virtually any chemical put into water where there are cultured fish (N.B. not only fish farmed for human consumption). The USA definition therefore includes algicides, herbicides, dyes to check water flows etc. It cannot be denied that such compounds may enter edible fish tissues, but this is a side effect, not the intention; they are outside the scope of this book.

3.1.3 SCOPE OF THE CONTROLS

In UK The Medicines Act 1968 was passed as a direct response to the thalidomide tragedy and it controlled the activities of the pharmaceutical industry - manufacture, distribution, warehousing, importation, exportation, labelling, sale or supply. It was not concerned with the end-user, and to this extent was more lax than some other national legislations, notably that of the USA.

Directive 90/676/EEC is still primarily concerned with "the placing on the market of a veterinary medicinal product"; but it does provide that, "No veterinary medicinal product may be administered to animals unless [an MA] has been issued ..." (Article 3).

3.2 Labelling

The MA issued by a government for a product is an extensive document. The material will vary between national legislations but will normally include detailed requirements for labelling, which term includes instructions-for-use leaflets and promotional material.

In addition to the name and address of the MA holder responsible for the medicine, the labelling may be expected to provide at least the following information:

3.2.1 QUALITY

- The composition of the product. The manner in which this is expressed may vary according to formulation (injection, premix, concentrate for use by immersion etc.); but the active ingredient must be specified using either its chemical name or better its "approved" (INN) name as would be found in a pharmacopoeia. The extent to which excipients must be mentioned varies between countries.
- The recommended storage conditions.
- The expiry date. This is the last date at which the safety and efficacy of the product may be expected assuming it has been stored as recommended. There may be a legal maximum shelf life (the time between manufacture and expiry); for example in the EU the latest expiry which may be claimed for any product is 5 years after manufacture.

3.2.2 EFFICACY

- The target species. National legislations vary in the closeness with which these may be defined, for example "fish", "salmonids" or "rainbow trout" may be specified.
- The indication - specification of the disease, parasitism or productivity enhancement claimed, and in the case of diseases whether the claim is for prophylaxis, therapy or both. Some national legislations require the association of indications with particular target species.
- Method of use - full details of: route of administration; dose; frequency of dosing; duration of dose regimen; any associated changes to husbandry and feeding routines.

3.2.3 SAFETY

- Contra-indications - species or circumstances in which the product should not be used;
- Precautions - actions to be taken to prevent mishaps, for example protective clothing to be worn, ages or conditions of the target species in which the product should not be used;
- Warnings - mention of possible unavoidable adverse effects, and where possible the action to be taken if they do occur;
- Recommendations for environmentally safe disposal of empty or part-used containers and discharge of medicated water;
- Withdrawal periods - for each target species.

3.3 Drug usage outwith MA provisions

3.3.1 OFF-LABEL USE

Off-label use (or, in the USA, "extralabel" use) means the use of a medicament with an MA, for a species or indication not mentioned on the label and hence by implication not included in the MA. In EU and North America so few products have MAs mentioning any fish species (At the time of writing there are eight in UK and four in USA of which one in each country is not actually on the market) that inevitably this usage is extensive.

The diversity of animal species (and especially of fish species) is such that most countries with developed medicines law have had to make legislative provision for off-label use. In the USA the Centre for Veterinary Medicines (CVM) of the Food and Drug Administration (FDA) at one time published an 'Enforcement Guide' indicating conditions under which FDA would not normally object to extra-label use (Geyer, 1992). This Guide had no statutory authority and in 1994 it was replaced by the Animal Medicinal Drug Use Clarification Act (AMDUCA) which, in effect, decriminalized extra-label use by veterinarians. It nevertheless makes veterinarians responsible for ensuring that there are no consumer safety violations. When AMDUCA was first enacted it was feared that it would act as a disincentive to pharmaceutical companies investing in extension of their NADAs to additional species. In the event the flow of such applications has been maintained for mammals and birds although there is little sign of interest in fish.

In EU off-label use is controlled by the "cascade" laid down in Article 4 of Directive 90/676/EEC.

Literal adherence to this would impose severe restrictions on the medication of farmed fish, and hence create serious welfare problems, because:

1. It applies only to "an animal or a small number of animals" - which excludes 25,000 salmon in cage!
2. It specifies that the medicinal product "contains only substances to be found in a veterinary medicinal product authorized for such animal ..." - and for fish there are hardly any such substances.

When the provisions of the directive were incorporated into UK law in 1995 the regulatory body, the Veterinary Medicines Directorate (VMD), issued an interpretation recognizing this problem. This publication, AMELIA 8, differs from others in the VMD's AMELIA series in that it is addressed not to the pharmaceutical industry but to practising veterinarians; it is in fact subtitled 'Guidance to the Veterinary Profession'. The entire document is required reading for veterinarians working with farmed fish in UK, and arguably also in the rest of EU. Points in it which deserve particular mention are:

1. "Where a veterinarian is required to treat an infection in, say, farmed fish, he or she may need to proceed on the assumption that all individuals in one cage in contact with one another are all equally and identically at risk, and the interpretation of 'a small number' may reflect this." (Paragraph 24)
2. Where no specifically authorized drug exists, the order of preferences (the "cascade") for alternative medicatons is:

a medicine authorized for a different use in the same species;
a medicine authorized for a different species;
a medicine authorized in UK for human use.

However where fish farmed for food are concerned a medicine authorized for a different species must be for a different *food-producing* species.

3. For any medication of fish farmed for food which is not specifically authorized, a withdrawal period of 500 degree-days applies.

3.3.2 ILLEGAL USE

In addition to off-label use within its strict meaning, there is also considerable medicinal use in fish of substances which are not the subject of any MA at all. Although fumagillin has been studied for its possible use against proliferative kidney disease in rainbow trout, the standard drug for this widespread disease is malachite green. Artificial breeding is practised in virtually all species of fish husbanded intensively (see Chapter 17) and this nearly always involves the use of hormones. Gonadotrophins and gonadotrophin-releasing hormones are in widespread use in EU member states, the latter usually being imported from North America.

These uses of drugs in fish are probably not hazardous to the consumer. Proliferative kidney disease affects mainly juvenile rainbow trout and any malachite green used in them would, despite its persistence, have ample time to depurate before the fish grow to harvestable weight. Artificial breeding is, by definition, conducted in broodfish and these are never suitable for harvesting for food. Nevertheless these uses are illegal. In effect the current EU medicines law is widely flouted where fish are concerned because it is restrictive and inadequately policed.

3.4 Legal control of in-feed medication

Medication of feed is an important method of drug administration not only for fish but also for most other farmed animals. It has therefore been necessary for governments to control the manufacture and supply of medicated feedstuffs to ensure that the controls exerted through MAs are not evaded by the supply of feeds already containing drugs.

3.4.1 THE MEDICATED FEEDINGSTUFFS DIRECTIVE

In EU this control is through Directive 90/167/EEC, and some of its provisions are of considerable significance for aquaculture. It decrees that a medicated feedingstuff may not be held, marketed *or used* unless it is produced in accordance with the directive. Requirements for legal production are that:

- Only authorized premixes (*i.e.* those the subject of MAs) are used. This can be a problem for aquaculture because fish feeds have much higher energy content per unit weight than, for example, ruminant feeds; they are fed at low percentages of bodyweight, and therefore need a high concentration of drug to achieve an adequate dose. "Premixes" with MAs for fish are often 100% drug; premixes with MAs for mammals may be of too low concentration for use in fish feeds.
- Manufacture must be on approved premises with staff competent at mixing - although there is an exception allowing on-farm mixing;
- The feedstuff used must be capable of producing a homogeneous mix - this should not be a problem with fish feeds; it would be with some sizes of cattle nuts;
- The MA conditions for the premix, including shelf life of the medicated feed, must be observed;
- Supply may only be on the prescription of the veterinarian who is treating the animals.

The directive also places restrictions on the nature of the prescription which the veterinarian can write:

- A prescription may only be for a single (course of) treatment. If prolonged medication is necessary only 1 month's requirement may be prescribed at a time.
- The daily dose must be mixed into a quantity of feed corresponding to at least half the daily ration. The reason for this requirement is not explained but it is presumably to obviate inadvertent overdosing through the provision of too much medicated feed to the animals;
- The prescription must be dated and is valid for only 3 months from that date.

3.4.2 THE FEED ADDITIVES DIRECTIVE

There is another directive, No. 70/524/EEC, which controls feed additives. The distinction between a premix and an additive may not be obvious but it is clear: whereas a premix is added for a limited period to the feed of animals suffering from or at risk from disease, an additive is added *routinely* to the feed of *normal, healthy* animals. The definition of an additive includes many substances which are not medicinal such as colouring agents, preservatives and anti-oxidants, gelling agents, minerals and vitamins etc. It also includes certain medicinal substances which are fed routinely (and therefore prophylactically), notably coccidiostats for poultry; and it includes growth promoters.

The Feed Additives Directive has had limited impact on aquaculture so far, although it seems probable that growth promoters for fish will be developed in due course. There is nevertheless one aspect of fish feed compounding which is covered by the directive, the addition of agents to colour the flesh of salmonids. It is stipulated in the Annex to the directive that levels of canthaxanthin, astaxanthin and mixtures of them may not exceed 100 mg per kg of complete feed. These compounds are in fact vitamin A precursors and act as vitamins for fish: deficiency causes poor appetite, slow growth and reduced fat and increased water content of the flesh. Fish cannot synthesize them, and the normal dietary source is crustacea which are absent from the compounded feeds normally fed to salmonids. However, the normal inclusion rates are far above the dietary requirements and the directive sets limits on their use as colouring agents.

3.5 Environmental safety legislation

It was noted in Section 3.1.1 that there is a distinct kind of legislation controlling the use of drugs in fish with the object of protecting the environment. This is particularly important in freshwater conditions because the effluent water from fish farms may be used for drinking. This kind of legislation normally provides for a "competent authority" to control discharges of medicated water into watercourses. The competent authority is usually a local body which can act on the basis of knowledge of flow rates in the rivers.

Environmental pollution from drugs used on fish is not confined to freshwater, but monitoring of the quantities used on marine sites is much more difficult. Legislative control is usually achieved by requiring an environmental impact assessment before an

MA is issued and the total banning, rather than quantitative restriction, of drugs which might cause permanent environmental damage.

3.5.1 EU DANGEROUS SUBSTANCES DIRECTIVES

EU legislation on the discharge of dangerous substances into water is complex, largely because of the wide range of types of substance which needs to be covered. Those whose use comes within the legal definition of medicinal probably represent only a small proportion of the total.

(a) Directive 67/548/EEC

This directive defines dangerous substances; requires the notification of all new ones; requires risk assessments of them, and requires them to be appropriately labelled. Substances are designated as dangerous if they are explosive, oxidizing, flammable, toxic, harmful, corrosive or irritant.

Of the characteristics of a substance leading to a designation as dangerous, most are chemical but "harmful" and "toxic" may be regarded as pharmacological. For many compounds it may be difficult to determine whether they are pharmacologically dangerous or not. ("The difference between a medicine and a poison is the dose." - Paracelsus).

(b) Directive 76/464/EEC

This directive divides dangerous substances into two lists: List I contains the more hazardous substances and the aim is to eliminate pollution by them; for List II the aim is to reduce pollution.

List I includes:
- Organo-phosphorus compounds;
- Carcinogens;
- Persistent synthetic substances which may float, remain in suspension or sink and which may interfere with any use of the water.

List II includes:
- Substances in List I for which limit values have not been determined;
- Copper, selenium, arsenic and boron and their compounds;
- Biocides;
- Substances having a deleterious effect on the taste and/or smell of products for human consumption derived from the aquatic environment, and substances likely to give rise to such substances in water;
- Inorganic compounds of phosphorus and elemental phosphorus; substances which have an adverse effect on the oxygen balance.

These lists are not exhaustive but include the main substances likely to be discharged from fish farms.

For List I substances the Council of EU sets limit values for emission, these being based on the toxicity, persistence and bioaccumulation of the compounds; different limits may be set for salt and freshwater. From these limit values the local competent authority sets an 'emission standard' for each applicant to discharge; this will include a maximum concentration and a maximum weight in a given period of time. The authorizations associated with the emission standards are for a limited period but are renewable.

For List II substances the competent authority sets emission standards but these are not dependent on limit values for emission set by the EU commission.

(c) Directive 93/67/EEC

This directive (enacted 26 years after Directive 67/548/EEC!) specifies how the risk assessments required in the earlier directive are to be done.

It provides that the risk assessment must be conducted by a competent authority and be based on a comparison of the potential adverse effects of a substance with the reasonably foreseeable exposure of man and the environment to that substance. The risk assessment must include:
1. Hazard identification;
2. Dose/effect assessment;
3. Exposure assessment - conducted separately for the foreseeably exposed human population and the foreseeably exposed environmental compartment and consisting of quantitative determinations of the emissions, pathways and rates of movement of a substance and its transformation or degradation;
4. Risk characterization - an estimate of the incidence and severity of adverse effects.

(d) Directives covering fish medicines

The main directives covering veterinary medicines other than vaccines are 81/852/EEC as amended by 92/18/EEC. These provide that an environmental risk assessment must be provided as part of an application for an MA. There are 2 significant differences from the Dangerous Substances directives: the assessment is provided by the applicant (normally a pharmaceutical manufacturing company), not by the "competent authority", and they apply to medicines for all animals, not just fish. In practice for fish medicines the applicant's environmental risk assessment is examined by the regulatory authority, and if it is approved it becomes adopted as the competent authority's assessment under the Dangerous Substances directives.

The EU Committee for Veterinary Medicinal Products (CVMP) has issued guidelines for the preparation of environmental risk assessments to be submitted in MA applications (CVMP, 1997). This gives 8 categories of substances for which there are exemptions from testing, those of interest for fish medicines being:
1. Physiological substances such as vitamins, electrolytes, natural amino-acids and herbs;
2. Substances intended for administration to companion animals (*not* including horses);
3. Substances intended for individual treatment of a small number of animals (as opposed to mass medication);
4. Substances that have a predicted environmental concentration (PEC) in ground water below 0.1 µg/L.

It may be presumed that categories 2 and/or 3 would include medicines for ornamental fish. There are further exemptions for medicines for terrestrial animals where there is good reason to believe their normal use would not present an environmental hazard, but there are no other exemptions for medicines for "direct entry into the aquatic environment".

3.5.2 COMPETENT AUTHORITIES IN THE UNITED KINGDOM

In the UK there is a National Rivers Authority for England and Wales and a separate Scottish Environmental Protection Agency. While an MA entitles the holder to sell a product, these bodies have jurisdiction over its use by farmers. They may demand a considerable amount of chemical and toxicological data before they will issue discharge consents. This creates ill-will with the pharmaceutical industry on two counts: since a large amount of such data has to be provided for the grant of an MA it is held that the MA itself is a guarantee of safety; and these authorities may demand the right to disclose the MA holder's intellectual property although this is illegal for the scrutineers of the MA application (The Medicines Act, Section 118).

3.5.3 ENVIRONMENTAL SAFETY PROCEDURES IN GERMANY

In a pessimistic survey of aquaculture in Lower Saxony, Schlotfeldt *et al.* (1990) draw attention to the preponderance of "hobbyists". They have some trenchant comments on hobbyists' standards of husbandry and see little prospect of improvement in the near future - largely because a small industry has little political clout with which to obtain legal controls over itself, and this is a self-perpetuating situation.

> *"From an environmental viewpoint, the most serious threats of* (sic) *fish health control effectiveness are the hobby-farmers; they remain the weakest part of any disease risk management scheme in Germany. This risk level has several reasons. Most (if not all) hobby farmers release pond water directly into receiving waters. These receiving waters are often those with very low flow in dry weather or in late summer and are located in the upper reaches of creeks and streams. These natural waters are the most vulnerable to organic enrichment because they are usually the least polluted."*

3.6 Consumer safety legislation in the EU

3.6.1 REGULATION 2377/90/EEC

The whole philosophy of the EU is that marketing conditions should be harmonized; and clearly the adoption of widely differing withdrawal periods by different member states would distort both the aquacultural and pharmaceutical industries. This regulation, which lays down a procedure for standardizing maximum residue limits (MRLs) throughout EU, was introduced as a contribution to the harmonization of the market for veterinary medicinal products. It provides that all new active ingredients of veterinary medicines must have MRLs determined at the Community level, before the withdrawal periods for different formulations can be assigned, and MAs issued, by national regulatory authorities. Active ingredients for which MAs already existed when this regulation came into force might continue to be used, but the sponsoring manufacturers or suppliers had to submit MRL applications with supporting toxicological data before a given date. Substances for which no MRL had been determined by a later given date (extended, at the time of writing, to 1999) must be taken off the market.

The regulation is concerned exclusively with consumer safety, having no regard to operator, target species or environmental safety.

3.6.2 THE ANNEXES

Regulation 2377/90/EEC provides for assigning substances to one of four Annexes, *viz.*

(a) Annex I

Substances for which MRLs have been established at the community level. Entries in this annex include not only the substance and MRL but also the animal species and/or tissue and the marker residue (which may be a metabolite of the drug) to which the MRL applies. For some substances there are multiple entries.

(b) Annex II

Substances for which it has been determined that no MRL is necessary.

(c) Annex III

Substances for which provisional MRLs have been determined. This annex initially included all pharmacologically active substances which were on the market when the regulation came into force, and additions of new substances can be made "in exceptional circumstances". Provisional MRLs are valid for a limited period and their expiry dates are specified in the annex.

(d) Annex IV

Substances for which MRLs cannot be established because residues at whatever limit constitute a hazard to the health of the consumer. The administration of substances in this annex to food-producing animals is prohibited.

3.6.3 MRL DETERMINATIONS RELEVANT TO FISH MEDICATION

(a) Annexes I and III

Determinations relevant to fish medication which had been made at the time of writing are listed in Table 3.1. For flumequine, azamethiphos, teflubenzuron and sarafloxacin these determinations are specified as applying to salmonids. However, it has been agreed that the same MRLs may be used for other finfish species, presumably on the reasonable assumption that the more a consumer eats of one finfish species the less he will eat of all others (see Section (d)). All the other determinations are relevant to fish by virtue of the animal species being listed as "All food producing species". The flumequine entry is notable in that the target tissue is shown as "muscle/skin" in contrast to virtually all earlier entries where commas are inserted between the tissues. For the last three entries the target tissue is given as "muscle and skin in natural proportions", and presumably this is what is intended for flumequine. Half of the pharmacologically active substances listed have been given provisional MRLs, *i.e.* they are in Annex III; those for which no expiry date is given are in Annex I.

Table 3.1. MRL determinations in the E.U.

Regulation number	Pharmacologically active substance	Marker residue	MRLs	Target tissues	Other provisions
2701/94	Sulfonamides	Parent drug	100μg/kg	muscle, liver, kidney, fat	The combined total residues of all substances within the sulfonamide group should not exceed 100μg/kg.
	Ampicillin	Parent drug	50μg/kg	muscle, liver	
	Amoxycillin	Parent drug	50μg/kg	muscle, liver	
1441/95	Febantel	Combined residues of oxfendazole, oxfendazole sulfone and fenbendazole	1000μg/kg 10μg/kg	liver muscle	The MRLs cover all residues of febantel, fenbendazole and oxfendazole.
	Fenbendazole	Combined residues of oxfendazole, oxfendazole sulfone and fenbendazole	1000μg/kg 10μg/kg	liver muscle	The MRLs cover all residues of febantel, fenbendazole and oxfendazole.
	Oxfendazole	Combined residues of oxfendazole, oxfendazole sulfone and fenbendazole	1000μg/kg 10μg/kg	liver muscle	The MRLs cover all residues of febantel, fenbendazole and oxfendazole.
281/96	Trimethoprim	Parent drug	50μg/kg	muscle, liver	
	Tetracycline Oxytetracycline Chlortetracycline	Sum of parent drug and its 4-epimer	300μg/kg 100μg/kg	liver muscle	
17/97	Flumequine	Flumequine	150μg/kg	muscle/skin	Provisional MRL expires on 1.1.2000
Procedure No. EU/96/037/TRW	Teflubenzuron	Teflubenzuron	500μg/kg	muscle and skin in natural proportions	Provisional MRL expires on 1.7.1999
1140/96	Azamethiphos	Azamethiphos	100μg/kg	muscle and skin in natural proportions	
Procedure No. 91/6/013/ABB	Sarafloxacin	Sarafloxacin	30μg/kg	muscle and skin in natural proportions	

(b) Annex II

The following pharmacologically active substances relevant to fish medication have been assigned to Annex II, *i.e.* no MRLs are considered necessary:

Follicle stimulating hormone
(natural FSH from all species and their synthetic analogues)
Human chorionic gonadotrophin (natural HCG and its synthetic analogues)
Luteinizing hormone (natural LH from all species and their synthetic analogues)
Gonadotrophin releasing hormone
Formaldehyde
Glutaraldehyde
Hydrogen peroxide
"Organic iodine compounds" (*i.e.* iodophors)
Sodium chloride

(c) Annex IV

The following pharmacologically active substances have been assigned to Annex IV and are therefore prohibited for use in food-producing species in EU:

All nitrofurans;
Chloramphenicol;
Dimetridazole.

(d) Extrapolations

The EU regulatory authority, The European Agency for the Evaluation of Medicinal Products (EMEA), recognizes that a requirement for data for every conceivable food species is impractical and would inhibit the development of both livestock production and the veterinary pharmaceutical industry. It has accordingly published specific extrapolations which it is prepared to make between species, and these are relevant to fish.

1. *Salmonidae* are regarded as a (single) major species.
2. Other fin fish species may be accorded the same MRLs as *Salmonidae*.
3. If an MRL has been established for a substance in muscle in a major mammalian species it may be applied to *Salmonidae* and other finfish as well (EMEA, 1998).

3.6.4 MONITORING RESIDUES

For several years compliance with Regulation 2377/90/EEC has been monitored by a system of sampling mammalian meat and offal at the point of sale and analysing it for drugs and contaminants. Directive 96/23/EC has extended the system to poultry and fish with effect from 1st January 1998. The minimum sampling rates specified for fish are:

1. One fish per 100 tonnes production; and
2. Samples taken from 10% of all production sites in the member state.

Requirements are laid down for the types of analyses to be made and for this purpose drugs are divided into two groups:

(a) Group A

This group contains prohibited drugs and so samples are to be taken from fish at all stages of farming up to the point of harvest. One third of all samples are to be tested for this group which consists of:

Stilbenes
Antithyroid agents
Steroids
Resorcylic acid lactones
Beta-agonists
Annex IV drugs

(b) Group B

This group includes drugs for which MAs exist and so the sampling has to be of fish ready for consumption. The Directive provides that:

> *"The sampling should be carried out a) preferably at the farm, on fish ready to be placed on the market for consumption; b) either at the processing plant, or at wholesale level, on fresh fish, on condition that tracing-back to the farm of origin, in the event of a positive result, can be done."*

Two thirds of all samples are to be tested for this group which consists of:

1. Antibacterials
2. Other veterinary drugs
   Anthelmintics, anticoccidials and nitroimidazoles
   Carbamates and pyrethroids
   Sedatives
   Non-steroidal anti-inflammatory drugs (NSAID)
   Other pharmacologically active substances
3. Other substances and environmental contaminants.

3.7 Market Authorizations

It must be emphasized that under most national legislative systems authorization for the use of a medicinal product for a particular indication in a particular species does not imply authorization for other formulations of the same active ingredient, for use in other species or even for other indications in the same species. This is perhaps more restricting in the case of chemotherapeutic agents (primarily anti-microbial and anti-parasitic drugs) than other types of drug, since the latter tend to have more defined indications. However the economic problem of affording to obtain MAs for these other drugs (mainly anaesthetics and hormones in the case of fish) is greater because the drugs are used on fewer individuals and hence have a smaller total potential market.

It should also be remembered that an MA is granted to a manufacturer or supplier. The use of an alternative brand of what is apparently the same medicinal substance will probably infringe medicines law.

The following sections review the MAs which were extant at the time of writing for pharmaceutical products (*i.e.* excluding vaccines) relevant to fish.

3.7.1 THE EUROPEAN UNION

Table 3.2 shows the MA's granted at the time of writing for the use of medicinal products in fish in the EU member states with significant aquacultural industries. Furazolidone and dimetridazole used to be authorized in some member states, but now that these drugs have been consigned to Annex IV of Regulation 2377/90 these authorizations are automatically cancelled. Potentiated sulfonamides are always authorized as specific formulations but some published lists mention trimethoprim and sulfadimethoxine, a combination which is not normally available and for which there are virtually no published pharmacokinetic data.

Table 3.2. Market authorizations in major E.U. member states for drugs for use in fish

Active ingredient	Denmark	France	Germany	Greece	Italy	United Kingdom
Amoxycillin						X
Chlortetracycline			X(t)		X	
Flumequine		X			X	
Oxolinic acid	X	X				X
Oxytetracycline		X	X(t)	X		X
Sarafloxacin						X
Sulfamerazine	X				X	
Tetracycline					X	
Trimethoprim & sulfonamide	X		X	X		X
Azamethiphos						X
Di-N-butyl tin oxide			X(t)			
Hydrogen peroxide						X
Trichlorphon		X	X			
MS-222						X
Chloramine-T	X					
Formalin	X	X			X	
Iodophors		X				
Malachite green	X			X	X	
Potassium permanganate		X				
Quaternary ammonium salts		X				

(a) Germany

The data in Table 3.6 are from Schlotfeldt and others (1990). Items marked X(t) were temporarily authorized.

(b) United Kingdom

The Atlantic salmon and the rainbow trout are the only fish species farmed to any extent and are the only species mentioned in any MA. The problem of obtaining authorization for veterinary use of antibacterial drugs in a range of different infections is often overcome (and not only in the case of fish products) by a wording such as "diseases caused by sensitive bacteria". Since only one dose regimen will be recommended the wording presupposes that all pathogenic species of bacteria showing any sensitivity at all to the drug will have closely comparable sensitivities to it!

The points made in Section 3.1.2 deserve repetition here: in the UK eggs are not legally animals, and substances are legally medicines only if they are administered to animals, whatever the intention behind their use. This exempts from legislative restrictions a considerable range of uses of chemicals, notably medication of hatchery water to prevent or treat fungal infections of eggs and medication of tank or pond water to prevent the transmission of protozoan parasites.

All the MAs for fish medicines issued in the UK categorize the products as POM (Prescription Only Medicine); such medicines may only be supplied by, or on the prescription of, a veterinarian who has the relevant animals "under his care". Since all vertebrates come under the legal definition of animals in The Medicines Act 1968 this restriction is perfectly valid. Anomalously the right to diagnose disease in animals is restricted to the veterinary profession in a different Act of Parliament, The Veterinary Surgeons Act 1966; and in the latter Act fish do not come within the legal definition of animals. As a result, diagnosis of disease in fish may be, and routinely is, conducted by fish pathologists who do not hold veterinary qualifications. If they conclude that medication with a POM drug is indicated a veterinarian must be called in; and he must make his own examination of the fish for them to be legally considered "under his care".

3.7.2 NORWAY

(a) Drug distribution law

In addition to being the world's largest producer of Atlantic salmon, Norway has unique features in its legislative control of the drugs used in fish farming. In any country an enactment requiring MAs for individual drugs is concerned with quality and efficacy as well as all aspects of safety; however the overriding concern is always consumer safety. Norway has a system of MAs but it achieves consumer safety less by controlling the range of drugs available than by close monitoring of all drug usage in farmed fish.

Drugs may only be prescribed by veterinarians; they are supplied by pharmacies or, in the case of medicated feed, by feed mills. These suppliers obtain stocks from a centralized Norwegian Medicinal Depot which, until the beginning of 1994, had a legal monopoly (Grave *et al.*, 1996). The veterinarian and the supplier must each, independently, send a copy of every prescription to the National Fish Inspection and Quality Control Service. These submissions are checked against each other and so act as a check on compliance by the veterinarians and suppliers. Withdrawal times must be quoted on the labels of all drugs for use in fish weighing over 100 g; but the system provides for a further check on residues in that before a batch of fish is harvested samples must be sent to the National Fish Inspection and Quality Control Service laboratories who will know from their prescription records exactly which drugs to look for.

Since 1992 the on-farm medication of feed has been allowed only on farms using wet feed mixed on the farm (Grave *et al.*, 1996).

(b) Authorized compounds

Antibacterial drugs

oxytetracycline	
furazolidone	
trimethoprim/sulphadiazine	powders for administration in feed
oxolinic acid	
flumequine	
erythromycin	powder or solution for oral administration to broodfish
dihydrostreptomycin sulfate	liquid suspension for administration to fry or fingerlings by immersion

The Norwegian government has advised against the use of furazolidone because little is known about the identity and pharmacology of its metabolites. Although the drug remains theoretically authorized this advice has clearly been followed because there have been no further prescriptions for it (Grave *et al.*, 1996).

Parasiticides

praziquantel	- powder
albendazole	- suspension
levamisol	- solution
trichlorfon	- for administration in baths for sea-lice
dichlorvos	
salt	- for protozoan ectoparasites
formalin	

Anti-fungal agents

- natamycin
- malachite green

Anaesthetics

- chlorbutanol
- tricaine methane-sulfonate

(c) Withdrawal periods

With only a single exception withdrawal periods are standardized for all antibiotics and chemotherapeutic agents. They are 40 days for water temperatures above 9°C and 80 days below 9°C. The exception is trimethoprim/sulphadiazine for which the withdrawal periods are 40 days for water temperatures above 12°C, 40-90 days for temperatures between 8 and 12°C and 90 days for temperatures below 8°C.

3.7.3 USA

The all-embracing legal definition of a medicine was explained in Section 3.1.2; and Geyer (1992) has given an important exposition of the approach to aquacultural medicines taken by the Food and Drug Administration (FDA) (or perhaps foisted onto FDA "by an ever-watchful Congress and by aggressive and well-informed consumer advocacy groups"). This paper also explains why all aquacultural medicines (including ice!) are legally new.

(a) New Animal Drug Approvals

Greenlees (1997) has published Table 3.3 showing the current approvals for food species of fish in the USA. Nifurpirinol is additionally approved only for aquarium fish. He notes that:

> *"There are no approved drugs for such normal hatchery management techniques as the induction and enhancement of spawning ... There are entire sections of the aquaculture industry with no approved therapeutic relief - molluscan shellfish, sturgeons, striped bass Morone saxatilis, tilapias, common bait fish species and ornamental fish species all lack approved therapeutants at this time."*

(b) 'Low regulatory priority'

Table 3.4 lists seventeen substances routinely used with medicinal intent in aquaculture which FDA has classified as of low regulatory priority (LRP). They cannot be NADA drugs because no data have been submitted in support of applications. Since from the nature of the substances it seems highly improbable that anyone ever will so submit, LRP appears to carry the implication of permanently shelved regulatory action.

(c) Investigational New Animal Drug

In 1993 the American Tilapia Association obtained an INAD for the (theoretically experimental) use of 17-α-methyl-testosterone (which is Methyltestosterone Ph Eur).

(d) An interim statement on HCG

In 1996 the FDA made an "interim statement" permitting the use of a brand of human chorionic gonadotrophin (HCG) on aquatic broodstock regardless of species. The brand is one already approved for use in some mammalian species and it may be used in brood-fish under the supervision of a veterinarian. Work to prepare a dossier in application for an NADA is in progress and the interim statement is based on data already submitted.

3.7.4 JAPAN

(a) Market authorizations

Japan is perhaps the only country in the world where the aquaculture industry is so important as to create an economic imperative for MAs for a wide range of drugs. Okamoto (1991) has published the MAs in Table 3.5 with the key to fish species in Table 3.6.

Table 3.3. Drugs approved by the U.S. Food and Drug Administration for use in aquaculture and their indications

Trade name	NADA[a] number	Sponsor	Active drug	Species	Uses
Finquel (MS-222)	42-427	Argent Chemical Laboratories	Tricaine methane -sulfonate	Ictaluridae, Salmonidae, Esocidae, and Pericidae (in other fish and cold-blooded animals, the drug should be limited to hatchery or laboratory use)	Temporary immobilization (anesthetic)
Formalin-F	137-687	Natchez Animal Supply	Formalin	Trout, salmon, catfish, largemouthbass, and bluegill	Control of external protozoa and monogenetic trematodes
				Salmon, trout, and esocid eggs	Control of fungi of the family Saprolegniacae
Parasite-S	140-989	Western Chemical Inc.	Formalin	Trout, salmon, catfish, largemouth bass, and bluegill	Control of external protozoa and monogenetic trematodes
				Salmon, trout, and esocid eggs	Control of fungi of the family Saprolegniacae
				Cultured penaeid shrimp	Control of external protozoan parasites
Romet 30	125-933	Hoffman-Laroche, Inc.	Sulfadimethoxine and ormetoprim	Catfish	Control of enteric septicemia
				Salmonids	Control of furunculosis
Sulfamerazine in Fish Grade[b]	033-950	American Cyanamid Co.	Sulfamerazine	Rainbow trout, brook trout, and brown trout	Control of furunculosis
Terramycin for fish	038-439	Pfizer, Inc.	Oxytetracycline	Catfish	Conttrol of bacterial hemorrhagic septicemia and pseudomonas disease
				Lobster	Control of gaffkemia
				Salmonids	Control of ulcer disease, furunculosis, bacterial hemorrhagic septicemia, an pseudomonas disease
				Pacific salmon	Marking of skeletal tissue

[a] New animal drug application.

[b] According to sponsor, this drug is not presently being distributed.

Table 3.4. Unapproved New Animal Drugs of Low Regulatory Priority for FDA[1]

Common Name	*Permitted Use*
Acetic acid	Used as a dip at a concentration of 1,000-2,000 milligrams per liter (mg/L) for 1-10 minutes as a parasiticide for fish.
Calcium chloride	Used to increase water calcium concentration to ensure proper egg hardening. Dosages used would be those necessary to raise calcium concentration to 10-20 mg/L calcium carbonate. Also used to increase water hardness up to 150 mg/L to aid in maintenance of osmotic balance in fish by preventing electrolyte loss.
Calcium oxide	Used as an external protozoacide for fingerling to adult fish at a concentration of 2,000 mg/L for 5 seconds.
Carbon dioxide gas	Used for anesthetic purposes in cold, cool, and warmwater fish.
Fuller's earth	Used to reduce the adhesiveness of fish eggs in order to improve hatchability.
Garlic (whole)	Used for control of helminth and sea lice infestations in marine salmonids at all life stages.
Hydrogen peroxide	Used at 250-500 mg/L to control fungi on all species and at all life stages of fish, including eggs.
Ice	Used to reduce metabolic rate of fish during transport.
Magnesium sulfate (Epsom salts)	Used to treat external monogenetic trematode infestations and external crustacean infestations in fish at all life stages. Used in freshwater species. Fish are immersed in a solution of 30,000 mg/L magnesium sulfate and 7,000 mg/L sodium chloride for 5-10 minutes.
Onion (whole)	Used to treat external crustacean parasites and to deter sea lice from infesting external surface of fish at all life stages.
Papain	Used as a 0.2% solution in removing the gelatinous matrix of fish egg masses in order to improve hatchability and decrease the incidence of disease.
Potassium chloride	Used as an aid in osmoregulation to relieve stress and prevent shock. Dosages used would be those necessary to increase chloride ion concentration to 10-2,000 mg/L.
Povidone iodine compounds	Used as a fish egg disinfectant at rates of 50 mg/L for 30 minutes during water hardening and 100 mg/L solution for 10 minutes after water hardening.
Sodium bicarbonate (baking soda)	Used at 142-642 mg/L for 5 minutes as a means of introducing carbon dioxide into the water to anesthetize fish.
Sodium chloride (salt)	Used as a 0.5-1% solution for an indefinite period as an osmoregulatory aid for the relief of stress and prevention of shock. Used as a 3% solution for 10-30 minutes as a parasiticide.
Sodium sulfite	Used as a 15% solution for 5-8 minutes to treat eggs in order to improve hatchability.
Urea and tannic acid	Used to denature the adhesive component of fish eggs at concentrations of 15 g urea and 20 g NaCl/5 L of water for approximately 6 minutes, followed by a separate solution of 0.75 g tannic acid/5 L of water for an additional 6 minutes. These amounts will treat approximately 400,000 eggs.

[1]FDA is unlikely to object at present to the use of these low regulatory priority substances if the following conditions are met:

1. The drugs are used for the prescribed indications, including species and life stage where specified.
2. The drugs are used at the prescribed dosages.
3. The drugs are used according to good management practices.
4. The product is of an appropriate grade for use in food animals.
5. An adverse effect on the environment is unlikely.

Table 3.4. (continued)

FDA's enforcement position on the use of these substances should be considered neither an approval nor an affirmation of their safety and effectiveness. Based on information available in the future, FDA may take a different position on their use.

Classification of substances as new animal drugs of low regulatory priority does not exempt facilities from complying with other federal, state, and local environmental requirements. For example, facilities using these substances would still be required to comply with National Pollutant Discharge Elimination System requirements.

Table 3.5. The status of approved fish drugs in Japan. After A.Okamoto.

Drug	*a*	*b*	*c*	*d*	*e*	*f*	*g*	*h*	*i*	*j*	*k*	*l*	*m*
Ampicillin	o	x	x	x	x	x	x	x	x	x	x	x	x
Erythromycin	o	x	x	x	x	x	x	x	x	x	x	x	x
Alkyltrimethylammonium calcium oxytetracycline	o	x	x	x	x	x	x	x	x	x	x	x	x
Oxytetracycline HCl	o	o	o	x	o	o	x	o	x	x	x	x	x
Oxolinic acid	o	x	o	o	o	o	o	o	o	o	o	x	x
Oxolinic acid (bath)	o	x	x	x	x	x	x	x	x	x	x	x	x
Oxolinic acid (liquid)	x	x	x	x	o	x	o	x	x	x	x	x	x
Kitasamysin	o	x	x	x	x	x	x	x	x	x	x	x	x
Spiramycin embonate	o	x	x	x	x	x	x	x	x	x	x	x	x
Sulphadimethoxine or its sodium salt	x	x	x	x	x	o	x	x	x	x	x	x	x
Sulphamonomethoxine or its sodium salt	o	x	o	x	o	o	o	x	o	o	x	x	x
Sulphamonomethoxine or its sodium salt (liquid)	x	x	x	x	x	o	x	x	o	x	x	x	x
Thiamphenicol	o	x	x	x	x	x	x	x	x	x	x	x	x
Florfenicol	o	x	x	x	x	x	x	x	x	x	x	x	x
Sulphamonomethoxine & ormetoprim	x	x	x	x	x	x	x	o	x	x	x	x	x
Amoxicillin	o	x	x	x	x	x	x	x	x	x	x	x	x
Colistin sulphate	x	x	x	x	x	x	x	o	x	x	x	x	x
Sulphisoxazole or its sodium salt	o	x	x	o	x	o	o	x	x	x	x	x	x
Tetracycline HCl	o	x	x	x	x	x	x	x	x	x	x	x	x
Doxycycline HCl	o	x	x	x	x	x	x	x	x	x	x	x	x
Lincomycin HCl	o	x	x	x	x	x	x	x	x	x	x	x	x
Josamycin	o	x	x	x	x	x	x	x	x	x	x	x	x
Nalidixic acid	x	x	o	x	x	o	o	x	o	o	o	x	x
Sodium nifurstyrenate	o	x	x	x	x	x	x	x	x	x	x	x	x
Sodium novobiocin	o	x	x	x	x	x	x	x	x	x	x	x	x
Piromidic acid	x	x	x	x	o	x	x	x	o	x	x	x	x
Flumequine	o	x	x	x	x	x	x	x	x	x	x	x	x
Sodium polystyrene sulphonate	o	x	x	x	x	x	x	x	x	x	x	x	x
Miloxacin	x	x	x	x	o	x	x	x	x	x	x	x	x

x = authorized o = not authorized

Table 3.6. Key to Table 3.5

a yellowtail	*Seriola quinqueradiata*	*g* ayu	*Plecoglossus altivelis*
b sea-bream	*Sparus aurata*	(*h* kuruma prawn)	
c coho salmon	*Oncorhynchus kisutch*	*i* amago salmon	*Oncorhynchus masou*
d common carp	*Cyprinus carpio*	*j* yamame salmon	*Oncorhynchus masou*
e Japanese eel	*Anguilla japonica*	*k* other salmonids	
f rainbow trout	*Oncorhynchus mykiss*	*l* crucian carp	*Carassius carassius*
		m other finfish	

The authorizations are of two types, *viz.*

- Drug/species usages which are controlled by the drug laws: for each authorized fish species the dose regimen and withdrawal period are specified and compliance is mandatory; these are shown in bold type in Table 3.5;
- Other authorized usages where advice on use must be obtained from the local (*i.e.* prefectural) government officers; these are in normal type in Table 3.5.

(b) Withdrawal periods

Withdrawal periods are determined separately for each drug and each species in which it is used. No provision is made for variation according to temperature.

Table 3.7a. Japanese withdrawal periods for orally administered antibacterial drugs in yellowtail

Drug	Dosage (mg/kg/day)	Withdrawal period (days)
Alkyltrimethyl ammonium calcium oxytetracycline	50	20
Amoxycillin	40	5
Ampicillin	20	5
Birozamycin benzoate	10	27
Doxycycline HCl	50	20
Erythromycin	50	30
Florfenicol	10	5
Flumequine	20	8
Josamycin	50	20
Kitasamycin	80	20
Lincomycin	40	10
Nifurstyrenate sodium	50	2
Novobiocin	50	15
Oleandomycin	25	30
Oxolinic acid	30	16
Oxytetracycline	50	20
Spiramycin	40	30
Sulfamonomethoxine	200	15
Sulfisoxazole	200	10
Tetracycline	110	10
Thiamphenicol	50	15

Table 3.7a and 3.7b show the withdrawal periods specified in the Japanese drug laws for four fish species for the authorized antibacterial agents.

Table 3.7b. Japanese withdrawal periods for antibacterial drugs in rainbow trout, eels and ayu

Drug	Route of Administration	RAINBOW TROUT		EELS		AYU	
		Dosage (mg/kg/day)	Withdrawal period (days)	Dosage (mg/kg/day)	Withdrawal period (days)	Dosage (mg/kg/day)	Withdrawal period (days)
Colistin	Immersion					1.3 ppm	3
Florfenicol	Oral	10	14				
Miloxacin	Oral			30	20		
Nalidixic acid	Oral					200	7
Oxolinic acid	Oral	20	21	20	20	20	14
" "	Immersion			5 ppm	25	10 ppm	14
Oxytetracycline	Oral	50	30	50	30		
Piromidic acid	Oral			20	20		
'Romet 30'	Oral					50	15
Sulfadimethoxine	Oral	100	30				
Sulfamonomethoxine	Oral	150	30	200	30	100	15
Sulfisoxazole	Oral					200	15

3.7.5 PHILIPPINES

In the Philippines MAs exist for a wide range of drugs (see Table 3.8). Fish species are not named but the MAs do specify whether the drug may be used only in ornamental species or also in edible ones. Withdrawal periods appear to be assigned only to anti-infective drugs.

Table 3.8. Market authorizations in Philippines

Only for ornamental fish	For edible fish	Withdrawal period (days)
	Anti-infective drugs – injectable	
Dihydrostreptomycin sulphate		
Gentamycin sulphate		
Kanamycin		
	Anti-infective drugs – oral	
Erythromycin phosphate	Oxolinic acid	30
Furazolidone	Oxytetracycline	21
Nifurpyrinol	Ormetoprim/Sulphadimethoxine	25
	Sulphamerazine	25
	Sulphisoxazole	25
	Co-trimazine	25
	Anti-infective drugs – immersion	
Isoniazid	Oxytetracycline	21
Neomycin sulphate		
Nifurpyrinol		
Nitrofurazone		
	Parasiticides – oral	
Ivermectin		
Metronidazole		
	Parasiticides – immersion	
Chloramine-T	copper sulphate	
Malachite green oxalate	Potassium permanganate	
Methylene blue	Sodium chloride	
Praziquantel		
Quinacrine HCl		
Trichlorfon		
	Disinfectants for ponds etc.	
Calcium oxide	Sodium hypochlorite	
	Dodecyl-dimethyl-ammonium chloride	
	Formalin	
	Antiseptics – immersion	
Calcium carbonate	Calcium hypochlorite	
Grapefruit juice	Potassium permanganate	

Table 3.8 (continued)

Only for ornamental fish	For edible fish	Withdrawal period (days)
	Disinfectants for eggs	
	Polyvinyl-pyrrolidone	
Antifungal drugs–immersion		
Griseofulvin	Chlorhexamine	
Malachite green oxalate	Copper sulphate	
	Formalin	
	Potassium permanganate	
	Trifuralin	
Anaesthetics–immersion		
Quinaldine sulphate	Tricaine methane-sulphonate	

3.7.6 PEOPLE'S REPUBLIC OF CHINA

Vaccines are widely used in fish in P.R. China but authorization of pharmaceuticals appears to be limited. This is partly due to the difficulty of monitoring compliance in so large a country and partly because the species commonly farmed are mostly fastidious feeders which will not take medicated pellets. One report states:

> *"... fish farmers in various regions have their own recognized prescriptions using herbs. But it is not easy to make clear the effective ingredient and mechanisms of the herbs. So very few of them were approved ..." (Anon., 1996)*

The report observed that farmers took the absence of official action as tacit approval of their herbal remedies. A list of commonly used herbs was appended but less than half of the plants had any recognizable parallels in European herbals. Nevertheless in 1995 there were 69 factories producing drugs for use in aquaculture; most of these were formulators (and presumably herb processors) rather than synthesizers of active ingredients.

3.7.7 TAIWAN

There is limited regulatory control over veterinary medicines in Taiwan, but the following drugs are used in aquaculture:

Antibiotics

- Erythromycin - orally or as a dip
- Chloramphenicol - orally or as a dip
- Florfenicol - orally
- Oxytetracycline - orally or, for eels, as a dip
- Streptomycin - as a dip

Other antibacterial drugs

Oxolinic acid	- orally
Flumequine	- orally
Furazolidone	- orally or as a dip
Nitrofurazone	- orally or as a dip
Nifurpirinol	- orally or as a dip

Disinfectants

Povidone-iodine
Sodium hypochlorite
Chlorinated lime

3.7.8 NEW ZEALAND

In its consumer safety policy New Zealand is the antithesis of the EU. Apart from nutritional (vitamin and/or mineral) premixes only three products have MAs specifically mentioning fish. These are all anaesthetics or sedatives, *viz.*

benzocaine	-	withdrawal period not specified
MS-222	--	withdrawal period 10 days
Aqui-S®	-	zero withdrawal period

Any other medicinal product used in fish is therefore “off-label”. This does not constitute a legal limitation on fish medication, and consumer safety is ensured by an efficient National Residue Monitoring and Surveillance program. This is conducted on samples of fish put on the market with severe sanctions on farmers whose produce is found to contain unacceptable residues.

The MRL applied to off-label use of “approved” products, *i.e.* those with MAs for other food species, is 0.1 ppm. A zero tolerance theoretically applies to unapproved medicinal substances but in practice 0.1 ppm is applied here too. A veterinary prescription is necessary for antibacterial drugs and live vaccines, and the veterinarian is therefore responsible for advising a withdrawal period. Non-veterinarians can buy other animal medicines but are entirely responsible if any problems arise such as detection of residues in their produce.

3.8 Future developments

3.8.1 THE POLITICAL SCENE

Virtually all legislation in the field of fish medicines is designed primarily to ensure consumer safety and secondarily to ensure environmental safety.

Of the national legislations surveyed in Section 3.7 those with MAs for a substantial range of drugs are in countries with developed fish farming industries.This situation is likely to continue for the foreseeable future. Data requirements for an MA are usually based on the limited range of species of mammals and poultry which are farmed; and terrestrial farming is perceived as posing less of a threat to the environment than aquaculture. (This may be an illusion: over the centuries animal farming has gradually become the norm for the terrestrial environment). Where the aquaculture industry is small it does not have the political power to secure changes to legislation developed with farmed mammals and poultry in mind, and it cannot economically justify pharmaceutical industry investing in generating data for inappropriate legislative requirements. Conversely there are countries with developed aquaculture industries, for example Norway, where it is politically and economically expedient for the government to facilitate the further development of the industry.

Legislation often owes less to science than to the outcome of a power struggle between the aquaculture industry on the one hand and consumer and environmental lobbies on the other.

3.8.2 STANDARDIZED WITHDRAWAL PERIODS

Section 2.4.4 surveyed the problem of setting withdrawal periods for fish even after a maximum residue limit (MRL) has been determined for a drug. A degree of pragmatism is essential but national legislations vary in the constraints they apply to the regulatory bodies. In a number of countries the withdrawal periods assigned to medicinal products used in fish are standardized. The following list is not exhaustive but illustrates the approach taken. No account is taken of either the chemical species or fish species and in some cases no account is taken of temperature, but the system makes MAs possible where the pharmaceutical industry cannot afford to generate the necessary data.

Austria - 42 days for all antibiotics and malachite green
Denmark - 40 days at water temperatures above 10°C;
80 days at water temperatures below 10°C.
Finland - 40 days at water temperatures above 10°C;
60 days at water temperatures below 10°C.
Israel - 14 days for antibiotics
Norway - 40 days for water temperatures above 9°C;
80 days for water temperatures below 9°C.
Poland - 180 days
Sweden - 30 days for water temperatures above 9°C;
60 days for water temperatures below 9°C.

3.8.3 USA

(a) Federal initiatives
In recognition of the political realities described in Section 3.8.1, in 1980 Congress passed the National Aquaculture Act which required *inter alia* that federal agencies should take steps to reduce the regulatory burden on aquaculture. It was re-enacted in 1985 with the proviso that the US Department of Agriculture should be designated the lead federal agency dealing with aquaculture. The legal provisions listed in Sections 3.7.3 (b), (c) and (d) are examples of actions taken in accordance with these Acts.

The latter act also established a federal Joint Subcommittee on Aquaculture (JSA) with a remit to implement the reduction in the regulatory burden. The JSA has "recognized the necessity for additional approved safe and effective aquaculture drugs to manage production and aquatic animal health ..." Among other initiatives JSA has produced the list in Table 3.9.

Table 3.9. JSA List of priority aquaculture drugs for approval through the compassionate INAD process

A. Therapeutants and Anesthetics

Amoxicillin - Antibacterial
Benzocaine - Anesthetic
Chloramine-T - Control of bacterial gill disease and flexibacter infections
Copper Sulfate - Antifungal, control of ectoparasites and surface bacteria
Cutrine-Plus - Antifungal, control of ectoparasites and surface bacteria
Diquat - Antifungal, control of surface bacteria
Erythromycin - Antibacterial
Formalin - Antifungal, control of ectoparasites
Hydrogen Peroxide - Antifungal, control of ectoparasites and surface bacteria
Neomycin Sulfate - Antibacterial for control of vibriosis
Oxytetracycline - Antibacterial and fish marking agent
Potassium Permanganate - Ectoparasites and control of surface bacteria
Praziquantel - Anthelminthic for trematodes and cestodes
Sea Lice Control therapeutants
Trichlorfon - Control of ectoparasites

B. Spawning and Gender Manipulation Aids

Common Carp Pituitary - Spawning aid
Human Chorionic Gonadotropin - Spawning aid
Luteinizing Hormone-Releasing Hormone Analogue - Spawning aid
17α-Methyltestosterone - Gender manipulation aid

The list has since been updated slightly because, as noted in Section 3.7.3(d), human chorionic gonadotrophin is now permitted for use in broodstock of all farmed fish species; additionally since the list was published 'Aqui-S' has been developed and is now under consideration as a possible improvement on benzocaine. The "compassionate" INAD procedure was developed by FDA because of the lack of pharmaceutical sponsors for aquaculture products. The procedure facilitates the generation of efficacy data by accepting data where there has been no specific identified pathogen, and by requiring safety data only from the most sensitive target species, this latter concession being called "crop grouping".

(b) Non-governmental initiatives

In 1984 a co-operative plan was developed by the National Biological Service, the US Fish and Wildlife Service and the International Association of Fish and Wildlife Agencies. This plan was to fund and carry out the research required for NADAs for high priority aquaculture drugs. Funding has been approximately half from states (about 40 of the 50 states in the Union) and half from the National Biological Service. Progress has been made, particularly with chloramine-T, copper sulfate and hydrogen peroxide.

An office has been created in USA to coordinate attempts to obtain NADAs for fish medicines. This is associated with Michigan State University and is called The Office of the National Coordinator for Aquaculture New Animal Drug Applications. It is sponsored by some four fish farmers' associations (two of them Canadian), a dozen pharmaceutical companies, and two national and two Canadian provincial governmental organizations. The Centre for Veterinary Medicines (CVM) of the FDA is cooperating with the coordinator's office as far as it is statutorily able to do and, indeed, helped to establish the office.

Further reading

Council Directive 67/548/EEC of 27th June 1967 on the approximation of laws, regulations and administrative provisions relating to the classification, packaging and labelling of dangerous substances. *Official Journal of the European Communities* No. L 196/1.

Council Directive 70/524/EEC of 14th December 1970 concerning additives in feedstuffs *Official Journal of the European Communities* No. L 270/70.

Council Directive 76/464/EEC dated 18th May 1976 on pollution caused by certain dangerous substances discharged into the aquatic environment of the Community. *Official Journal of the European Communities* No. L 129/23.

Council Directive 81/851/EEC of 28th September 1981 on the approximation of laws of the Member States relating to veterinary medicinal products *Official Journal of the European Communities* No. L 317/81.

Council Directive 81/852/EEC of 28th September 1981 on the approximation of laws of the Member States relating to analytical, pharmacotoxicological and clinical standards and protocols in respect of the testing of veterinary medicinal products *Official Journal of the European Communities* No. L 317/81.

Council Directive 90/167/EEC of 26th March 1990 laying down the conditions governing the preparation, placing on the market and use of medicated feedingstuffs in the Community *Official Journal of the European Communities* No. L 92/42.

Council Directive 90/676/EEC of 13th December 1990 amending Directive 81/851/EEC on the approximation of the laws of the Member States relating to veterinary medicinal products *Official Journal of the European Communities* No. L 373/15.

Commission Directive 92/18/EEC of 20th March 1992 modifying the Annex to Council Directive 81/852/EEC on the approximation of laws of the Member States relating to analytical, pharmacotoxicological and clinical standards and protocols in respect of the testing of veterinary medicinal products *Official Journal of the European Communities* No. L 97/1.

Commission Directive 93/67/EEC of 20th July 1993 laying down the principles for assessment of risks to man and the environment of substances notified in accordance with Council Directive 67/548/EEC *Official Journal of the European Communities* No. L 227/93.

Council Directive 96/23/EC of 29th April 1996 on measures to monitor certain substances and residues thereof in live animals and animal products and repealing Directives 85/385/EEC and 86/469/EEC and Decisions 89/187/EEC and 91/664/EEC *Official Journal of the European Communities* No. L 125/10.

Council Regulation (EEC) No 2377/90 of 26th June 1990 laying down a Community procedure for the establishment of maximum residue limits of veterinary medicinal products in foodstuffs of animal origin *Official Journal of the European Communities* No. L 224/1.

European Agency for the Evaluation of Medicinal Products (1998) Committee for Veterinary Medicinal Products note for guidance on the establishment of maximum residue limits for *Salmonidae* and other fin fish. *Veterinary Medicines Evaluation Unit* Document No. EMEA/CVMP/153b/97-Final Anon. (1996) in *Proceedings of the SEAFDEC/FAO/CIDA Meeting on the Use of Chemicals in Aquaculture in Asia* Aquaculture Department, Southeast Asian Fisheries Development Centre, Tigbauan, Iloilo, Philippines.

Geyer, R.E. (1992) FDA regulation of drugs for use in Aquaculture: Law and Policy. *Salmonid* **17**, 5-9.

Grave, K., Marjestad, A. and Bangen, M. (1996) Comparison in prescribing patterns of antibacterial drugs in salmonid farming in Norway during the periods 1980-1988 and 1989-1994. *Journal of Veterinary Pharmacology and Therapeutics* **19**, 184-191.

Greenlees, K.J. (1997) Laboratory studies for the approval of aquaculture drugs. *Progressive Fish-Culturist* **59**, 141-148.

van Houtte, A. (1996) Preliminary review of the legal framework governing the use of chemicals in aquaculture in the Asian Region in *Proceedings of the SEAFDEC/FAO/CIDA Meeting on the Use of Chemicals in Aquaculture in Asia* Aquaculture Department, Southeast Asian Fisheries Development Centre, Tigbauan, Iloilo, Philippines.

Meyer, F.P. (1989) Solutions to the shortage of approved fish therapeutants. *Journal of Aquatic Animal Health* **1**, 78-81.

Okamoto, A. (1991) Restrictions on the use of drugs in aquaculture in Japan. In *Chemotherapy in Aquaculture: from theory to reality* (eds. Michel, C. and Alderman, D.J.) Office International des Epizooties, Paris. pp. 109-114.

Proceedings of the IR-4/FDA Workshop for Minor Use Drugs: Focus on Aquaculture, October 1990. *Veterinary and Human Toxicology* **33**, Supplement 1.

Schlotfeldt, H.J., Herbst, J., Rosenthal, H., Alvarado, V., Stanislawski, D. and Boehm, K.H. (1991) Freshwater fish production and fish diseases in Lower Saxony (Federal Republic of Germany): Fish Health Service experiences. *Journal of Applied Ichthyology* **7**, 26-35.

Schnick, R.A., Alderman, D.J., Armstrong, R., Le Gouvello, R., Ishihara, S., Lacierda, E.C., Percival, S. and Roth, M. (1997) World wide aquaculture drug and vaccine registration progress. *Bulletin of the European Association of Fish Pathologists* **17**, 251-260.

Veterinary Medicines Directorate (1995) *AMELIA 8: The Medicines (Restrictions on the Administration of Veterinary Medicinal Products) Regulations 1994 (SI 1994/2987)* VMD, Addlestone, UK.

Wilder, M.N. (1996) Government Regulations Concerning the Use of Chemicals in Aquaculture in Japan in *Proceedings of the SEAFDEC/FAO/CIDA Meeting on the Use of Chemicals in Aquaculture in Asia.* Aquaculture Department, Southeast Asian Fisheries Development Centre, Tigbauan, Iloilo, Philippines.

PART TWO

ANTI BACTERIAL DRUGS

4. COMPARATIVE ASPECTS OF ANTIBACTERIAL DRUGS

4.1 Introduction

4.1.1 SPECTRUM OF ACTIVITY

As in other fields of veterinary medicine, antibacterial drugs are the most extensively used of all types of medicinal products in fish. In a number of respects these drugs need to be considered as one group rather than several separate groups because despite their chemical diversity they are used for a common purpose - either to kill, or to inhibit the multiplication of, bacteria.

In the rest of this part of the book limited attention will be given to the range of bacteria or diseases for which the drugs are active. In general newly introduced agents are active against most bacteria pathogenic to fish, but as a result of their use over time bacterial resistance develops. As a broad generalization the activity of an antibacterial agent is in inverse proportion to the time it has been available. An exception to this rule must be made for some important narrow- and medium-spectrum antibacterial agents: a majority of bacteria pathogenic to fish are Gram-negative, and for this reason alone agents such as benzylpenicillin and the macrolides have limited use in fish. The only important Gram-positive pathogens are *Renibacterium salmoninarum*, for which erythromycin is being used experimentally, and a non-haemolytic streptococcus which is found mainly on the coast of South Africa and on the Pacific rim.

4.1.2 WITHDRAWAL PERIODS

In the following chapters in this part of the book considerable emphasis will be given to pharmacokinetics and especially to the elimination phases as these determine withdrawal periods. However only rarely will any pronouncement be made as to what the withdrawal period should be. This is because the MRL is a politically determined parameter which will vary between countries. The elimination half-life for a fish species/drug combination will vary with temperature, and countries differ in the legislative provisions made for this variation (see for example Section 3.8.2). The withdrawal period after oral administration will vary with the initial concentration of drug absorbed; this will depend both on the dose regimen and, to an extent which does not occur in mammals or birds, on the formulation of the drug. Thus while an MRL is a characteristic of a drug, a withdrawal period following oral administration is dependent on fish species, formulation, dose regimen and temperature.

4.1.3 ASSAY METHODS

If it is accepted that the hazard to the environment of an antibacterial drug resides in its antibacterial action, then assay of that action - a bioassay - is the correct method of measuring it in the environment. This is in direct contrast to the consumer hazard which is rarely if ever related to antibacterial action. (While some orally administered but poorly absorbed agents may cause indigestion, the residue levels in food would never be high enough to do so). Thus bioassays, while being correct for environmental contamination, are to be avoided for tissue residues. For the latter purpose chemical methods, usually chromatography, should be used whenever possible.

4.2 Selection of antibacterial drugs

In selecting an antibacterial drug the clinician must ensure that the drug will be efficacious against the bacterial pathogen and that it will reach the right site in the animal. Knowledge of the efficacies of drugs on any particular farm is a matter of experience, but as with terrestrial animals use of antibacterial agents eventually leads to bacterial resistance. Antibiograms become necessary, but in aquaculture they need to be used with circumspection.

4.2.1 SEA-WATER ANTIBIOGRAMS

Divalent and trivalent metallic cations have a powerful chemical complexing action on some groups of antibacterial agents, notably quinolones (including fluoroquinolones) and tetracyclines. Seawater with a salinity of 3.5% contains about 54 mMol Mg^{++} and 10 mMol Ca^{++}; other divalent and trivalent cations are negligible. Chemical determinations of the complexing of oxytetracycline (OTC) with the Mg^{++} and Ca^{++} ion concentrations in seawater have shown that when concentrations of OTC normally inhibitory to bacteria are present only about 5% is free. Studies of magnesium and calcium separately show that the former reduces the antibacterial efficacy of OTC more strongly.

In vitro MICs of quinolones and tetracyclines have been shown to rise 40-60-fold following the addition of 50 mMol magnesium chloride to the agar. This is fully pertinent to in-feed medication of marine fish because, as noted in Section 1.1, they are continously drinking seawater and excreting salt. Their gut contents consist of a continuous phase of seawater with suspended food particles and the concentrations of Ca^{++} and Mg^{++} ions may be higher in the hind gut than in the seawater. Medicated feed formulations such as Aqualets® will protect the drug from leaching before the pellet is eaten by the fish, but they will not prevent chelation before the drug is absorbed from the gut. For drug usage at marine sites it is essential that antibiograms are conducted using media with appropriate concentrations of divalent cations.

4.2.2 SITE OF ACTION

If, as is normal on fish farms, the only economic method of administration is *per os* the clinician must bear in mind that he is probably dosing only prophylactically and that ideally the drug should concentrate at the portal of entry of the pathogen. Obvious portals to consider are the gut mucosa, for which a non-absorbed drug is desirable, and the gills, for which high blood levels and hence high bioavailability is requisite.

4.2.3 IDENTIFICATION OF THE POTENTIAL PATHOGEN

There is a limited number of primary bacterial diseases of fish and antibiograms may be useful for controlling them. However fish live in a medium, water, which probably contains a higher concentration of bacteria than the air in which terrestrial animals live. A large and entirely unidentified proportion of these bacteria may potentially become opportunist secondary invaders of fish. Furthermore, since these bacteria are adapted to an aqueous medium, it is thought that possibly less than 1% will develop colonies in air on the surface of an agar gel. In consequence it is unlikely that a conventional plate antibiogram will examine the sensitivities of the real potential secondary pathogens.

4.2.4 EFFECTS ON THE BACTERIAL FLORA OF FISH GUT

Since the gut of a fish will contain a very high concentration of food compared to the surrounding water it will normally develop a dense bacterial flora. Fish gut contents have been estimated to contain 10^8 aerobic and 10^5 anaerobic bacteria per gramme, the latter being largely confined to the upper intestine. The faeces of fish contain very high bacterial counts and on farms will profoundly affect the flora of the water.

Oral administration of oxolinic acid, oxytetracycline or a sulphonamide has been shown to cause a significant increase in bacterial counts in the gut within 24 hours. This increase is maintained during a course of treatment and falls steadily over the following two weeks. In contrast the administration of the narrower spectrum antibiotics, erythromycin or benzylpenicillin, caused a reduction in the aerobic bacterial count. With penicillin the oesophagus and stomach became sterile during the course of treatment and were afterwards recolonized; with erythromycin the fall in count was progressive over the period of treatment and there was subsequent gradual recovery. With erythromycin the main fall in count was in Gram-positive bacteria and there was a slight compensatory rise in *Pseudomonas* and *Aeromonas*, but these were later replaced by enterobacteria.

Erythromycin caused total resistance to itself in the faecal flora within 10 days; there was also some cross-resistance to oxolinic acid and sulfonamides. Oxolinic acid caused increased resistance to erythromycin and sulfonamides but a reduced level of resistance to oxytetracycline and penicillin. Penicillin had no effect on the level of resistance to oxytetracycline or oxolinic acid, but it caused a high level of resistance to sulfonamides; its apparent effect on resistance to itself and erythromycin was to increase and later reduce it.

In contrast to this it has been found that in the goldfish oxytetracycline has no significant effect on bacterial counts in the intestines but resistance to it does develop. It has been suggested that the difference from the increased counts mentioned above may be related to the fact that the goldfish has no stomach.

In an investigation of the effect of some antibacterial agents on the gut flora of yellowtail, *Vibrio* was found to be the predominant genus both before and after administration of ampicillin or the macrolide, josamycin. In contrast, after oxolinic acid administration streptococci predominated. The only genus to be reduced by josamycin was streptococci and on that evidence the new antibiotic was adjudged to be safe for use in yellowtail.

4.2.5 THE ILLUSION OF IMMERSION THERAPY

As a broad generalization, antibacterial agents are not absorbed from water by fish.

> *"While the bath method is indicated to reduce the level of microflora in the water and to reduce the number of microorganisms in skin lesions, it is generally useless in the control of systemic infections. The popularity of administering virtually every antibacterial by the bath method is a result of convenience and traditional usage rather than efficacy. The home aquarist in particular is usually unaware that the antibacterial bath method of treatment for systemic infections of freshwater fishes is relatively useless and expensive and sometimes hazardous."* (Blasiola *et al.*, 1980).

There are exceptions, notably nifurpirinol, a few sulfonamides, lincomycin, spectinomycin and the sodium salts of quinolones, but where they are available at all they are usually not in a formulation suitable for administration by immersion. Furthermore Michel (1986) notes that where absorption through the gills occurs the rate is affected by the pH, the content of chelating divalent cations and the temperature of the water; it is consequently unpredictable. He recommends that in the case of antibacterial drugs immersion administration should be used only for superficial infections.

4.3 Environmental safety

Particular emphasis has been given to sediments and to antibacterial drugs in published studies on environmental depuration not only because antibacterials are the most extensively used class of drugs but also because:

- Dispersal is very much slower in sediments than in water;
- In-feed medication is the usual method of administering systemically acting antibacterial drugs;
- The solubility of antibacterial drugs used in medicated fish feed is such that they usually pass into sediments rather than into water.

The important hazards to the environment which are common to all antibacterial drugs are items 3 and 4 in Redshaw's (1995) list quoted in Section 2.5.1, that is, inhibition of microbiological activity in sediments and the induction of antibiotic resistance.

4.3.1 HOW DO SEDIMENTS DEPURATE?

Hektoen *et al.* (1995) studied boxes of spiked sediment placed on the sea-bed with differential determinations of the depuration of the various layers. The medicinal substances tested were:

Co-trimazine (TMP/SDZ)
Oxytetracycline (OTC)
Oxolinic acid (OXA)
Flumequine (FLQ)
Florfenicol (FF)
Sarafloxacin (SFX)

Florfenicol was depurated rapidly from all layers; a little FF-amine was found but this represented less than 3% of the parent drug used. All the other drugs were depurated faster from the surface than the deep layers. TMP/SDZ was less persistent than the other (OTC and quinolone) drugs. These latter three drugs are known to be photo-labile but this is obviously of no significance at the depths normal for marine fish farms. In the lowest layers the quinolones were still near their initial concentrations after 180 days; in the case of OXA this was not surprising as it is known to adsorb strongly onto sediment. It was concluded that leaching (outwashing) was the mechanism for the depuration of all except florfenicol. The variations in half-lives would have been due to differences in water solubility and to reversible chemical reactions with sediment. Florfenicol may be assumed to have been chemically unstable or to have been metabolized by sediment micro-flora.

4.3.2 HALF-LIVES IN SEDIMENTS

Hansen *et al.* (1992) made a laboratory study of artificial marine sediments spiked with either OTC hydrochloride (which is soluble in water), OXA or FLQ (which are sparingly soluble), and with a through flow of water. They found that initially all disappeared rapidly - to the extent of 35-40% in 3 weeks. Since no metabolites were found this was attributed to diffusion in the sediments and dispersal ("outwashing") from the surface. Subsequent depuration was slower with drug half-lives of the order of 150 days. The hazards of these compounds relate to their antibacterial activities and it was notable that while all caused an initial 50% decrease in bacterial populations, oxytetracycline lost its activity in 25 days. This was attributed to chemical complexing with Ca^{++} ions and binding to clay, humic substances and proteins.

In one survey of the literature it was noted that the half-life of OTC in sediments had been variously determined at 9 — 419 days! OTC has been the most extensively studied antibacterial agent because it is the most extensively used; it has been on the market for a long time, and it is cheap. Were other antibacterial agents to be as extensively studied there is no reason to suppose that some of them would not show half-lives with similar variability. In considering the factors causing this variation the change in half-life with time is important. It is also important to consider the method used to assay the drug concentration: a chemical method will show a much longer half-life than a microbiological method in the marine environment and probably also in hard freshwater. Since the identified hazards of antibacterial agents to the environment are both microbiological it is microbiological half-lives that are relevant. Chemical assays of OTC and probably also of quinolones give misleading estimates of their environmental impact.

4.3.3 DECOMPOSITION IN SEDIMENT

Since the main mechanism of marine sediment depuration is thought to be outwashing, Samuelsen *et al.* (1994) used a closed system to study the chemical decomposition of antibacterial drugs. Since they could not find any farms with no history of antibacterial drug use from which to obtain sediments, they prepared an artificial sediment by mixing mud from a non-farmed marine site with detritus from a farm not then using antibacterial agents. OTC, OXA, FLQ, SDZ, TMP, ormetoprim and sulphamethoxazole were studied.
1. All were stable at -8°C.
2. Ormetoprim and trimethoprim were unstable at 8°C, disappearing in 1 and 2 months respectively;
3. After 180 days at 8°C sulphamethoxazole had decreased by 20%;
4. After 180 days at 8°C, the other four drugs showed no decrease; nevertheless the chemical extraction procedure would have re-activated the otherwise inactive oxytetracycline.

To these points may be added the previously mentioned finding that florfenicol is rapidly depurated from sediment. How fast this would happen at -8°C has not been examined. It is also known that furazolidone is rapidly metabolized by sediment bacteria.

4.3.4 DECOMPOSITION IN WATER

It has been noted that drugs administered in-feed tend to accumulate in sediments; only the fraction excreted in urine or leached from faeces will enter the water compartment. In contrast drugs given by water medication (in baths, dips or immersion) stay in the water, and a higher proportion of the dose used will remain than is normally the case with in-feed medication.

Compounds in water are subject to photo-degradation near the water surface and hydrolysis at all levels. Lunestad *et al.* (1995) conducted degradation tests on the antibacterials commonly used in fish farming - SDZ, TMP, sulphamethoxazole, ormetoprim, OXA, FLQ, OTC and furazolidone (FZD). Daylight at sea-level caused a significant change in the absorption spectra of OXA, FLQ, OTC and FZD. There was less change for the sulfonamides; and the two pyrimidine potentiators, ormetoprim and TMP, were stable. In the dark, solutions of all the drugs were stable. At 1 metre depth in sea-water all gave stable bioassay results except OTC and FZD; this may be attributable to chelation and binding to organic matter in the water in the case of OTC, and bacterial degradation in the case of FZD.

4.3.5 FACTORS AFFECTING ENVIRONMENTAL IMPACT

The environmental impact of antibacterial drugs is microbiological and depends on whether the concentration in any environmental compartment exceeds a microbiological activity threshold. Only if this happens will the rate of depuration be of any significance. The time for which the threshold will be exceeded will depend on the peak concentration.

OTC, although widely used, has a low bioavailability in fish; the majority of the drug used in aquaculture reaches the sediment either in uneaten food or in faeces. Studies have been made at marine salmon farms of the peak concentrations of OTC in sediments after courses of treatment, and no significant correlation has been found between peak concentrations and weights of OTC used. The peak concentration in a sediment obviously will depend on the weight used but it will also be affected by:

- The proportion reaching the sediment - which will depend on bioavailability
- The area over which deposition occurs - which will depend on currents;
- The depth to which the drug penetrates the sediment - which will depend on the physical characteristics of the sediment and its bioperturbation (*i.e* activity of invertebrates).

4.4 Bacterial resistance

4.4.1 INDUCTION AND MAINTENANCE OF RESISTANCE

Bacteria pathogenic to fish have often been found in the sediments under cages so bacterial resistance is relevant to the selection of therapeutic agents. Unmedicated food alone has been found to lead to an increase in the proportion of bacteria in marine sediments which are resistant to OTC. It has also been found that use of OTC leads to a rise in the proportion of bacteria resistant to it even in sediments at sufficient distance from the cage site for the drug itself to be undetectable. On one marine site where the sediments were examined at 19 days and again at 33 days after the end of medication, it was found that at 19 days the reduction in resistance was not significant but at 33 days the proportion of resistant bacteria was significantly less than at the end of medication and at 19 days.

A similar phenomenon has been found in freshwater - at channel catfish farms in southern USA. Comparisons were made between the percentages of bacteria in sediments resistant to antibacterial agents in rivers, in ponds where antibacterial agents had not been used for a long time and in ponds where they had been used recently. In the sediments from rivers, which were in effect the controls, 57% of all bacteria were resistant to only one drug, usually ampicillin; less than 1% of bacteria were resistant to the tetracyclines. In the sediments from ponds where no antibacterial agents had been used for a long time 26% of bacteria were resistant to two drugs, usually ampicillin and either tetracyclines or kanamycin, and a further 52% were resistant to one drug; in the sediments from ponds where drugs had been used 44% of bacteria were resistant to three drugs, usually the ones already mentioned. The authors of the report remarked on the high degree of resistance in the pond sediments where no antibacterial agents had been used. It seems possible that a similar mechanism is operating as at the marine farm where food alone increased the proportion of OTC-resistant bacteria.

In a case where *sub-therapeutic* levels of OTC were inadvertently added to catfish feed it was found to have had a profound effect on the fishes' intestinal flora and on the water-borne bacteria. The latter continued to show a high proportion of resistant strains for 21 days.

One site where antibacterial drug usage might be expected to cause the development of bacterial resistance is the lumen of the fish intestine. However in an experiment where goldfish were monitored during 7 days oral medication with oxolinic acid at 20 mg/kg/day their intestinal flora were found to change only slightly in terms of either total counts or resistance to the drug used. Insofar as bacterial resistance is an environmental hazard of antibacterial drug use it would appear to be confined to the sediment compartment.

4.4.2 CROSS-RESISTANCE

In the study on artificial sediments spiked with OTC, OXA or FLQ mentioned in Section 4.3.2 it was found that resistance to all three drugs developed in two stages, a rapidly developed low level resistance and a slowly developed high level resistance. The following points relating to the individual drugs were noted:

- Although OTC was undetectable after 25 days, resistance to it continued to develop. Eventually 35% of bacteria developed high level resistance to it.
- OTC did not cause cross-resistance to the other two drugs.
- OXA caused cross-resistance to all three drugs.
- Resistance to FLQ developed only slowly. It caused a lower level of cross-resistance to OXA than OXA caused to FLQ.

In another experiment where sediments were artificially spiked with either OTC or OXA, and after 12 months the proportions of bacteria resistant to those two antibacterial agents or to FZD were determined, it was found that each drug had caused the proportion of bacteria resistant to itself to triple. In the OTC-spiked sediment the proportion of OXA-resistant bacteria had risen from 7% to 20%; in the OXA-spiked sediment the proportion of OTC-resistant bacteria had risen from 5 % to 22 %. Cross-resistance to FZD had risen in the OTC-spiked sediment but not significantly, whereas in the OXA-spiked sediment it had risen from 5% to 22%. Unlike the previously mentioned findings with OTC, it was concluded that both drugs induced resistance not only to themselves but also to some other antibacterial agents not necessarily of the same chemical grouping. It was however found that OXA induced this cross-resistance to a greater extent than OTC. The results showed that it would be inadvisable to alternate treatments between OXA and OTC.

In a survey of bacterial resistance on freshwater (mainly trout) farms in northern Germany, Schlotfeld and others (1985) noted that at that time there were six antibacterial drugs with market authorizations (MAs) for use in fish, *viz.* chloramphenicol (CLP), chlortetracycline (CTC), oxytetracycline (OTC), furazolidone (FZD), nifurprazine (NFP) and trimethoprim-sulfadiazine (TMP/SDZ). Pathogenic bacteria isolated over a 14-month period in 1983/4 were submitted to antibiograms for these antibacterial agents and also for five agents, ampicillin (AMP), erythromycin (ERY), oxolinic acid (OXA), flumequine (FLQ) and Linco-Spectin® (L-S), which were not authorized. The percentages of resistance found are shown in Table 4.1.

Table 4.1. Percentages of resistant bacterial isolates from freshwater fish farms in Northern Germany

Authorized drugs		Unauthorized drugs	
Trimethoprim-sulphadiazine	60%	Ampicillin	76%
Nifurprazine	44%	'Linco-Spectin'	19%
Chloramphenicol	39%	Oxolinic acid	3%
Furazolidone	28%		
Oxytetracycline	24%		
Chlortetracycline	14%		

The authors remarked on the low level of resistance to the tetracyclines, but the high level of resistance to ampicillin is also notable. It seems unlikely that this degree of resistance would have resulted from illicit use of ampicillin; it must be attributed either to uneaten food or cross resistance from chemically unrelated drugs.

Further reading

Blasiola, G.C., Dempster, R.P. and Morales, M. (1980) Uptake studies of ampicillin, nitrofurazone and oxytetracycline in carp, *Cyprinus carpio*, with special reference to use of antibacterials in aquaria and notes on heparin as an anticoagulant. *Journal of Aquariculture* **1**, 1-5.

Hansen, P.K., Lunestad, B.T. and Samuelsen, O.B. (1992) Effects of oxytetracycline, oxolinic acid and flumequine on bacteria in an artificial marine fish farm sediment. *Canadian Journal of Microbiology* **38**, 1307-1315.

Hektoen, H., Berge, J.A., Hormazabal, V. and Yndestad, M. (1995) Persistence of antibacterial agents in marine sediments. *Aquaculture* **133**, 175-184.

Lunestad, B.T. and Goksøyr, J. (1990) Reduction in the antibacterial effect of oxytetracycline in sea water by complex formation with magnesium and calcium. *Diseases of Aquatic Organisms* **9**, 67-72.

Lunestad, B.T., Samuelsen, O.B., Fjelde, S and Ervik, A. (1995) Photostability of eight antibacterial agents in seawater. *Aquaculture* **134**, 217-225.

Michel, C. (1986) Practical value, potential dangers and methods of using antibacterial drugs in fish. *Revue Scientifique et Technique de l'Office International des Epizooties* **5**, 659-675.

Redshaw, C.J. (1995) Ecotoxicological risk assessment of chemicals used in aquaculture: a regulatory viewpoint. *Aquaculture Research* **26**, 629-637.

Samuelsen, O.B., Lunestad, B.T., Ervik, A. and Fjelde, S (1994) Stability of antibacterial agents in an artificial marine aquaculture sediment studied under laboratory conditions. *Aquaculture* **126**, 283-290.

Smith, P., Hiney, M.P. and Samuelsen, O.B. (1994) Bacterial Resistance to Antimicrobial Agents Used in Fish Farming: a Critical Evaluation of Method and Meaning. *Annual Review of Fish Diseases* **4**, 273-313.

Schlotfeldt, H.J., Neumann, W., Fuhrmann, H., Pfortmueller, K. and Boehm, H. (1985) Remarks on an increasing resistance of fishpathogenic and facultative-fishpathogenic bacteria in Lower Saxony. *Fish Pathology* **20**, 85-91.

5. TETRACYCLINES

5.1 Group characteristics

5.1.1 THE RANGE OF DRUGS

The tetracyclines are broad spectrum bacteriostatic drugs, three of which are natural fermentation products and have been available since about 1950. There are also several semi-synthetic derivatives. Two of the natural tetracyclines, oxytetracycline (OTC) and chlortetracycline, have been used in aquaculture; OTC in particular has been widely used because in most markets it is not only available but also cheaper than other broad-spectrum antibacterial drugs. Of the semi-synthetic derivatives tetracycline and doxycycline have been used in aquaculture but to only a limited extent; they are more expensive than OTC and whatever their pharmacokinetic merits the long usage of OTC has meant that many bacterial pathogens are resistant to the whole group.

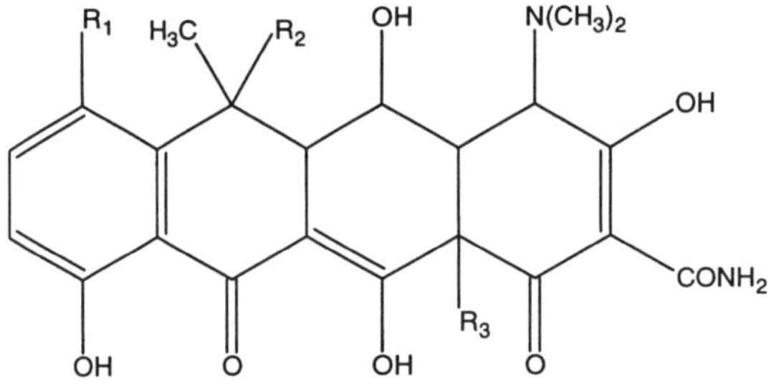

	R_1	R_2	R_3
Oxytetracycline	H	OH	H
Chlortetracycline	Cl	OH	H
Tetracycline	H	OH	OH
Doxycycline	H	H	OH

Figure 5.1. Antibacterial tetracyclines.

5.1.2 CHEMISTRY

The tetracyclines are yellow crystalline compounds. They have low octanol/water partition coefficients and varying solubilities in water, but all are soluble in both acids and alkalis. They are stable as powders but unstable in solution, particularly alkaline solution. Injections are therefore often formulated as hydrochlorides in non-aqueous carriers.

An aspect of the chemistry of the tetracyclines which is pharmacologically important is that they are chelating agents - they form complexes with divalent cations. These complexes are microbiologically inert and, because they are electrically charged, cannot easily cross lipid-rich biological membranes. The tetracyclines will also complex with organic material, especially proteins, and with clay.

5.2 Uses of oxytetracycline

5.2.1 SPECTRUM OF ACTIVITY

Tetracyclines are broad spectrum antibiotics and OTC has been used as a first choice drug for nearly all bacterial diseases of fish. For example since 1980 there have been reports of its activity when given in feed for the following diverse range of infections:
1. Cold water vibriosis (*V. salmonicida* infection) in Atlantic salmon (*Salmo salar*) in seawater;
2. Enteric redmouth (*Yersinia ruckeri* infection) in Atlantic salmon in seawater;
3. Rosette disease of chinook salmon (*Oncorhynchus tschawytscha*) in seawater;
4. Flavobacteriosis in bream (*Sparus aurata*) in seawater and common carp (*Cyprinus carpio*) and grass carp (*Ctenopharyngodon idella*) in freshwater;
5. Furunculosis (*Aeromonas salmonicida* infection) in coho salmon (*Oncorhynchus kisutch*) in freshwater;
6. Columnaris disease (*Flexibacter columnaris* infection) in rainbow trout (*Oncorhynchus mykiss*);
7. Strawberry disease in rainbow trout;
8. Streptococcosis in rainbow trout.

5.2.2 DEMERITS OF OXYTETRACYCLINE

This widespread use has led to the development of bacterial resistance in most countries with major aquacultural industries. Nearly all cases of furunculosis are now resistant to it and other bacterial diseases are becoming increasingly resistant.

Quite apart from the resistance aspect, which has resulted from use, the extent to which OTC has been used is remarkable because its complex formation with Ca^{++} and Mg^{++} ions makes it generally an unsuitable antibiotic for fish. In all fish it has a low bioavailability. In marine fish it is a requirement for osmoregulation that they drink continually and hence the contents of their alimentary tracts are modified seawater; in this medium the complexing of OTC creates a requirement for high dose rates and exacerbates the low bioavailability. As the drug is hardly metabolized at all by fish virtually the whole of the large dose is excreted or defaecated into the environment. It is persistent in the environment, although, again due to complex formation, it may well be inactive (see Section 4.3.2).

The effect is less in freshwater than in seawater not only because of the lower concentrations of cations but also because freshwater fish do not drink. On grounds of environmental safety the minimum effective dose should be used and this means that the doses recommended for freshwater use should be lower than those for seawater. However in the vast majority of market authorizations worldwide they are the same.

5.2.3 DOSE REGIMENS

(a) Injection

Because injection is a labour-intensive method of administering drugs it would normally be done only once with antibacterial agents. It is essential for fish which are not feeding and are too large for immersion treatment. Its value is in the rapid achievement of therapeutic plasma levels; and for a given dose of OTC these have been shown in adult sockeye salmon (*Oncorhynchus nerka*) to be higher following intraperitoneal injection than following intramuscular. As the intraperitoneal route causes less necrosis at the injection site this route should always be used. The dose rate normally recommended is 20 mg/kg which is almost certainly more than necessary. Complexing in seawater does not apply to this route of administration so the dose should be the same as in freshwater. Since the drug is relatively persistent in the fish body maintenance dosing after this initial dose is unnecessary, and many market authorizations specify a longer withdrawal period than for other broad-spectrum antibacterial drugs.

(b) In feed

The normally recommended dose rate is 75 mg/kg/day; this may be appropriate for marine fish in which case it represents an overdose for freshwater fish due to the effects considered in Section 5.2.2. In groups of rainbow trout kept in either freshwater or seawater and given the same dose regimen of oral OTC, it was found that the tissue concentration in the seawater group was only 30% of that in the freshwater group (Lunestad and Goksøyr, 1990).

The usual regimen is daily dosage at this rate for 10 days. The rate of elimination of the drug from most fish species is so slow that if the daily dose is any more than the minimum requirement it should not be necessary to repeat it in less than 2-3 days; alternatively a lower daily maintenance dose rate should be adequate.

(c) Immersion

Concentrations of OTC ranging from 5 to 120 mg/l have been recommended, with the appropriate concentration varying with the hardness of the water. This very wide range illustrates the absence of critical dose titration studies of this widely used drug. In fact although OTC could be used to prevent water-borne transmission of bacterial disease it would not be therapeutic (see Section 4.2.5).

5.3 Pharmacokinetics of oxytetracycline

5.3.1 RAINBOW TROUT

(a) Palatability

In one study mixing OTC into a moist diet at 10 g/kg led to a mean reduction in intake by rainbow trout of 61%, although a wide variation between individual fish was noted. At this medication rate the normal dosage of 75 mg/kg bodyweight per day would be achieved at a feeding rate of 0.75% per day. Conversely to allow for reduced feeding of this order the feed medication rate should be 65% above that calculated for the rate of feed consumption before medication.

(b) Bioavailability

In a study on rainbow trout kept at 14°C, feeds medicated with either 0.1% or 0.5% OTC were given for 2 weeks. Faecal analyses showed the bioavailability of OTC to be 7.1% for the 0.1% medication rate and 8.6% for the 0.5% rate. Only four fish were studied at each medication rate and the variation in the results was such that the differences in bioavailabilities was not statistically significant. In view of the findings in Atlantic salmon mentioned in Section 5.3.2(a) it is regrettable that the form of OTC used was not recorded. The higher of the two feed medication rates used was within the normal range, so it would provide the normal dose rate of 75 mg/kg/day to fish feeding at 1.5% per day.

These findings may be compared with another study at 7°C using a single oral dose of 150 mg/kg given in a gelatine capsule. Here at the time of maximum absorption only 2.6% of the dose had been absorbed. It is important to note the different methods of determining bioavailability in these two studies: in the latter absorption may have continued after the time of peak absorption (T_{max}), but this would have been at a low level and the total absorption would not have reached the 7-9% found at 14°C. Temperature obviously has an important effect on bioavailability and should theoretically affect dose rates.

In comparison with other antibacterial agents the oral bioavailability of OTC is very low in all fish species. Similar bioavailabilities to that for rainbow trout are found in most other other salmonids although it may be 14-15% in the amago salmon (*Oncorhynchus masou*).

(c) Absorption

Several studies of the pharmacokinetics of OTC in rainbow trout have been published and some cases of conflicting results have so far not been resolved. The different results are in part due to different vehicles in which the drug was administered but they do show the profound significance of temperature. Björklund and Bylund (1990) standardized factors other than temperature and determined absorption at 16, 10 and 5°C using an HPLC method. The observations shown in the middle three columns of Table 5.1 were made after a single oral dose of 75 mg/kg. These findings are of interest in that the highest concentrations were achieved at the intermediate temperature. A possible explanation lies in the very rapid excretion found at 16°C in the same study.

In contrast the results in the first column were obtained by a different worker in rainbow trout kept at 15°C and using the slightly higher dose of 100 mg/kg. These results are notable for the rapid absorption (low T_{max}) and low concentrations found. Yet another group has reported the results in the fifth column from a dose of 150 mg/kg. They noted two maxima on the concentration/time curves for both muscle and liver. They attributed the delayed T_{max} (serum) to the vehicle used for drug administration.

(d) Distribution

All the results in Table 5.1 show that of the organs assayed the highest concentrations were in the liver. The study of the 150 mg/kg dose showed therapeutic concentrations in skin, mucus and vertebrae within 48 hours at 7°C. T_{max} for skin was about 72 hours but the concentration in vertebrae rose continuously for the first week after medication.

The OTC in blood is bound to plasma proteins to the extent of 52-55%.

The absorption and distribution findings strongly suggest that the normal oral dose regimen of 75 mg/kg/day is excessive. 100 mg/kg every 48 hours should be fully efficacious for any sensitive systemic infection.

Table 5.1. Absorption of OTC in rainbow trout following oral administration of single doses at different temperatures

	Temperature	15 °C	16 °C	10 °C	5 °C	7 °C
	Dose (mg/kg)	100	75	75	75	150
SERUM	T_{max} (hrs, days)	9 h	1 h	12 h	24 h	1 d, 3 d
	C_{max} (µg/ml)	1.14	2.1	5.3	3.2	-
MUSCLE	T_{max} (hrs, days)	2 h, 3 d	2 d	4 d	9 d	4 d, 7 d
	C_{max} (µg/g)	0.56	2.9	4.0	2.6	-
LIVER	T_{max} (hrs, days)	3 h	2 d	12 h	9 d	3 d, 5 d
	C_{max} (µg/g)	13.3	24.1	45.8	20	44.7

(e) Elimination

The various reported studies of the pharmacokinetics of OTC in rainbow trout have considered elimination from serum or plasma, liver and muscle. However in salmonids generally, and certainly in rainbow trout, the liver is not considered an edible tissue; thus while it is clearly an important storage compartment for OTC the precise rate of depletion from it may be considered somewhat academic. The same consideration applies to the serum or plasma.

Table 5.2 also relates to Björklund and Bylund (1990) and shows half-lives in plasma calculated from "the linear terminal part of the excretion curve". Other data in the report establish that a regression line on a semi-logarithmic plot was calculated, but it is not clear how much data was used in the regression calculation. The profile of OTC absorption into rainbow trout vertebrae means that determination of the terminal half-life at low temperatures should not use data generated until at least a week after the end of dosing. The half-life in liver at 10°C looks anomalous but exactly the same half-life may be calculated from the data generated by other workers at 7°C.

A rather different result for the terminal half-life was obtained by Grondel *et al.* (1989) in a detailed comparative study of the pharmacokinetics of OTC in the plasma of rainbow trout (Table 5.3) and African catfish following intramuscular injection of 60 mg/kg. It will be noted that $t^{1/2}\gamma$ is given a mean value of 89.5 hours; 136 hours is the half-life in serum which may be interpolated from Björklund and Bylund's data, and more recently the same group has published a plasma half-life of 60 hours at 16°C.

Table 5.2. Elimination of OTC in rainbow trout

	Half-lives ± s.e.m. (days)			Withdrawal period (days)
	Serum	Muscle	Liver	
5°C	8.9 ± 2.3	8.8 ± 3.9	9.5 ± 2.1	92
10°C	6.1± 1.6	5.9 ± 1.7	4.2 ± 0.5	48
16°C	4.8 ± 1.3	5.1 ± 0.8	4.7 ± 0.9	37

Table 5.3. Pharmacokinetic values for oxytetracycline administered i.v. to rainbow trout (mean ±SD, n = 4); three-compartment open model. After Grondel *et al*, 1989, with permission.

Parameter	Units	Value
Temperature	(°C)	12
Dose	(mg/kg)	60
Weight	(g)	323 ± 9
C_P^0	(μg/ml)	753.3 ± 291
A	(μg/ml)	601 ± 247
B	(μg/ml)	135 ± 86
C	(μg/ml)	17.5 ± 3.6
α	(h^{-1})	1.35 ± 0.63
β	(h^{-1})	0.123 ± 0.043
γ	(h^{-1})	0.0078 ± 0.0008
$t_{1/2\alpha}$	(h)	0.6 ± 0.2
$t_{1/2\beta}$	(h)	6.3 ± 2.6
$t_{1/2\gamma}$	(h)	89.5 ± 8.7
K_{el}	(h^{-1})	0.175 ± 0.013
K_{12}	(h^{-1})	0.648 ± 0.367
K_{21}	(h^{-1})	0.383 ± 0.239
K_{12}/K_{21}		1.78 ± 0.41
K_{13}	(h^{-1})	0.230 ± 0.092
K_{31}	(h^{-1})	0.018 ± 0.003
K_{13}/K_{31}		12.4 ± 3.6
AUC_{tot}	(μg h/ml)	3759 ± 306
Cl_B	(ml/min/kg)	0.27 ± 0.02
$V_{d(area)}$	(l/kg)	2.1 ± 0.3
V_c	(l/kg)	0.09 ± 0.04

Grondel *et al.* (1989) calculated that a three-compartment mathematical model gave the best representation of their results, and A, B, C, α, β and γ in Table 5.3 are the parameters of this model. The K parameters are the calculated transfer constants between the three compartments. Only plasma was studied so no attempt was made to identify the other compartments; other work quoted above seems to indicate that liver is one.

(f) Withdrawal periods

Björklund and Bylund (1990) calculated withdrawal periods from their data for each of the three tissues, using Salte and Liestøl's method (see Section 2.4.5); those shown in Table 5.2 are for the only edible tissue, *i.e* muscle. They also conducted a field trial using 100 mg/kg/day for 10 days, and then studied elimination. The water temperature rose from 17 to 20°C after the treatment; it may be assumed to have been 20°C during the terminal elimination phase. From their data they calculated a terminal half-life of 1.7 days and a withdrawal period of 9 days. In calculating withdrawal periods they used the limit of detection of their assay, 50 μg/kg, as the MRL; the EU MRL is 100 μg/kg for OTC in muscle so the EU withdrawal period should be shorter by one half-life in muscle. Using the degree-day concept the EU withdrawal periods work out at 415, 420 and 512 degree-days for 5°, 10° and 16°C respectively. This apparently validates the concept for temperatures within the range 5-16°C, but 500 degree-days should not be taken as an adequate withdrawal period because although the dose was 75 mg/kg it was given only once.

The concept fails for the results at 20°C. Although a higher dose rate was used and it was given for 10 days instead of once, the withdrawal period works out at about 150 degree-days. This result cannot be lightly dismissed as an outlier however because it was obtained under field trial conditions. The same group later reported another field trial on rainbow trout in brackish water (0.6% salinity) at 15°C using 85 mg/kg/day for 10 days. They found muscle residues to fall below the level of detection by 25-30 days and a terminal half-life of 5.6 days. From these results an EU withdrawal period in the range 300-360 degree-days may be calculated. There is no obvious reason why the salinity of the water should affect drug elimination, and for a higher dose rate to produce a shorter withdrawal period is inexplicable.

There seems to be a need for better understanding of the biochemical processes underlying drug elimination before a rational formula for calculating withdrawal periods can be devised.

Jacobsen (1989) made a very detailed study by HPLC of residue depletion at 6, 12 and 18°C. Using a dose regimen of 50 mg/kg/day for 8 days, which was that approved in Denmark at the time. Assays were conducted on whole gutted fish at 6 and 18°C and the four tissues, skin, muscle, kidney and blood at 12°C. The limit of detection was 0.05 ppm in all except skin where it was 0.01 ppm. The EU MRL for muscle is twice the former value; none has been determined specifically for skin. 24 hours after the end of the dose regimen the highest concentrations of OTC were in the gutted fish at 18°C (1.4 ppm) and the skin at 12°C (1.9 ppm). At 6°C OTC was still detectable in gutted fish 47 days after the end of medication, but not subsequently. At 12°C it was detectable in muscle at 22 days and in skin at 52 days. At 18°C it was detectable in gutted fish at 22 days. Jacobsen concluded that:

1. The maximum absorption (highest bioavailability) of OTC in rainbow trout was at 18°C;
2. The withdrawal period should be based on whole gutted rainbow trout since the skin may be eaten;
3. OTC excretion is a linear function of temperature;
4. Considerable variation in the dosage of OTC leaves the excretion time roughly constant.

5.3.2 ATLANTIC SALMON

(a) Absorption

OTC hydrochloride has been shown to have a bioavailability of only 2% in seawater at 7-8°C. After a single dose of 50 mg/kg given by force feeding medicated pellets C_{max} (plasma) was 0.42 μg/ml at T_{max} 12 hours and the plasma concentration was found to remain constant for the first 40-50 hours. This was taken to indicate that absorption was not only at a low level but also slow. These findings were attributed to OTC complexing with Ca^{++} and Mg^{++} ions and the complex molecules being unable to pass through lipid membranes.

Profound pharmacokinetic differences have been shown between OTC dihydrate and hydrochloride. Pre-smolts held in seawater, and given either drug in feed at 75 mg/kg/day for 10 days and then bioassayed, showed a sixfold higher muscle concentration after the dihydrate than after the hydrochloride. Similarly fish held in freshwater and given a single dose by oral intubation showed a three to four times higher muscle concentration after the dihydrate. The effect is presumably associated with the pH of the two compounds. The hydrochloride is normally used in commercially available premixes but it would appear that the dihydrate is the more economical form of OTC for Atlantic salmon, and hence also it would potentially present less risk to the environment.

(b) Elimination

A comparative study has been made of the depletion of OTC following a single intra-peritoneal injection of 20 mg/kg in groups of cultured salmon of four different states of maturity and health:

1. Sexually mature fish which had been returned to freshwater at 4°C; they were therefore not feeding. They had become infected with furunculosis and this was controlled by the OTC treatment. At 3 weeks after treatment the serum concentration was 3.4 μg/ml and it had scarcely fallen any further by 8 weeks.
2. Similarly mature fish in freshwater at 7°C which were not infected. At 3 weeks after treatment these fish had lower serum concentrations (2.5 μg/ml) than the infected ones and the levels continued to fall reaching 1.3 μg/ml by 8 weeks after treatment.
3. Fish kept in a similar way to the second group but which had not matured. They had nevertheless been transferred to freshwater. Their serum levels at 1 week were little more than those of the mature fish at 8 weeks, and the levels fell to 0.5 μg/ml by 8 weeks
4. Post-smolts in seawater. Their mean serum level at 1 week was only 1.2 μg/ml and was undetectable by 5 weeks.

Liver concentrations were not measured in the infected fish; in the uninfected ones they were eight to ten times the serum concentrations. The difference in water temperatures at which the two groups were kept would probably not account for more than a twofold difference in residue levels; the major part of the difference must be attributed to the infection. Residues were not measured in the main edible tissue, the musculature, and hence the results are of no value in the determination of a withdrawal period. Nevertheless they show the difficulty of making such a determination. Assuming that the muscle concentrations are proportional to the serum concentrations and if the withdrawal period is to be based on the worst case, residue assays should be conducted on broodstock which are the most expensive fish; and they should be infected with furunculosis. However this is an unrealistic approach: fish grown for the table are medicated in feed; broodstock do not feed and they are never sold for the table. Infected fish are never used for residue studies but this also does not matter because they also do not feed; normal in-feed medication is prophylaxis for healthy fish.

The field trial in Björklund and Bylund (1990) involved two pens of Atlantic salmon as well as the rainbow trout. The salmon were also dosed at 100 mg/kg/day and the water temperature fluctuated between 13.5 and 17°C. The residue levels were determined separately for the two pens and showed the same means and the same half-lives of 5.3 days in muscle. However the standard error of the mean (s.e.m.) was higher in one case and this led to a longer withdrawal period calculated by Salte and Liestøl's method. The two withdrawal periods were 27 and 37 days, or, using the EU MRL of 100 µg/kg, 22 and 32 days. The latter would give degree-day results in the range 335-500, or slightly less than that for rainbow trout.

5.3.3 SOCKEYE SALMON

Strasdine and McBride (1979) studied absorption of OTC by adult sockeye salmon returning from seawater to freshwater to breed. They tested single intraperitoneal and intramuscular injections of approximately 5 mg/kg, immersion (5 ppm in freshwater for 2 hours) and oral administration (approximately 5 mg/kg). Because fish at this stage of maturity do not feed the orally administered drug was in gelatine capsules. Serum OTC was determined by bioassay.

No OTC was detected at any time in the serum of fish treated by immersion. intraperitoneal injection produced greater and more prolonged serum concentrations than intramuscular injection, although in the fully mature ("ripe") fish this difference did not reach statistical significance. Absorption from oral administration increased with increasing maturity of the fish. In view of the low oral bioavailability which OTC normally has in fish it may be considered surprising that as small a dose as 5 mg/kg produced any detectable serum level at all. The experimenters hypothesized that their results were due to disintegration of the biological barriers of the intestinal tract leading to the leakage of OTC into the peritoneal cavity.

The object of the study was to find a method of preventing deaths, apparently due to microbial infection, shortly before spawning; and intraperitoneal injection was clearly the route of choice for administration of OTC.

5.3.4 CHINOOK AND COHO SALMON

Namdari *et al.* (1996) studied the distribution and elimination of OTC in chinook and coho salmon in seawater. They were interested in determining a withdrawal period for chinook salmon, and in the absence of a legally determined MRL in Canada they were looking for absence of residue at a limit of detection of 100 µg/kg (= 0.1 ppm). The chinook salmon were kept at either 9 or 15°C, and the coho salmon at 10°C. The feeding rates were 75 mg/kg/day for 21 days and 100 mg/kg/day for 42 days respectively. These treatment periods are much longer than would ever be recommended in a medication regimen but were used with the aim of attaining steady state concentrations in the tissues.

Table 5.4. Steady stateconcentrations and elimination times of OTC in chinook salmon and coho salmon in seawater

	Chinook salmon 9°C		Chinook salmon 15°C		Coho salmon 10°C
	Steady state conc.	Elimination to <0.1 ppm	Steady state conc.	Elimination to <0.1 ppm	Steady state conc.
Liver	47.6 ppm	57 days	50.5 ppm	44 days	9.0 ppm
Kidney	11.2 ppm	50 days	8.6 ppm	36 days	3.8 ppm
Muscle	4.5 ppm	50 days	4.4 ppm	41 days	0.95 ppm
Skin	10.6 ppm	57 days	9.9 ppm	41 days	2.6 ppm
Bone	11.3 ppm	> 72 days	10.3 ppm	> 72 days	0.22 ppm

If bone is excluded, the order of steady state concentrations in the different tissues studied was the same for the two species, *viz.*

Liver>kidney>skin>muscle

Nevertheless although the coho salmon had been given a higher dose rate for twice as long as the chinook salmon their steady state tissue concentrations were significantly lower. This was attributed primarily to the shorter half-life in coho salmon, but it was noted that over the period for which the observations were continued growth of the fish would have had some effect on drug concentration.

5.3.5 AMAGO SALMON

It was noted in Section 5.3.1(b) that the bioavailability of OTC in amago salmon may be 14-15%; this result was obtained at 15°C, and it must be assumed that as in the previously mentioned salmonid species the parameter varies with temperature. Other results in amago salmon are shown in Table 5.5 and it is notable that two maxima were found for all three tissues. This effect was commented on by one group as occurring in muscle and liver of rainbow trout although their graphs show it also in serum. The possibility cannot be excluded that it is a consistent feature for all salmonids; whether it is seen or not may depend on the frequency with which samples are taken for assay.

Table 5.5. Absorption parameters of OTC in amago salmon

	Serum		Muscle		Liver	
C_{max} (μg/g)	0.89	2.05	0.19	0.29	10.9	9.12
T_{max} (hours)	1	24	0.5	9	0.5	9

The initial maximum in liver occurred before that in serum, and may reasonably be attributed to "first pass" absorption, that is, active absorption into the liver from the portal bloodstream. This too is a feature in rainbow trout and probably in Atlantic salmon.

5.3.6 COMMON CARP

(a) Absorption

The bioavailability of OTC following oral administration in the carp is extremely low - of the order of 0.6%. In consequence the drug is not given orally for systemic infections, although it is used as a growth promoter in some cyprinid species. It is given by injection to large and valuable individuals such as koi and by dipping to juveniles. The exceptionally low oral bioavailability may be associated with the absence of a stomach, and hence of acid digestion, in the carp.

The normal dose by injection is 60 mg/kg; injectable formulations should not be given intraperitoneally to carp. When it is given intravenously, there is an immediate rapid fall in plasma concentration (half-life 3.5 hours at 20°C) due to distribution to a peripheral and a deep compartment. Distribution to the peripheral has been estimated to be 29 times that to the deep compartment. After intramuscular injection the half-life of absorption from the injection site is about 12 hours, but because of the rapid onward distribution from the plasma T_{max} (plasma) is only about 14 hours. C_{max} by bioassay is 57 μg/ml.

(b) Excretion

48 hours after injection the highest concentration is in the liver, and by 96 hours equilibrium is attained with the liver:plasma concentrations in the ratio 2:1. There is some accumulation in the scales and bone, and slow excretion. The terminal half-life in the plasma appears to vary according to the route of injection, being of the order of 140 hours following intravenous injection and 78 hours following intramuscular. This could be due to the high plasma level following intravenous injection producing a higher deposition in scales and bone. At no stage is the concentration in muscle particularly high and by 500 hours after injection it is undetectable by bioassay, even at the injection site.

5.3.7 CATFISH

(a) African catfish (Clarias gariepinus)

Grondel *et al.* (1989) have published the data in Table 5.6 for African catfish. They comment that while rainbow trout and African catfish have comparable terminal half-lives, the volume of distribution was larger in the trout. In consequence when the two species are given the same dose, at any given time point in the elimination phase the catfish contains twice the plasma OTC concentration of the rainbow trout. The corollary of this is that there are tissues in catfish into which OTC does not penetrate, but these have not been identified.

(b) Blue and channel catfish (Ictalurus furcatus and *I. punctatus)*

As in salmonids and carp, at equilibrium the highest concentrations of OTC are found in the liver, and in consequence in the elimination phase this tissue takes longer to deplete than others.

Residue studies have been reported for the two species kept at either 16.6-21.1 or 22.2-25°C and dosed at 50, 100 or 200 mg/kg/day for 10 days. Using a bioassay with

Table 5.6. Pharmacokinetic values for oxytetracycline administered i.v. to African catfish (mean ±SD, *n*=4); two-compartment open model After Grondel *et al*, 1989, with permission.

Parameter	Units	Value
Temperature	(°C)	25
Dose	(mg/kg)	60
Weight	(g)	293 ± 45
C_p^o	(μg/ml)	86.4 ± 10.9
A	(μg/ml)	42.5 ± 8.5
B	(μg/ml)	43.9 ± 9.1
α	(h^{-1})	0.184 ± 0.121
β	(h^{-1})	0.0087 ± 0.0004
$t_{½α}$	(h)	5.2 ± 3.2
$t_{½β}$	(h)	80.3 ± 3.9
K_{el}	(h^{-1})	0.16 ± 0.003
K_{12}	(h^{-1})	0.084 ± 0.071
K_{21}	(h^{-1})	0.092 ± 0.048
K_{12}/K_{21}		0.69 ± 0.49
AUC_{tot}	(μg h/ml)	5369 ± 1102
Cl_B	(ml/min/kg)	0.19 ± 0.03
$V_{d(area)}$	(l/kg)	1.33 ± 0.22
V_c	(l/kg)	0.70 ± 0.09

a limit of detection of 0.25 ppm in muscle, residues were only found at all in the muscle of channel catfish kept at the lower temperature and dosed at the rate of 200 mg/kg/day for 10 days, and these residues were detected for only 3 days after the end of medication; none were found in blue catfish muscle. For serum the limit of detection was 0.09 ppm and no residues were found after 2 days at any temperature or dose rate. In liver the residues were more persistent at the higher temperature which, it was noted, is the converse of the position in salmonids. The limit of detection was 0.15 ppm and at 22.2-25°C residues fell below this in 10 days for the 50 mg/kg/day dose rate and in 21 days for the two higher dose rates.

5.3.8 YELLOWTAIL (*SERIOLA QUINQUERADIATA*)

The bioavailability of OTC in yellowtail is similar to that in rainbow trout. After a single oral dose of 100 mg/kg was given to fish kept at 22°C, maxima were seen as in Table 5.7.

Table 5.7. Absorption parameters of OTC in yellowtail

	Serum		Muscle			Liver
C_{max} (μg/g)	0.89	0.53	0.11	0.12	0.50	35.5
T_{max} (hrs, days)	3 h	2 d	0.5 h	9 h	2 d	0.5 h

The data for both serum and muscle should be treated with caution because no observations were made between 3, 6 and 9 hours, 24 hours and 2 days, or between 2 and 3 days. The concentration data suggest that in serum the first T_{max} was probably about 4 hours and the second about 36 hours; and in muscle the second at 8 hours and the third about 36 hours. The one maximum in liver must be attributed, as in salmonids, to first pass absorption from the portal bloodstream.

The terminal half-life in serum was estimated to be 1.2 days. That in muscle was not determined: the terminal elimination phase did not appear to start until about 4 days and residue levels fell below the level of detection (0.05 ppm) soon afterwards.

5.3.9 MARINE SPECIES

(a) Sea bass (Dicentrarchus labrax)
A study has been conducted in Italy where OTC is approved for use in "fish" (*i.e* all species) at 35-75 mg/kg/day for 7-14 days. Sea bass and sea bream were fed a diet containing 7.5 g/kg OTC at 1% per day for 14 days, that is, the maximum permitted dose regimen. The water temperature was 19-25°C until 20 days after the end of treatment and then rose to 24-28°C. Tissues were assayed with limits of quantitation 0.02 ppm in muscle, 0.01 ppm in skin and bone, and 0.20 ppm in liver.

In sea bass only limited observations were made because OTC has very low bioavailability. In muscle it was detectable only during treatment.

(b) Sea bream
An interesting feature of the pharmacokinetics of OTC seen in sea bream was its concentration in the vertebrae and to a lesser extent in the skin. The AUC value for vertebrae was 17 times and for skin five times that for muscle. C_{max} for skin was actually higher than for vertebrae; it was reached by half way through treatment and the skin concentration had fallen back to a low level before the end of treatment. Falls of a much lower extent occurred simultaneously in muscle and liver; they may have been a palatability effect.

The terminal half-lives were estimated to be 3.3 days in skin and 52 days in muscle. So although the concentration in muscle at the end of treatment was considerably lower than that in skin, the drug was undetectable in skin by 30 days afterwards, at which time there was still 0.04 ppm in muscle.

5.4 Pharmacodynamics of oxytetracycline

5.4.1 IMMUNO-SUPPRESSION

(a) Cellular immunity
Although OTC is primarily used for its antimicrobial action, a chemotherapeutic action, it does have pharmacodynamic effects on fish, the most important of which is immunosuppression.

That antibiotics would delay the rejection of allografts in fish had been reported as early as 1963, and the time to rejection of allografted scales was the measure used in one of the first systematic studies of immuno-suppression. This was by Rijkers *et al.* (1980) who used the common carp as their subject species and OTC "because of its frequent use in fish culture" as the antibiotic. In addition to unmedicated controls they studied fish fed at 2.5% of bodyweight with a diet containing 2000 ppm OTC (equivalent to 50 mg/kg/day), and fish given an intraperitoneal injection of OTC every 3 days at either 60 or 180 mg/kg. They noted that OTC was not detectable in serum until the medicated diet had been fed for 48 days and that C_{max} was only 1.25 mg/ml; this work was published before that (ironically from the same Dutch research institute) showing the very low bioavailability of oral OTC in carp.

They concluded that "orally-administerd OTC had no effect upon the rejection time of transplanted scales, but injection with OTC significantly increased the survival time of allogenic scales and thus suppressed the cellular immune response." This was with repeated injections of OTC; other work has shown that a single injection delays rather than inhibits the cellular immune response. It was also found that polyvinylpyrrolidone (PVP), the normal excipient for injectable OTC, has delaying action.

(b) Humoral responses

Rijkers *et al.* (1980) also studied humoral responses using rabbit erythrocytes as antigen and counting rosette forming cells (RFC) in the carp spleen. Some fish were given a single injection of antigen either 15 or 35 days after the beginning of medication and some of each were given a second injection of antigen after an interval of 24 days. There was a four-fold increase in RFC after the first injection in unmedicated controls and an equal or greater but short-lived response after the second. There was little, or in some cases no, response in any of the medicated fish. Because there was no humoral response in orally medicated carp although the cellular response appeared to be unaffected, it was concluded that the humoral response is more sensitive to low levels of antibiotic.

Subsequent work on serum proteins of carp given the same dose regimens of OTC showed that by either route of administration the drug produced a 20% fall in total protein but a profound drop in immuno-globulin. This work was done with the OTC injections beginning some time before the antigen; significant inhibition of antibody production also occurs with a single injection a day before the antigen, but not if the OTC and antigen are given simultaneously.

(c) Immunological memory

The effect of OTC on the development of immunological memory was also investigated using sheep erythrocytes as antigen and counts of antibody forming cells as measure. A first injection of antigen produced a lower response in medicated carp but after the second injection there was no significant difference between medicated and control fish. It is hypothesized that it is interference of OTC with normal monocyte activity which results in the low response to the primary antigenic stimulus; but that with the secondary stimulus the antigen is trapped more efficiently, monocyte activity is less important and the anamnestic response in the medicated fish would need to be considerably lower to reach statistical significance.

An important mode of action of OTC on monocytes is probably inhibition of mitosis. It has been shown *in vitro* that a concentration of 4-6 μg/ml OTC will reduce the initial response of carp leucocytes to mitogens to 50% of unmedicated control cells; however in concentrations of OTC up to 15 μg/ml the effect is a delay rather than an inhibition. Rijkers *et al.* (1980) made differential blood cell counts of their control fish and those given injections of 180 mg/kg OTC every 3 days. They did not find any differences in the granulocyte or platelet counts, but the total number of lymphocytes and monocytes was significantly reduced by the medication.

5.4.2 INHIBITION OF ERYTHROPOIESIS

Rijkers *et al.* (1980) found a reduction in erythrocyte counts as a result of OTC injections. The mean count in the control carp was 135×10^7 and in the treated fish 110×10^7 per ml., a reduction of 18% from what was a fairly high dose of OTC even bearing in mind that it was given only every 3 days.

5.4.3 GROWTH PROMOTION

The growth promoting effect of oral OTC has been recognized for many years. It is used routinely in aquaculture in a number of countries - particularly in the culture of Indian carp species. Dose regimens used are largely empirical.

Rijkers *et al.* (1980) noted that their common carp dosed orally at 50 mg/kg/day had greater weight gains than unmedicated controls; small fish (weight 35-39 g) showed a greater response than larger ones (75-77 g). They tested only the one dose regimen, which was selected as potentially therapeutic, in their investigation of immuno-suppression; it was almost certainly too high for economic growth promotion. They also noted a dose-related inhibition of growth in fish given regular injections of OTC.

5.4.4 EFFECTS OF OVER-DOSAGE

In a search for a suitable anti-Chlamidial drug for lake trout *(Salvelinus namaycush)* the effects of high doses of OTC hydrochloride have been studied.

(a) Immersion

LC_{50} values were calculated to be

840 mg/l at 1 hour

<200 mg/l at 24 hours

The symptoms of toxicity were attributed to the effect of the hydrochloride on the pH of the water.

(b) Injection

55 mg/kg was tolerated satisfactorily when injected into the dorsal sinus once weekly for 4 weeks. This route of administration necessitated the use of the same injection site on each occasion; three times the dose each week caused injection site lesions after the third dose and five times caused lesions after the second. It is not clear whether the effects derived from the OTC, the pH, the excipient (presumably polyvinylpyrrolidone), or simply the volume of inoculum. The last alternative is unlikely because the five times dose was also tested by

intraperitoneal injection where it produced more severe effects than in the dorsal sinus.

(c) In-feed
Food containing 2.2 g/kg OTC hydrochloride was eaten readily by the lake trout. Higher inclusion rates were still usable therapeutically with most of the food being eaten but they caused slower growth rates.

5.5 Oxytetracycline in the environment

5.5.1 CONTAMINATION OF WILDLIFE IN THE WATER COLUMN

The bioavailability of oral OTC is low; so although excreted OTC is in solution, in which it is unstable, a majority of the drug used in aquaculture enters the environment in solids (food and faeces). As the carrier solid descends through the water column a proportion is subject to ingestion by wildlife. Studies at marine fish farms in both Norway and Finland have shown wild fish in the vicinity to contain OTC in their musculature for up to a week after the end of a course of medication. In one case bacteria were cultured from these wild fish and of 139 isolates which could be cultured *in vitro* without OTC only 123 could be cultured on a medium containing OTC.

There is some evidence that where OTC-resistant organisms are found in wildlife, particularly mussels, under a cage the medicated food may be the source of the organisms as opposed to it being the source of the OTC producing resistance in the normal marine flora. Such food-borne resistant bacteria may not be pathogenic and probably do not survive long in the marine environment. However if they do get into wildlife their resistance may be transmissible.

5.5.2 CONTAMINATION OF SEDIMENTS

The major proportion of any OTC given in feed on fish farms ends up in the sediment below the water column, and its depletion has been the subject of a series of studies by a team led by Smith at the University of Galway in Ireland. An important issue to which they have drawn attention is the meaning of the word sediment: it is frequently used in the literature, often without adequate definition. It can mean the mud, sand or other material which would form the sea-bed whether a fish farm were above it or not; and it can mean the low mounds of detritus, mainly food and faeces, produced by and existing only below farm sites.

One report concerns 5-10 and 10-30 cm thick "organic sediments under fish pens" which were studied after therapeutic use of OTC for 10 days. Clearly these sediments did not develop exclusively during the medication but OTC concentrations ranging up to 16 ppm were found in them. Assay was by HPLC and the farms were in brackishwater at 0.5% salinity so the figure will exaggerate antimicrobial activity. Nevertheless a biologically significant concentration of OTC probably does get into the organic deposits from farms where it is persistent (see Section 2.5.1(d)).

5.5.3 INACTIVATION IN SEDIMENTS

(a) In freshwater

OTC sinks through the water column in organic detritus sediments but after arriving at the bottom it does diffuse into muds. In one study (in seawater) it was found to increase in sediments at depths greater than 2 cm for 19 days after the end of a course of therapy; thereafter it decreased. Where OTC diffuses into mud it probably has very little biological activity. Smith and Samuelsen (1996) have shown that it binds to humic acid, peat and clays forming antimicrobially inactive complexes. Vaughan and Smith (1996) have shown that although a river bed may contain divalent cations at a high enough concentration to be significantly inhibitory to OTC *in vitro*, the cations probably cannot interact with OTC. Nevertheless the organic sediment may reduce the activity of OTC 100-fold. This may be compared with an estimate that the cations in seawater reduce its activity to 7% of that in freshwater.

It must be assumed that outwashing of OTC from a river bed into freshwater would restore its antimicrobial activity.

(b) In seawater

Smith (1996) has made an analysis of a number of determinations of OTC concentrations and half-lives in marine sediments. He showed that the mean weight of OTC accounted for was as little as 4.3% of that used; theoretically it should have been nearly 100%! Possible causes of the discrepancy were:

1. Food eaten by other animals - bottom-dwellers including invertebrates to perhaps a greater extent than pelagic fish;
2. Leaching into the water column;
3. Inhibition by divalent cations and organic sediment.

(c) Minimum inhibitory concentrations

The hazards of antibacterial agents in the environment are the killing of beneficial bacteria, especially in this instance bacteria which oxidize sulphide ions to sulphate ions, and the induction of resistance, particularly transmissible resistance. The profound inactivation of OTC by both sediments and divalent cations therefore raises the question of what constitutes resistance. Pursell *et al.* (1996) have discussed this and suggested that for *in vitro* studies of marine bacteria breakpoints between sensitivity and resistance as high as 100-200 ppm may be appropriate.

5.5.4 DEPURATION OF SEDIMENTS

The depuration of OTC from sediments is believed to be almost exclusively by dissolution into the water column. Thus its half-life in a sediment depends on the physical and chemical characteristics of the sediment rather than any inherent property of OTC. It has been found that covering a sediment containing OTC by further, uncontaminated, sediment can halve the rate of depuration. Depuration will also be accelerated by perturbation of the sediment by invertebrates, and this in turn will depend on the oxygen tension in the sediment. However it has been calculated that after a normal course of treatment for a marine cage of Atlantic salmon, if all the OTC lost from the sediment were to remain in the bottom 1 cm of the water column then at a worst case the concentration in that water would be only 0.016 µg/ml. On the criteria of Pursell *et al.* (1996) this is negligible.

5.6 Other tetracyclines

5.6.1 TETRACYCLINE

Tetracycline has a higher solubility in water at neutral pH than other antibiotics in the group and is therefore of particular value for formulation as an injection. It is available as a premix for in-feed administration to mammals but this has no particular advantages over OTC for fish. In channel catfish its oral bioavailability is only 1-2% at 27°C; furthermore in-feed administration lowers the rate of absorption - T_{max} (plasma) is 4 hours compared with only 2 hours after gavage.

In channel catfish tetracycline concentrates in the liver and biliary system. Muscle contains the lowest concentration of any tissue but because of its total mass it contains a similar weight of drug to the liver - in both cases about 33% of the dose at 72 hours. The terminal half-life of tritium-labelled tetracycline in plasma has been estimated to be 16.5 hours which is significantly less than OTC.

Tetracycline in addition to OTC was tested for tolerance by lake trout in the study mentioned in Section 5.4.4. The LC_{50} values by immersion were generally similar; tetracycline was considered unsuitable for use as an injection because of the low pH of the aqueous solution, but in feed it was readily taken at all concentrations tested and the weight gains of the fish were as good as those of untreated controls.

5.6.2 IMMUNO-SUPPRESSION

In separate studies tetracycline and doxycycline have both been shown to be immuno-suppressive in *in vitro* measures of cellular immune responses. Different measurement techniques and fish species were used so the results are not comparable but in both cases comparisons were made with OTC. Tetracycline appeared to have a similar activity to OTC. Doxycycline was significantly more immuno-suppressive than OTC and this was attributed to its greater solubility in lipids.

Further reading

Björklund, H. and Bylund, G. (1990) Temperature-related absorption and excretion of oxytetracycline in rainbow trout (*Salmo gairdneri* R.). *Aquaculture* **84**, 363-372.

Elema, M.O., Hoff, K.A. and Kristensen, H.G. (1996) Bioavailability of oxytetracycline from medicated feed administered to Atlantic salmon (*Salmo salar* L.) in seawater. *Aquaculture* **143**, 7-14.

Grondel, J.L., Nouws, J.F.M., Schutte, A.R. and Driessens, F. (1989) Comparative pharmacokinetics of oxytetracycline in rainbow trout (*Salmo gairdneri*) and African catfish (*Clarias gariepinus*). *Journal of Veterinary Pharmacology and Therapeutics* **12**, 157-162.

Jacobsen, M.D. (1989) Withdrawal times of freshwater rainbow trout, *Salmo gairdneri* Richardson, after treatment with oxolinic acid, oxytetracycline and trimethoprim. *Journal of Fish Diseases* **12**, 29-36.

Lunestad, B.T. and Goksøyr, J. (1990) Reduction in the antibacterial effect of oxytetracycline in sea water by complex formation with magnesium and calcium. *Diseases of Aquatic Organisms* **9**, 67-72.

Namdari, R., Abedini, S. and Law, F.C.P. (1996) Tissue distribution and elimination of oxytetracycline in seawater chinook and coho salmon following medicated feed treatment. *Aquaculture* **144**, 27-38.

Pursell, L., Dineen, T., Kerry, J., Vaughan, S. and Smith, P. (1996) The biological significance of breakpoint concentrations of oxytetracycline in media for the examination of marine sediment microflora. *Aquaculture* **145**, 21-30.

Rijkers, G.T., Teunissen, A.G., van Oosterom, R. and van Muiswinkel, W.B. (1980) The immune system of cyprinid fish. The immunosuppressive effect of the antibiotic oxytetracycline in carp (*Cyprinus carpio* L.). *Aquaculture* **19**, 177-189.

Rogstad, A., Ellingsen, O.F. and Syvertsen, C. (1993) Pharmacokinetics and bioavailability of flumequine and oxolinic acid after various routes of administration to Atlantic salmon in seawater. *Aquaculture* **110**, 207-220

Smith, P. (1996) Is sediment deposition the dominant fate of oxytetracycline used in marine salmonid farms: a review of the available evidence. *Aquaculture* **146**, 157-169.

Smith, P. and Samuelsen, O.B. (1996) Estimates of the significance of outwashing of oxytetracycline from sediments under Atlantic salmon sea-cages. *Aquaculture* **144**, 17-26.

Strasdine, G.A. and McBride, J.R. (1979) Serum antibiotic levels in adult sockeye salmon as a function of route of administration. *Journal of Fish Biology* **15**, 135-140.

Vaughan, S. and Smith, P. (1996) Estimation of the influence of a river sediment on the biological activity of oxytetracycline hydrochloride. *Aquaculture* **141**, 67-76

6. PENICILLINS

6.1 Group characteristics

The penicillins are part of the larger antibiotic group, the beta-lactams, which includes the cephalosporins. However none of the latter are used in aquaculture and for practical purposes the penicillins form a distinct group. They are weak acids and are chemically rather unstable, being decomposed by heat, light, oxidizing and reducing agents. They are rendered microbiologically inactive by heavy metals. They are sensitive to alkaline solutions although their crystalline sodium and potassium salts are reasonably stable in cool, dark conditions. Most penicillins are sensitive to acids but a few are stable and therefore active by mouth.

Penicillins are sensitive to hydrolysis by bacterial beta-lactamase enzymes but some can be potentiated by beta-lactamase inhibitors, notably clavulanic acid.

Benzylpenicillin (Penicillin G) is a natural antibiotic produced by the fungus, *Penicillium notatum*; it has a narrow spectrum of action, mainly against Gram-positive bacteria, and hence is of little use in aquaculture. Other penicillins are produced by chemical treatment of benzylpenicillin and hence are known as semi-synthetic. Two of them, ampicillin and amoxycillin, have similar spectra of activity, which are broader than that of benzylpenicillin, and they are widely used in aquaculture. They do not form complexes with divalent cations and therefore their cost-benefit comparison with oxytetracycline is better in seawater than in freshwater.

Penicillins are sparingly soluble in water but their sodium and potassium salts are soluble, as are the trihydrate forms of semi-synthetic penicillins. The latter are the normally available forms.

6.2 Uses

Of the two semi-synthetic penicillins mentioned above the one usually used in Europe and North America is amoxycillin. This is not because it has any inherent superiority over ampicillin but because it was more recently patented, and the exclusive rights to its manufacture and supply have justified the investment in obtaining a market authorization. Conversely in countries where little or no investment is necessary to obtain market authorization, the cheaper ampicillin is usually used.

Semi-synthetic penicillins are used in most species of farmed fish. The major indications are:

Furunculosis	- in salmonids
Pasteurellosis (pseudotuberculosis)	- in sea-bass, sea-bream, yellowtail and marine perches

Edwardsiellosis - in catfish, eels, tilapia and sea-bream
Streptococcosis - in Japanese eels, yellowtail and occasionally rainbow trout.

The drugs have a poor activity against *Vibrio* spp. and *Yersinia ruckeri*. From their activity against *Aeromonas salmonicida*, semi-synthetic penicillins might be expected to be active against other *Aeromonas* species; but amoxicillin is poor for *A. hydrophila* and in one study of 42 isolates of *A. sobria* in ornamental fish all were found to be insensitive to ampicillin.

6.3 Dose rates

6.3.1 IN-FEED ADMINISTRATION

While the literature contains numerous case reports of successful use of semi-synthetic penicillins there are few, if any, reports of satisfactory dose titrations in fish. Amoxycillin dose rates in the range 40-80 mg/kg/day have been recommended. For furunculosis the higher end of the range for 10 days is usually necessary if relapses are to be avoided. This is especially the case if twice daily feeding of half the daily dose is practised, because penicillins are rapidly excreted and low dose rates will not produce therapeutic blood levels in salmonids.

For *Pasteurella piscicida* and streptococcal infections in yellowtail a lower dose regimen of 40 mg/kg/day for 5 days has been found satisfactory; ampicillin is usually used.

There are problems with the mixing of amoxycillin into feed. The trihydrate is hygroscopic and liable to cake, so it does not form a homogeneous mixture when used under farm conditions. It should not be mixed into material to be pelleted at a mill because the pelleting temperature will decompose the drug; and if it is surface-coated onto pellets any length of time before use as much as 20% may be lost through photo-decomposition.

6.3.2 INJECTION

The recommended dose rate is 10 mg/kg/day. This is rarely used in food species however because of the labour costs involved. The route may be used in suitably large and valuable ornamental fish or brood fish of food species. For these it can be used as an initial dose to be followed by in-feed medication or, if the fish are not feeding, as a course of injections.

Brown and Grant (1992) injected adult male Atlantic salmon of mean bodyweight 3.6 kg with 45 mg amoxycillin trihydrate. They used the deep intramuscular route and monitored serum concentrations. T_{max} was about 24 hours with mean C_{max} 3.3 μg/ml, but wide variations between individual fish were noted. The authors concluded that using the deep intramuscular route, "Amoxycillin can produce an effective serum level which can be maintained for a significant period of time."

6.3.3 IMMERSION

The bacteria in aquarium filters are very sensitive to semi-synthetic penicillins so where the drugs are used it should always be by dipping (see Section 1.1.2) with the solution being discarded afterwards. However very little drug is absorbed through the gills and this route of administration is not recommended for therapy of systemic infections.

6.4 Pharmacokinetics

6.4.1 AMPICILLIN IN YELLOWTAIL

Table 6.1 shows the concentrations in yellowtail following a course of ampicillin at 40 mg/kg given once daily for 5 days. It illustrates the rapid absorption and elimination of the drug. The persistence in the liver relative to other tissues results from its high C_{max} and early T_{max}, the latter probably being due to first pass accumulation.

Table 6.1. Pharmacokinetics of ampicillin in yellowtail

Time after last dose (hours)	Concentrations (ppm)			
	Serum	Liver	Kidney	Muscle
s.b.	0	0.015	0.013	0
1	0.36	0.98	0.44	0.20
3	0.24	0.62	0.75	0.16
6	0.21	0.55	0.64	0.09
9	0.04	0.34	0.24	0.015
24	0	0.01	0.01	0
48	0	0.01	0	0
72	0	0	0	0

s.b. = shortly before the final dose

6.4.2 WITHDRAWAL PERIODS

Withdrawal periods obviously apply only to food species and these are always medicated through the feed. In EU the MRL for both ampicillin and amoxycillin in muscle is 50 ppm; in some member states withdrawal periods are quoted in degree-days although the temperature range, if any, for which this concept is valid has not been evaluated. In any case the withdrawal period should theoretically depend on C_{max}; for a rapidly excreted drug like amoxycillin C_{max} will depend on the rate of absorption, and this in turn will depend on the formulation, particularly particle size. In UK different brands of amoxycillin have authorized withdrawal periods (for salmonids) varying between 30 and 150 degree-days, all, apparently, for dose regimens of 80 mg/kg/day for 10 days. A short withdrawal period is an obvious advantage for fish about to be harvested, but for any others it will lead the wary prescriber to wonder to what extent and in what tissues therapeutic concentrations are maintained between doses.

Further reading

Brown, A.G. and Grant, A.N. (1992) Use of ampicillin by injection in Atlantic salmon broodstock. *Veterinary Record* **131**, 237

7. MACROLIDES

7.1 Group characteristics

7.1.1 CHEMICAL AND PHYSICAL PROPERTIES

Macrolides are antibiotics having a large ring in the molecular structure. The ring may be 12-, 14- or 16-membered but no 12-membered ring macrolides are used medicinally. Only three are used to any extent in aquaculture - erythromycin which has a 14-membered ring, and spiramycin and josamycin both of which have 16-membered rings. Nearly all macrolides are mixtures of a few (in the case of erythromycin, four) closely related compounds.

Macrolides form colourless crystals which are poorly soluble in water. They are bases and although they are soluble in both acids and alkalis they are chemically unstable at high and, particularly, at low pH. Erythromycin is decomposed very rapidly at pH 4 and is stable only in the range pH 6-8. In consequence it is frequently used as an ester, for example thiocyanate or ethylsuccinate. Microbiologically macrolides are more active in slightly alkaline conditions, and erythromycin is most active at close to the physiological pH of salmonids.

7.1.2 USES

Macrolides are medium-spectrum antibiotics active mainly against Gram-positive bacteria, *Chlamydia* and *Rickettsia*. Because a majority of fish bacterial pathogens are Gram-negative the indications for macrolides are limited and specific; the important ones are:

- Bacterial kidney disease (BKD) caused by *Renibacterium salmoninarum*
- Streptococcosis, especially in yellowtail
- *Chlamydia* spp. infections
- *Piscirickettsia* spp. infections

7.2 Pharmacokinetics of erythromycin

7.2.1 CHINOOK SALMON

(a) Palatability

Erythromycin is rather unpalatable, at least to salmonids, and this may be a problem especially at low temperatures where the feeding rate may be low and a high concentration of drug is consequently required. In a trial where chinook salmon were offered feed pellets containing concentrations of erythromycin varying from nil to 12% it was found that the fish first contacted the unmedicated pellets significantly more often than medicated ones and that there was no significant difference between the concentrations. However after the medicated pellets had been taken they were rejected at a rate dependent on concentration.

This trial used abnormally high concentrations of erythromycin. In chinook salmon feeding at 1.5% of bodyweight per day and given a diet containing 1.3% erythromycin (equivalent to 200 mg/kg/day or double the normally recommended rate) it was found that feed intake fell by not more than 20%.

It has been suggested that medicated feeds should be prepared using the highest concentration of premix available to minimize the proportion of carrier in the final mix. This is not to suggest that the carrier (usually maize flour) is positively unpalatable but it is unfamiliar to the fish.

There appear to have been no specific studies of the oral bioavailability of erythromycin.

(b) Absorption and distribution

Moffitt (1988) made a comparative study of the distribution of erythromycin following 21-day courses at 50 and at 100 mg/kg/day using a commercially available 11% thiocyanate premix. At 50 mg/kg/day the concentration in both plasma and tissues increased for the first few days and then reached a plateau. At 100 mg/kg/day a similar effect was seen in plasma but not in tissues. The plateau level in plasma was twice as high, that is, it was in proportion to the dose rate; in the tissues doubling the dose rate increased the concentration 4- to 10-fold.

At the lower dose rate the concentrations after 3 days were higher in the spleen and kidney than in liver, plasma or muscle. At 10 days the concentrations were all equal except that the kidney contained a higher concentration than the muscles. At the end of dosing, the order of concentrations was

kidney>spleen>liver>plasma>red muscle>white muscle.

At the higher dose rate the concentrations at 3 days were all the same. At 10 days the kidney, liver and spleen had higher concentrations than the plasma or muscles; and at 21 days the kidney still had the highest level. In the kidney, a target tissue for control of BKD in susceptible classes of fish, the concentrations at the end of medication were

using 50 mg/kg/day, 25 ppm
using 100 mg/kg/day, 107 ppm

(c) Elimination

In the same study erythromycin was undetectable in muscle or plasma 10 days after the end of dosing. It was nevertheless detectable in the other three tissues studied until at least 19 days after the end of dosing, always being highest in the kidney. This elimination profile makes it an ideal drug for the control of BKD.

Although erythromycin persists in yolk it is eliminated from juvenile (feeding) fish faster than from adults; an injection at the very high rate of 75 mg/kg was undetectable after only 7 days.

7.2.2 OTHER SPECIES

(a) Coho salmon

Brood fish of some salmonid species, particularly anadromous ones, do not feed, and medication of any sort has to be by injection. It has been shown that erythromycin injected into female coho salmon 28 days prior to spawning penetrates into the yolk of the eggs. A dose of 20 mg/kg injected into the dorsal sinus of the female will prevent the vertical transmission of BKD and the drug persists into the alevin stage at a concentration bactericidal to *R. salmoninarum in vitro*.

If injected only two weeks before spawning erythromycin is found in many tissues and organs at spawning, and is retained not only in the kidneys but also in the eggs for longer than elsewhere.

(b) Yellowtail

Absorption is very rapid: at the normal dose rate of 50 mg/kg/day T_{max} (plasma) is only 1-3 hours, and by only 1 hour after dosing concentrations in the blood and tissues vary between 10 and 100 times the MIC of streptococci. The dose rate is nevertheless necessary because elimination is also more rapid than in salmonids. The residual concentrations are only just above the MIC at 24 hours, and a 10-day course results in only slightly higher concentrations than a single dose. Elimination half-lives have been determined to be in the following ranges:

blood	5 - 7.15	hours
liver	8.25 - 9.15	hours
kidney	14.15 - 15.15	hours
spleen	14.25 - 16	hours
muscle	7 - 8.25	hours

By 7 days after dosing erythromycin is below the limit of detection of 0.03-0.08 ppm in yellowtail. Table 7.1 shows the results of a study conducted in both caged fish under field conditions and under more controlled conditions in tanks.

Table 7.1. Elimination of erythromycin in yellowtail

Time after last dose (hours)	Concentrations (ppm)									
	Blood		Liver		Kidney		Spleen		Muscle	
	cage	tank	cage	tank	cage	tank	cage	tank	cage	tank
1	2.05	1.31	15.9	7.53	10.3	7.51	13.1	9.70	1.07	1.69
3	2.23	1.88	9.73	6.08	16.9	10.5	20.9	7.55	3.55	2.75
6	1.42		5.88		14.7		12.1		3.15	
9		0.44		2.53		4.06		4.20		1.94
24	0.12	0.23	1.29	1.59	2.17	2.29	2.17	3.46	0.43	0.93
48	N.D.	N.D.	0.27	0.20	0.67	0.39	0.76	0.47	0.05	0.06
72			N.D.	N.D.	0.27	0.17	0.24	0.22	N.D.	N.D.
96					0.17	0.07	0.17	0.09		
120					0.09	0.04	0.07	N.D.		
144					0.02	N.D.	0.03			

N.D. = Not detectable

7.3 Toxicology of erythromycin

7.3.1 INJECTION

A safety study has been conducted in lake trout (*Salvelinus namaycush*) as part of a search for a suitable treatment for *Chlamydia* infection. Doses of 0, 9.9, 20 or 40 mg/kg were injected into the dorso-median sinus weekly for 4 weeks and injections of 40 mg/kg were similarly tested by the intraperitoneal route. 2 of 30 fish died as a result of the 40 mg/kg dorso-median sinus regimen, this being attributed to stress rather than any inherent toxicity of the erythromycin being used. The same dose regimen given intraperitoneally had no effect on either behaviour or feeding.

7.3.2 IMMERSION

Erythromycin phosphate is not toxic to lake trout at up to 800 mg/l for 6 hours. The 96-h LC_{50} is 410 mg/l.

7.3.3 IN FEED

The low palatability of erythromycin to chinook salmon has been noted and it applies equally to lake trout. Nevertheless lake trout given a diet medicated with 2.5% erythromycin showed weight gains of only 10% over a period when unmedicated fish nearly tripled in weight. This is obviously a toxicity rather than a palatability effect.

7.3.4 TOXIC EFFECTS

When erythromycin is given orally to rainbow trout at 110 mg/kg/day there are no changes in behaviour for the first two weeks. After 14-18 days medication partial or complete anorexia develops - fish necropsied at this stage have reduced gut contents. Since the anorexia does not develop until after the end of a normally recommended course erythromycin can be used satisfactorily for the treatment of an outbreak of BKD.

Erythromycin is selectively toxic to the cells of the renal tubules; it has no apparent effect on either the liver or gills although the drug accumulates in these organs to almost the same concentration as in the kidney. Lesions in the first proximal segment of the nephron are detectable within 24 hours. The cells lining the tubules develop a vacuolated cytoplasm with lysosome hypertrophy, the latter causing the displacement of the nucleus to the periphery. Residual bodies develop in the lysosomes after prolonged medication; these are believed to be evidence of repair processes.

The effects are reversible following the cessation of medication; hence the anorexia caused by erythromycin makes the antibiotic sub-lethal.

7.4 Control of bacterial kidney disease

7.4.1 PREVENTION OF VERTICAL TRANSMISSION

Bacterial Kidney Disease (BKD) affects all salmonid species. The reservoir of infection appears to be the brood female carrier and the disease is maintained by vertical transmission (through the egg). Although horizontal transmission has been shown to occur, prophylactic strategies focus medication on brood females, eggs and juveniles. Medication of brood males is not necessary because infected milt does not infect eggs.

Heavily infected brood females can be identified by microbiological examination of their coelomic fluid and this should be a routine procedure at premises where the disease has occurred previously. However the causative bacterium, *Renibacterium salmoninarum* is an intra-cellular parasite and not easy to detect, and such tests produce false negative results in lightly infected fish. It is normal practice to medicate apparently uninfected females and sometimes also the fry.

Brown *et al.* (1990) made a comparative study of five antibiotics for injection into maturing female coho salmon at 7°C, those used being erythromycin phosphate 20 mg/kg, benzylpenicillin 50 mg/kg, oxytetracycline 10 mg/kg, cephradine 25 mg/kg and rifampicin 5 mg/kg. In each case the fish spawned 14-26 days after injection. All five antibiotics were found in the egg yolks; and all were at active concentrations, reducing the prevalence of infection. However only erythromycin and rifampicin persisted long enough to be detectable in the alevins.

Another study compared the use of a course of three injections of either benzylpenicillin, dihydro-streptomycin, oxytetracycline or erythromycin. It was found that the first three antibiotics were teratogenic, causing abnormalities of the mandible and fins of the progeny of treated fish. This could only be avoided by completing the course of injections by 32 days before spawning, but this interval, while ensuring the absence of the drug from the eggs, gave no protection against BKD. Only erythromycin was persistent, and hence protective, but not teratogenic.

7.4.2 CONTROL OF OUTBREAKS

Because *R. salmoninarum* is an intra-cellular parasite it is generally easier to prevent infections than treat them with antibacterial agents. Treatment of established infections requires a drug which will penetrate cells; prevention requires a drug which will persist in the blood for a long time. When given in feed, clindamycin and the macrolides, erythromycin and spiramycin, have been shown to reduce mortality in outbreaks of BKD; cephradine, lincomycin and rifampicin are prophylactic but not therapeutic.

Moffitt (1989) conducted an artificial challenge study on chinook salmon smolts in seawater at 10°C. An in-feed regimen of 100 mg/kg/day for 21 days was used and groups of fish were challenged on the day before treatment started or 1, 11 or 29 days after the start. There was 80% survival in the fish challenged immediately before medication as against 7% in the controls, a relative percent survival of 78%. The treatment was only partially effective if the challenge was delayed until the 11th day, that is, during the medication, and there was no protection if the challenge was after the end of medication.

7.4.3 ERYTHROMYCIN DOSE REGIMENS

(a) Injection

In some countries erythromycin base is available as an injectable solution in polyethylene glycol, ethyl acetate and ethanol. Where this is not available it is usually possible to obtain erythromycin phosphate as a 30% soluble powder. This can be prepared, immediately before use, as a 4% or 5% solution in Water for Injection. (Higher concentrations cannot be produced because the salt has a solubility of only 6%).

In Pacific salmon species such solutions are injected at 10-20 mg/kg into brood females on one to three occasions. Within broad limits the timing of these injections is important: in coho salmon injections 70 or more days before spawing do not result in any erythromycin being deposited in the eggs; it is deposited when the injections are made between 56 and 14 days before spawning. With the injections made at at the correct time the erythromycin goes into the yolk, from which it does not leach to any significant degree (see Section 7.4.2(d)).

In wild Atlantic salmon doses of 11 mg/kg have been used in females first entering traps, with repeat doses at 3 week intervals until spawning.

The dorso- median sinus is the injection site normally used; the intramuscular route may cause site lesions. The evidence from lake trout mentioned above would suggest that the intraperitoneal route is safer but it may be difficult in ripe female fish.

(b) Immersion of fish

Immersion medication of Atlantic salmon has been investigated using a 4-5% solution of erythromycin phosphate in tanks. The technique is not very practical for a limited number of large fish and in any case either a two-step hyperosmotic infiltration procedure (see Section 1.1.3) or the addition of a surfactant was found necessary to achieve adequate blood levels.

(c) Immersion of eggs

Eggs of coho salmon, chinook salmon and steelhead trout (anadromous *O. mykiss*) have all been shown to absorb erythromycin. In the case of coho salmon and steelhead trout eggs, as little as 10 ppm for an hour at only 4°C led to the absorption of a higher concentration than the MIC of *R. salmoninarum*. However once the eggs are transferred to unmedicated water the erythromycin leaches out again very quickly - with a half-life of little more than 20 minutes.

This procedure is of little value in the control of BKD for two reasons. In the first place eggs are normally incubated at low temperatures, where bacterial metabolism will be slow; a drug acting on dividing cells will need a long contact time to be effective. In the second place the bacteria are located primarily in the yolk of infected eggs whereas erythromycin absorbed from the water goes into the perivitelline fluid, that being the only egg fluid equilibrating rapidly with the surrounding water.

(d) In feed

As noted in Section 7.2.2, in-feed medication of salmonid broodfish is rarely possible. For control of BKD outbreaks in growing fish the normal dose regimen is 100 mg/kg/day;

this is usually given for 21 days but sometimes for 14 days followed after an interval by a further 7 days.

Moffitt (1992) has conducted a dose titration of erythromycin in the control of an artificial BKD infection of chinook salmon. This was in freshwater unlike the artificial challenge study mentioned in Section 7.4.2. Erythromycin thiocyanate was used at 50, 100 or 200 mg/kg/day for 7, 14 or 21 days, in each case at 8 or 12°C - a total of 18 treatment groups apart from controls! Each group was challenge inoculated with *R. salmoninarum* intraperitoneally 3 days before medication began.

Table 7.2. Mean weekly percent survival in duplicate groups of 50 fish inoculated with *R. salmoninarum* and administered oral dosages of erythromycin at 50, 100, or 200 mg/kg body weight. Dosages were fed for 7, 14, or 21 consecutive days beginning 3 d after inoculation. Data for controls (fish fed no erythromycin) are shown in the middle panels. After Moffit C., 1992, with permission

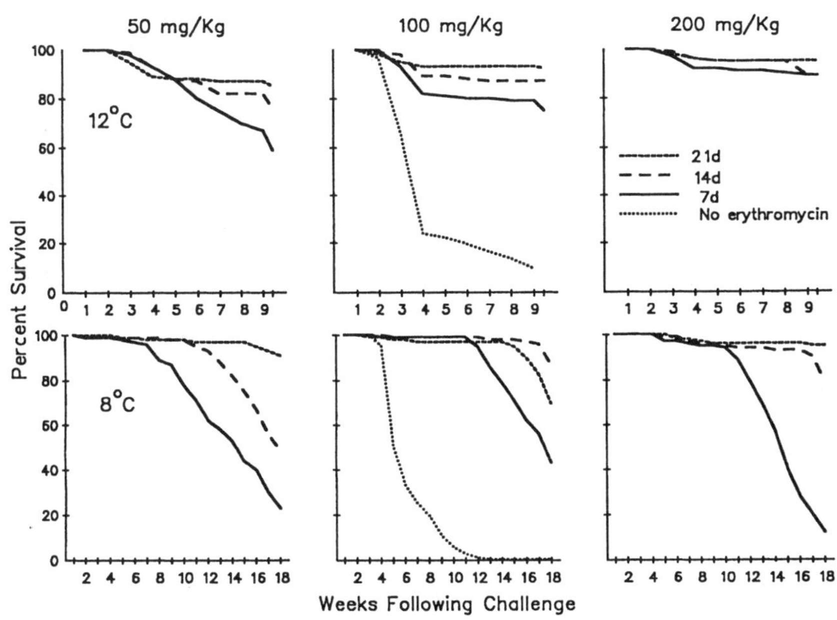

All the unmedicated controls had died by 13 weeks at 8°C, and 90% by 9 weeks at 12°C when the trial at this latter temperature was terminated. Survival was dose-related and duration-related, the best results being from using 200 mg/kg/day for 21 days. However anomalously with the 7 and 14 day regimens at 8°C, 200 mg/kg/day appeared to be less efficacious than 100 mg/kg/day.

Peters and Moffitt (1996) have pointed out that the commercially available erythromycin thiocyanate premix is intended for use in poultry and has shortcomings when used in fish. The coarse carrier does not mix well with fish feeds and can cause oesophageal damage. They tested a premix based on wheat flour in chinook salmon artificially infected with BKD. 50 mg/kg/day was inadequate; 100 mg/kg/day for 28 days was successful. At 200 mg/kg/day there were palatability problems: “Fish usually consumed all of the medicated rations on the first day of therapy, and feed refusal began on the second or third day of treatment.”

7.5 Other macrolides

In Japan both spiramycin and josamycin are authorized for use in yellowtail, sea bream, horse mackerel and tilapia. The indications are generally confined to streptococcosis.

Further reading

Brown, L.L., Albright, L.J. and Evelyn, T.P.T. (1990) Control of vertical transmission of *Renibacterium salmoninarum* by injection of antibiotics into maturing female coho salmon. *Diseases of Aquatic Organisms* **9**, 127-131.

Elliott, D.G., Pascho, R.J. and Bullock, G.L. (1989) Developments in the control of bacterial kidney disease of salmonid fishes. *Diseases of Aquatic Organisms* **6**, 201-215.

Evelyn, T.P.T., Ketcheson, J.E. and Prosperi-Porta, L. (1986) Use of erythromycin as a means of preventing vertical transmission of *Renibacterium salmoninarum. Diseases of Aquatic Organisms* **2**, 7-11.

Lee, E.G-H. and Evelyn, T.P.T. (1994) Prevention of vertical transmission of the bacterial kidney disease agent *Renibacterium salmoninarum* by broodstock injection with erythromycin. *Diseases of Aquatic Organisms* **18**, 1-4.

Moffitt, C.M. and Schreck, J.A. (1988) Accumulation and depletion of orally administered erythromycin thiocyanate in tissues of chinook salmon. *Transactions of the American Fisheries Society* **117**, 394-400.

Moffitt, C.M. and Bjørnn, T.C. (1989) Protection of chinook salmon smolts with oral doses of erythromycin against acute challenges of *Renibacterium salmoninarum. Journal of Aquatic Animal Health* **1**, 227-232.

Moffitt, C.M. (1992) Survival of juvenile chinook salmon challenged with *Renibacterium salmoninarum* and administered oral doses of erythromycin thiocyanate for different durations. *Journal of Aquatic Animal Health* **4(2)**, 119-125.

Peters, K.K. and Moffitt, C.M. (1996) Optimal dosage of erythromycin thiocyanate in a new feed additive to control bacterial kidney disease. *Journal of Aquatic Animal Health* **8(3)**, 229-240.

8. SULFONAMIDES

8.1 Group characteristics

8.1.1 CHEMISTRY

The sulfonamides are a large range of synthetic compounds which are chemically related, all being derivatives of sulfanilamide.

They are amphoteric but most are more soluble in alkalis than in acids; they are usually formulated, especially in injections, as their sodium salts. The solubility of the sodium salt of any one sulfonamide is usually unaffected by the presence of others and hence more concentrated solutions can be achieved by using mixtures of sulfonamides.

8.1.2 THERAPEUTIC GROUPS

The therapeutic uses of the different sulfonamides depend on their pharmacokinetic profiles. These in turn depend to some extent on their solubilities, which for many sulfonamides will depend on the pH and buffering capacity of the water. As a result, the pharmacokinetic profiles of some sulfonamides vary between fish in freshwater and seawater; and there can be profound differences in pharmacokinetics between fish and mammals (and in some cases the profiles in birds are different again). Another related factor is how polar the compound is, because non-polar compounds are more lipid-soluble and cross cell membranes more easily. Although it has limited therapeutic efficacy, sulfanilamide itself is rapidly absorbed; the tissue concentrations are high and are not significantly affected by the pH of the water.

1. A few sulfonamides are sparingly soluble in water and are used in mammalian (*e.g.* sulfathiazole) and avian (*e.g.* sulfaquinoxaline) medicine for intestinal infections. The only place for such comounds in fish medicine might be for topical application to skin wounds, abrasions and ulcers. However some sulfonamides (*e.g.* sulfaguanidine) which are not absorbed by mammals are absorbed to some extent by fish.

2. A majority of sulfonamides are moderately soluble. While they are absorbed from the gut at a moderate rate they persist in blood and tissues for sufficiently long to have significant therapeutic action on systemic infections.

3. A few sulfonamides are very soluble and hence rapidly excreted. Excretion is often before any metabolic detoxification has taken place and so the drug appears in the urine unchanged. Such drugs are useful for urinary tract infections. Among them sulfisoxazole has found an occasional place in fish medicine.

8.1.3 USES

The sulfonamides have a broad spectrum of activity; and since they were the first modern antibacterial drugs to be developed members of the group have been used for virtually all bacterial diseases of fish. However the doses required often leave little margin below toxicity and many bacteria develop resistance fairly easily. In consequence the group has little use now except in synergistic combination with pyrimidine potentiators (see Chapter 9).

One advantage which sulfonamides have over other antibacterial drugs is that they are absorbed through the gills. Administration by immersion is therefore feasible and is of particular value for juvenile fish. This route of administration has a further advantage in marine species in that they are drinking continuously and can absorb water medication through the gut mucosa. However Michel (1986) has pointed out that dose rates are profoundly affected by such variables as water pH and temperature.

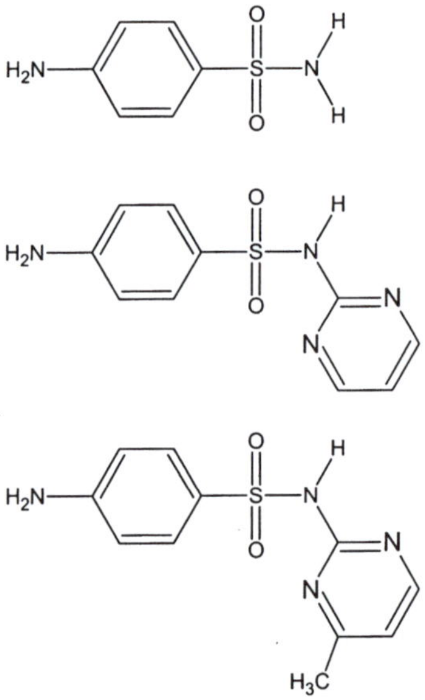

Sulfanilamide

Sulfadiazine

Sulfamerazine

Figure 8.1.

8.2 Pharmacokinetics

8.2.1 SULFAMERAZINE

Sulfamerazine was one of the earliest sulfonamides to be the subject of a market authorization for fish. This does not mean it is necessarily the best sulfonamide, only that a company manufacturing it took the trouble to study its use in fish. Experiments were conducted mainly on freshwater salmonids, notably brown, brook and rainbow trout.

Sulfamerazine has been approved in the USA for use in farmed fish with a dose regimen of 200 mg/kg/day for 14 days.

(a) Brook trout

Brook trout kept at 11-13°C and given sulfamerazine in the feed at 100, 200 or 300 mg/kg/day for 9 days showed a rise in tissue concentrations over the first 4 days, a steady level for the rest of the period of medication and a rapid fall after the end of medication. Only two samples, each a pool of three fish, were assayed at each time point and some wide variations in observations occurred but tissue levels were dose-dependent. It was noted that, for any one dose rate, tissue levels were higher in older fish; this was attributed to the older fish having a lower feeding rate as a percentage of bodyweight and hence requiring a higher concentration of drug in the feed in order to maintain the dose per unit bodyweight.

Yearling brook trout fed sulfamerazine at 20 mg/kg/day for 14 days showed T_{max} in blood, muscle, liver and kidney at 9 days, the relative concentrations being blood>liver> kidney>muscle. Concentrations fell over the next 5 days despite continued medication, the falls in the blood and liver being of the order of 65-70%.

Brook trout acetylate a high proportion of their tissue sulfamerazine, for example more than 60% of that in the liver and 33% of that in blood and kidney. These proportions remain approximately constant as the concentrations fall after withrawal of the medication. Elimination is initially very rapid after the end of medication with about 80% of the drug being lost in the first 5 days at 8°C. It proceeds more slowly over the following 2 weeks, and there is a slight rise in blood content with T_{max} at about 24 days which has been attributed to release of the drug from visceral fat. Terminal elimination is very slow. Elimination rates appear to be independent of temperature in the range 8-13°C.

Since for therapeutic efficacy a high tissue concentration is required as early as possible, and since sulfamerazine is apparently accumulated over the first few days, the possibility arises of using a "loading dose" as is commonly done with sulfonamides in mammalian medicine. Fish were given 300 mg/kg/day for 7 days and then one of the three dose rates as above. It was found that in the fish given the lower dose rates the levels rapidly fell to those observed in fish dosed at those rates. Field trials of sulfamerazine for furunculosis have found C_{max} resulting from 200 mg/kg/day to be appropriate so a loading dose of 300 mg/kg/day has been proposed for an initial 4 days, with a maintenance rate of 200 mg/kg/day for the remainder of the (usually 10 day) treatment.

(b) Rainbow trout

The finding that older brook trout had higher tissue concentrations of sulfamerazine was further studied using rainbow trout all of the same age. Fingerlings were medicated at 200 mg/kg/day but in diets fed at 2, 4 or 8% of bodyweight per day. Some medicated food given at 8% per day was uneaten; (this could not have been a palatability problem because this diet would have had the lowest concentration of sulfamerazine). As in the brook trout with different ages so in the rainbow trout at the same age the lower feeding rates produced the higher tissue concentrations of drug. It must be concluded that bioavailability of sulfamerazine is related to the concentration in the diet.

Dosing rainbow trout with sulfamerazine at 200 mg/kg/day for 14 days produces significantly different tissue concentrations from those in brook trout. The concentrations in blood, muscle, liver and kidney all rise steadily throughout the period. C_{max} for blood, liver and kidney are very similar; in one report it was highest in blood and in another in the liver.

While rainbow trout acetylate much smaller proportions of sulfamerazine than brook trout, their elimination profiles are very similar. The main difference is that rainbow trout do not have the C_{max} in plasma at 24 days but show a continuous reduction in sulfamerazine concentration.

(c) Brown trout

The distribution of sulfamerazine in brown trout is very similar to that in brook trout. However, like rainbow trout, brown trout acetylate much smaller proportions of sulfamerazine than brook trout. In brown trout elimination rates are temperature-sensitive.

(d) Tolerance by chinook salmon

In a search for suitable therapy for natural *Pseudomonas* infection sulfamerazine was found to be one of the safer sulfonamides for use in chinook salmon. It produced no mortality at 220 mg/kg/day in either infected or uninfected fish at 8°C. (The latter would be better able to tolerate a xenobiotic compound but would consume more medicated feed). Further investigation showed that over 330 mg/kg/day caused gastric symptoms; the stomachs were swollen with constrictions anterior to the gastric caeca. There was no evidence of crystal formation in the kidneys as occurred with other sulfonamides. At dose rates in excess of 660 mg/kg/day the feed was refused after 24 hours.

8.2.2 SULFADIMIDINE

This compound is called sulfamethazine in some pharmacopoeias; 'Sulphamezathine' was at one time used as a trade name for it.

(a) Chinook salmon

Blood levels of sulfadimidine (SDD) in juvenile chinook salmon have been studied after courses of 110 or 220 mg/kg/day for 6 days given in a pelleted diet. The 110 mg/kg/day regimen developed C_{max} 150 μg/ml whole blood at T_{max} 4 days; the 220 mg/kg/day results were C_{max} 210 μg/ml at T_{max} 2 days. After the end of medication the drug was eliminated from the blood with a half-life between 1 and 2 days.

In the safety trial described in Section 8.2.1(d) for sulfamerazine, SDD over-dosage produced the same gastric symptoms but three times the mortality. When 110 mg/kg/day was given for 30 days the food was taken readily and safely over the first 12 days. However the rate of loss of fish rose over the last 2 weeks of the trial and there was a 70% mortality over the period. The outstanding observed effect in moribund fish was fungal infection of the gills. The fish used would normally have migrated to the sea in April but were given the medicated diet in freshwater during July; it was hypothesized that the sulfadimidine was adversely affecting gill tissues which were already under osmotic stress.

(b) Rainbow trout

Van Ginneken *et al.* (1991) have studied the pharmacokinetics of SDD in rainbow trout at 10 and 20°C. For each temperature fish were given a single intravenous injection of sodium sulfadimidine at 100 mg/kg and serial blood samples were analysed for SDD and its known metabolites, N_4-acetyl-SDD, 5-hydroxy-SDD and 6-hydroxy-SDD (see Fig. 8.2).

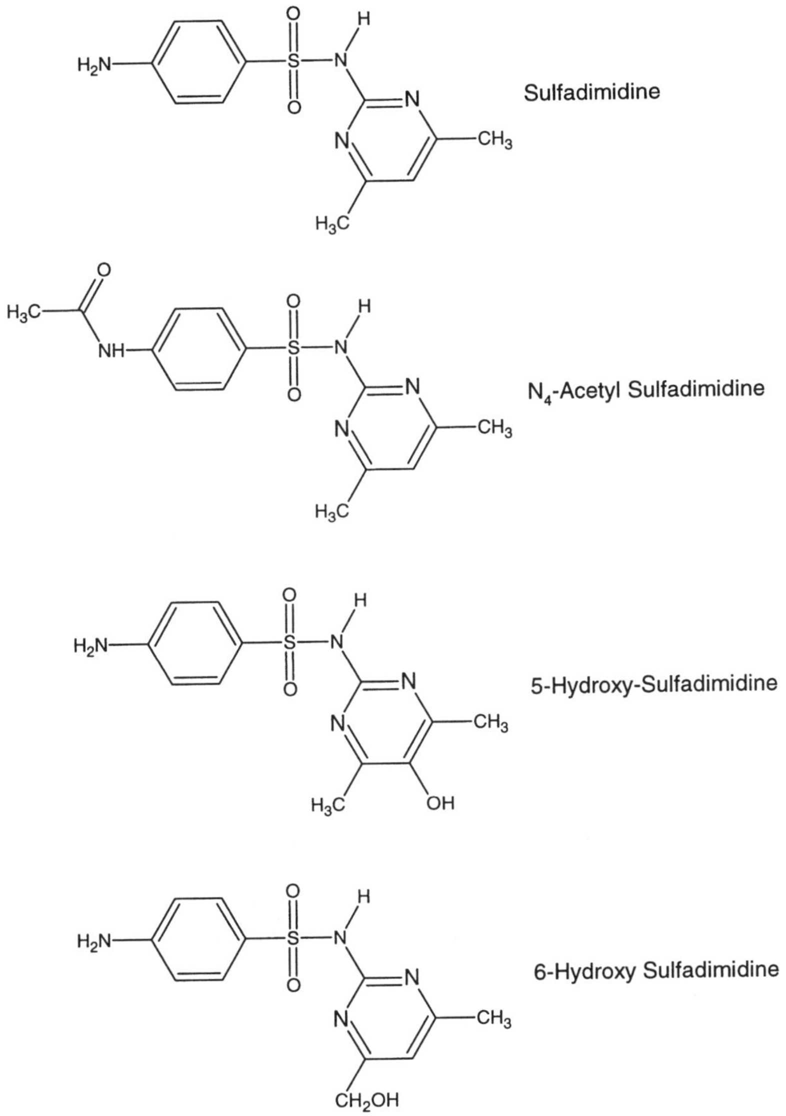

Sulfadimidine

N_4-Acetyl Sulfadimidine

5-Hydroxy-Sulfadimidine

6-Hydroxy Sulfadimidine

Figure 8.2.

Distribution of SDD into the tissues was complete within 1 hour of the injection. The volume of distribution (Vd) was significantly lower at 20 than at 10°C. The major metabolite was N_4-acetyl-SDD. Table 8.1 shows the important calculated pharmacokinetic parameters. C_{max} for N_4-acetyl-SDD at the two temperatures differed at the 5% level of significance.

Table 8.1. Plasma pharmacokinetics of sulfadimidine in rainbow trout

		10 °C	20 °C
Vd	(l / kg)	1.20	0.83
AUC	(μg.h/ml)	2508	2543
$t^{1/2}$ β	(hours)	20.6	14.7
Conjugation of N4-SDD		4.05%	3.17%
C_{max} (N4-SDD)	(μg/ml)	1.47	1.63
T_{max} (N4-SDD)	(hours)	24	14

Figure 8.3 shows the elimination of sulfadimidine (which the authors abbreviate to SDM) and N_4-acetyl-SDD at each of the two temperatures. Concentration is shown on a logarithmic scale; and assuming that after 2-3 hours the graph should be linear, elimination is faster at the higher temperature. This observation taken together with the similarity of the AUC determinations at the two temperatures implies that SDD is much more rapidly absorbed at higher temperatures. These findings confirm a much earlier study at 7 and 14°C. Here the terminal half-lives were determined to be 52 and 25 hours respectively, which means that the degree-day concept (see Section 2.4.4) is valid within this temperature range.

SDD should be expected to have higher antibacterial efficacy at lower temperatures but this would be offset by a longer withdrawal period.

(c) Common carp

Van Ginneken *et al.* (1991) also made the same study in carp. At 20°C the distribution was very rapid (as in rainbow trout), but at 10°C distribution took about 14 hours. Again similarly to the position in rainbow trout, the volume of distribution was lower at 20 than at 10°C; the difference was proportionately less in carp but it was nevertheless highly significant. N_4-acetyl-SDD was again the main metabolite and again C_{max} for it did not differ significantly between the two temperatures. More SDD was conjugated than in rainbow trout and T_{max} for the conjugate was 67% later in the carp at each temperature; whereas in rainbow trout a lower proportion was conjugated at the higher temperature, in carp the proportion was higher at the higher temperature.

The difference in elimination rates between the two temperatures was more marked in carp than in rainbow trout, and at both temperatures the elimination half-life was longer in the carp. The very slow elimination of N_4-SDD at 10°C may make the drug unsuitable for therapeutic use in carp at this temperature.

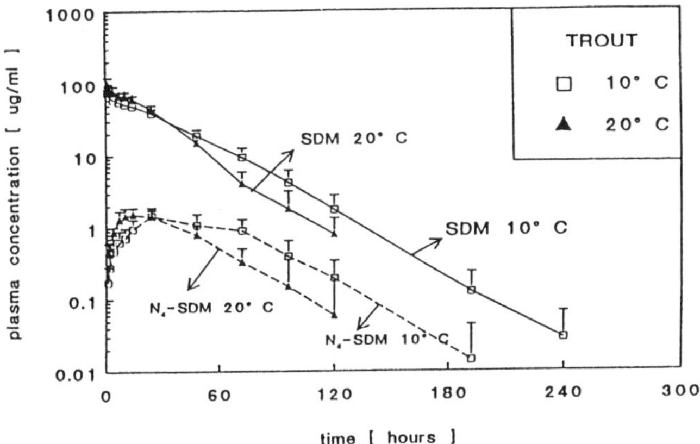

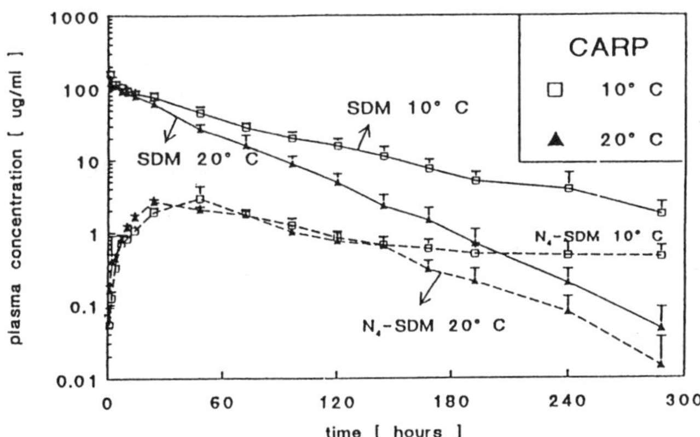

Figure 8.3. Top. Mean ± (S.E.) plasma concentration-time profiles of sulphadimidine (SDM) and its main metabolite N_4-acetylsulphadimidine (N_4-SDM) at two different temperature levels (10°C and 20°C) in trout (*Salma gairdneri Richardson*) following intravenous administration of SDM at a dosage of 100 mg/kg.
Figure 8.3. Bottom. Mean ± (S.E.) plasma concentration-time profiles of sulphadimidine (SDM) and its main metabolite N_4-acetylsulphadimidine (N_4-SDM) at two different temperature levels (10°C and 20°C) in carp (*Cyprinus carpio* L.) following intravenous administration of SDM at a dosage of 100 mg/kg.
After van Ginneken *et al*, 1991, with permission.

Figure 8.3.

Table 8.2 Plasma pharmacokinetics of sulfadimidine in common carp

		10 °C	20 °C
Vd	(l/kg)	1.15	0.90
AUC	(μg.h/ml)	6379	4140
$t^{1/2}$ β	(hours)	50.3	25.6
Conjugation of N4-SDD		4.18%	5.62%
C_{max} (N4-SDD)	(μg/ml)	2.35	2.69
T_{max} (N4-SDD)	(hours)	40	24

In carp dosed by the intraperitoneal route at 550 mg/kg C_{max} of 124 µg/ml was found at T_{max} 2 hours. T_{max} for N_4-SDD was less than 20 hours after the injection and from 20 hours the plasma concentrations of sulfadimidine and N_4-SDD were parallel. The terminal half-life of sulfadimidine was estimated to be 17.5 hours as against 25.6 hours following intravenous administration. After 48 hours only 64% of the dose was detected in the water, and of this only 2% was N_4-acetyl-SDD, but some may have hydrolysed.

(d) Immersion

The dependence of SDD absorption on the pH of the medicated water has been studied in rainbow trout and it may be assumed that the findings are applicable to other salmonids if not to all other fish species. Groups of rainbow trout were immersed for 96 hours in SDD at 650 mg/l in freshwater at 11°C with pH 4, 6, 8 or 10. The pKa of SDD is 7.4 and there was no significant difference in absorption between pH 6 and 8, presumably because there would have been little change in the ionization of SDD over this range. Absorption was poor at pH 10 whereas it was very rapid at pH 4 but the fish died. In a further experiment fish were immersed for 24 hours in SDD in either freshwater at pH 6.6 or seawater at pH 8. Although these pHs are in the range for which there was no significant difference in absorption from freshwater there was significanly higher absorption from the seawater. This was attributed to a contribution from intestinal absorption of medicated seawater which the fish had drunk.

Bergsjø (1974) investigated the possibility of medicating a marine species, cod, by immersion. Groups of cod weighing 80-100 g were kept for 4 days in seawater medicated with sodium sulfadimidine equivalent to 1000 mg/l SDD. Replicates were tested at pH 6.0 and 10.0. Samples were taken at intervals during the treatment and blood, liver, kidney and muscle were assayed. At pH 6 the concentration of sulfadimidine in all four tissues rose throughout the exposure period, exceeding the nominally therapeutic level of 100 µg/mg in all of them by 6-12 hours. At pH 10 all concentrations were lower and a C_{max} was reached at 48 hours in blood, liver and kidney; in all tissues the 100 µg/mg threshold was passed within 24 hours.

8.2.3 SULFADIMETHOXINE

(a) Rainbow trout

Sulfadimethoxine has been studied in rainbow trout by at least three separate research groups. They have used different temperatures and looked at different aspects of pharmacokinetics and so their results cannot be practically synthesized.

One group found that after oral administration of 42 mg/kg by gavage, the sodium salt showed considerably greater bioavailability than the free drug (63% as against 34%). This might be considered an anomalous result but at three times the dose the bioavailability of the salt was slightly reduced, possibly due to saturation of the water in the gut lumen.

Kleinow *et al.* (1992) compared a 'hot' study (*i.e* using radio-active drug; and measuring radioactivity, which may be drug or metabolite) with a cold one, where the drug was injected intravenously in both cases. The drug disappearance from blood followed a bi-exponential rate ($t^{1/2}\alpha = 0.4$ h; $t^{1/2}\beta = 16$ h); and it was essentially similar in the two studies implying that there was little if any metabolism in the blood. The volume of distribution (Vd) was 420 ml/kg; this indicates only limited distribution into the extra-vascular space, probably into extra-cellular but not intra-cellular fluids. Despite this the drug was found in virtually all tissues, the concentrations being:

bile>intestine>liver>blood>skin>kidney>spleen>gills>muscle>fat

Little drug was excreted through the gills or urine and the high concentration in the bile would result in recirculation, so slow elimination was to be expected.

Another group found that after a single oral dose of 200 mg/kg by gavage sulfadimethoxine was present in all tissues within 0.5 hour, although T_{max} for tissues varied between 1 and 5 days. Concentrations in five tissues studied were in the same order as above. They found the N_4-metabolite in liver at 1 hour and in all five tissues at 9 hours. As with the parent drug, the highest concentration of the metabolite was in the bile. The metabolite was shown to be bacteriologically inert and is believed to cause crystalluria.

Kleinow *et al.* (1992) found a terminal half-life of 17 hours after a single oral dose at 42 mg/kg but 36 hours after 5 daily doses. This implies a change in metabolism. It was attributed to acetylation of the drug because although this process accelerates elimination in man it is known to slow it down in goldfish.

Residues were undetectable in serum at 10 days but persisted in tissues until 4 weeks. The group commented that the Japanese withdrawal period is 30 days; and at 15°C, the temperature at which their work was done, this appears to be inadequate.

Table 8.3 Biological half-life and elimination time of sulphamonomethoxine, sulphadimethoxine and their N^4-acetyl metabolites. After Uno K. *et al*, 1993, with permission

		Serum	Muscle	Liver	Kidney	Bile
SMM[1]	$T_{1/2}$[a]	32.6	46.7	38.5	52.4	61.6
	Et[b]	14.4	17.9	17.0	20.0	26.4
AcSMM[2]	$T_{1/2}$	129	123	146	104	201
	Et	43.4	34.8	67.1	39.9	132
SDM[3]	$T_{1/2}$	24.5	41.0	67.9	52.2	46.2
	Et	10.7	13.6	23.8	18.6	27.1
AcSDM[4]	$T_{1/2}$	134	72.3	41.3	50.2	59.4
	Et	35.9	18.7	20.2	17.8	29.8

[1]Sulphamonomethoxine.
[2]N^4-Acetylsulphamonomethoxine.
[3]Sulphadimethoxine.
[4]N^4-Acetylsulphadimethoxine.
[a]Biological half-life (h).
[b]Elimination time (days).

(b) Channel catfish

The bioavailability of sulfadimethoxine in channel catfish has been estimated to be about 34% at 20°C, and this appears to apply also to the sodium salt, in contrast to the rainbow trout where the salt has a bioavailability of 63%. Despite this low bioavailability the fraction of the drug which is absorbed from the gut is absorbed rapidly. It is also distributed from plasma to the tissues rapidly, with determinations of the distribution half-life after intravenous dosage of 0.06 and 0.09 hours (= 3½ and 5½ minutes)!

However the volume of distribution is only about 400 ml/kg, a comparable figure to that for rainbow trout.

The major proportion of the absorbed sulfadimethoxine appears to go initially to the musculature with a T_{max} for that tissue of about 3 hours. It also declines rapidly from muscle with a half-life of 12-13 hours and by 24 hours the major fraction of the drug is in the bile (whence it will be recirculated). Muscle residues have been reported to be below the FDA MRL by 8 days, but this was after only a single dose at only 40 mg/kg.

With the exception of muscle and skin (20 hours) the half-lives of sulfadimethoxine in tissues appear to be over 24 hours. It should therefore be useful for visceral infections in channel catfish; however one report has suggested that as much as 90% may be in the form of the microbiologically inert N-acetyl conjugate.

8.2.4 SULFAMONOMETHOXINE

This drug is little used except in Japan and is not listed in most pharmacopoeias.

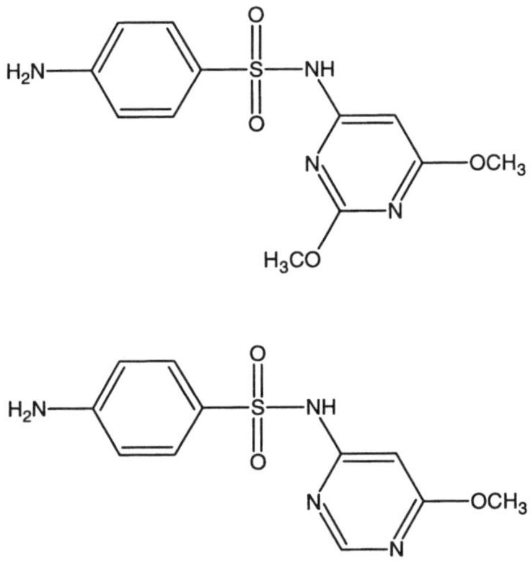

Sulfadimethoxine

Sulfamonomethoxine

Figure 8.4.

(a) Rainbow trout

The dose rate normally used in rainbow trout in Japan is 300 mg/kg/day. The absorption and elimination is similar to, but a little slower than, that for sulfadimethoxine. An important feature is the formation of the N-acetyl conjugate, which persists in all tissues for over 32 days following a normal dose regimen.

(b) Japanese eels (Anguilla japonica)

An investigation has been made of the medication of Japanese eels by immersion. Concentrations ranging from 100 to 600 ppm for 3 hours and temperatures ranging from

10 to 35°C were used. Absorption was shown to rise with temperature, and after immersion in 600 ppm for 3 hours at 25°C therapeutic concentrations in blood and tissues were maintained for over 4 days. The addition of 1% sodium chloride to the water increased absorption by factors varying from 11% (in blood after immersion in 100 ppm) to 44% (in liver after immersion in 200 ppm).

Concentrations in spleen, kidney, blood, liver and muscle were assayed. The order of the concentrations varied with the concentration in the water and to a lesser extent with temperature. However the spleen always contained the highest concentration and the muscle the lowest.

Further reading

Bergsjø, T. (1974) The Absorption of sulphadimidine in cod fish. *Acta Veterinaria Scandinavica* **15**, 442-444.

Bergsjø, T. and Bergsjø, T.H. (1978) Absorption from water as an alternative method for the administration of sulphonamides to rainbow trout, *Salmo gairdneri*: the significance of the pKa value of the sulphonamides and the pH and salt content of the water. *Acta Veterinaria Scandinavica* **19**, 102-109.

van Ginneken, V.J.Th., Nouws, J.F.M., Grondel, J.L., Driessens, F. and Degen, M. (1991) Pharmacokinetics of sulfadimidine in carp (*Cyprinus carpio* L.) and rainbow trout (*Salmo gairdneri* Richardson) acclimated at two different temperatures. *The Veterinary Quarterly* **13**, 88-96.

Kleinow, K.M., Beilfuss, W.L., Jarboe, H.H., Droy, B.F. and Lech, J.J. (1992) Pharmacokinetics, bioavailability, distribution and metabolism of sulfadimethoxine in the rainbow trout (*Oncorhynchus mykiss*). *Canadian Journal of Fisheries and Aquatic Science* **49**, 1070-1077.

Michel, C. (1986) Practical value, potential dangers and methods of using antibacterial drugs in fish. *Revue Scientifique et Technique de l'Office International des Epizooties* **5**, 659-675

Michel, C.M.F., Squibb, K.S. and O'Connor, J.M. (1990) Pharmacokinetics of sulfadimethoxine in channel catfish (*Ictalurus punctatus*). *Xenobiotica* **20**, 1299-1309.

Snieszko, S.F. and Friddle, S.B. (1950) Tissue levels of various sulfonamides in trout. *Transactions of the American Fisheries Society* **80**, 240-250.

Snieszko, S.F. and Friddle, S.B. (1951) Further studies on factors determining tissue levels of sulfamerazine in trout. *Transactions of the American Fisheries Society* **81**, 101-110.

Uno, K., Aoki, T. and Ueno, R. (1993) Pharmacokinetics of sulfamonomethoxine and sulfadimethoxine following oral administration to cultured rainbow trout (*Oncorhynchus mykiss*). *Aquaculture* **115**, 209-217.

9. POTENTIATED SULFONAMIDES

9.1 Group characteristics

9.1.1 MODE OF ACTION

(a) Pharmacodynamics

The potentiated sulfonamides are combinations of two antibacterial drugs, a sulfonamide and a pyrimidine potentiator. The combination is synergistic, that is, the antibacterial potency of the combination is greater than the sum of the potencies of the two separate drugs. This is because they act as competitive inhibitors of two successive steps in the synthetic pathway of folinic acid in bacteria.

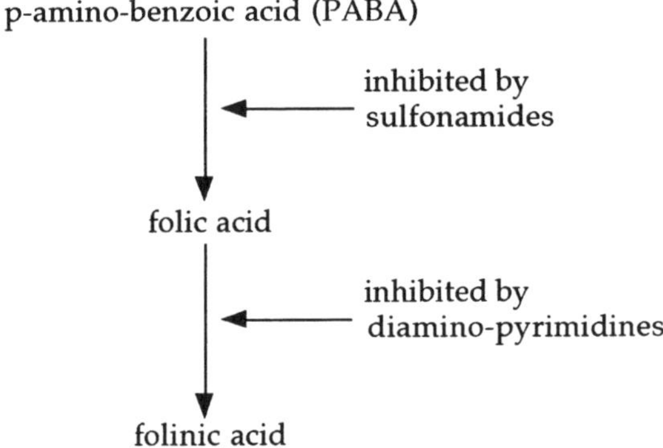

Figure 9.1. Mode of action of potentiated sulfonamides.

The combination delays, although it does not entirely prevent, the development of bacterial resistance, because the bacteria have to produce resistance factors for the two drugs simultaneously.

(b) Fractional inhibitory concentration

The fractional inhibitory concentration (FIC) is a measure of the *in vitro* antibacterial activity of an individual drug in a combination. It has been developed and used as a means of both demonstrating synergy and determining the optimum ratio of drugs in a synergistic mixture. The FIC of a drug is, strictly speaking, specific to both the other drug in the combination and the bacterial strain. For a given combination and bacterial strain the FIC of a drug is defined as:

$$\text{FIC} = \frac{\text{concentration in a mixture}}{\text{concentration alone giving the same antibacterial effect}}$$

The FIC index for a mixture is the sum of the FICs of all the active ingredients in it.

- A FIC index <1 indicates synergy
- A FIC index =1 indicates additive action
- A FIC index >1 indicates antagonism between the drugs

Table 9.1. Calculation of fractional inhibitory concentration (FIC) index for the combinations sulphachlorpyridazine/trimethoprim (SCP/TMP) (5:1) against five type or referrence strains of fish pathogenic bacteria*. After Oppegaard H. *et al*, 1995, with permission

	MIC (μg ml^{-1})								
	SCP alone:	TMP alone:	SDZ alone:	SCP-TMP combination		SDZ-TMP combination		SCP-TMP FIC index:	SDZ-TMP FIC index:
Organism	A_0	B_0	a_0	A	B	a	b	$A/A_0 + B/B_0$	$aa_0 + b/B_0$
A. salmonicida subsp. *salmonicida* ATCC 14174	12·8	6·4	32·0	0·5	0·1	1·0	0·2	0·06	0·06
A. salmonicida subsp. *achromogenes* NCMB 1110	8·0	6·4	12·8	0·5	0·1	0·5	0·1	0·08	0·06
V. anguillarum ATCC 14181	2·0	3·2	12·8	0·125	0·025	1·0	0·2	0·07	0·14
V. salmonicida NCMB 2262	1·0	0·8	6·4	0·0625	0·0125	1·0	0·2	0·14	0·41
Y. ruckeri ATCC 29473	2·0	0·2	6·4	0·25	0·5	0·25	0·05	0·15	0·05

*All values are obtained at 15 °C.

Table 9.2. Calculation of fractional inhibitory concentration (FIC) index for different concenttration rations of sulphachlorpyridazine (SCP) *versus* trimethoprim (TMP) against *Aeromonas salmonicida* subsp. *salmonicida* ATCC 14174*. After Oppegaard H. *et al*, 1995, with permission

	MIC (μg m^{-1})					
	SCP alone :	TMP alone:	SCP-TMP combination		FIC index:	Ratio in mixture
	A	B	A_0	B_0	$A/A_0+B/B_0$	(SCP:TMP)
A. salmonicida subsp. *salmonicida* ATCC 14174	12·8	6·4	0·2	0·2	0·05	1:1
			0·2	0·1	0·03	2:1
			0·3	0·1	0·03	3:1
			0·4	0·1	0·04	4:1
			0·5	0·1	0·05	5:1
			1·2	0·2	0·12	6:1
			0·4	0·05	0·04	8:1

* All values are obtained at 15 °C.

Oppegaard *et al.* (1995) used this principle in an investigation of the activity against five fish pathogens of two sulfonamides, sulfachlorpyridazine (SCP) and sulfadiazine (SDZ), and the pyrimidine potentiator, trimethoprim (TMP). Table 9.1 shows the *in vitro* MICs for each of the three drugs and the two TMP-sulfonamides; also tabulated are the calculated FIC indices for the two combinations. This led to the initial conclusion that SCP alone was better than SDZ alone, and that TMP-SCP was better than TMP-SDZ for *Vibrio* spp. and just as good for the other species tested. They therefore selected SCP for an investigation of the optimum ratio with TMP, with the results shown in Table 9.2. Overall they concluded that *in vitro*:

1. TMP was the best of the three drugs when used alone.
2. SCP was better than SDZ, and as good as TMP for some bacteria.
3. Both sulfonamides synergize with TMP.
4. Synergy occurs against sulfonamide-resistant strains.
5. There was little difference in activity between TMP:SCP ratios of 1:1 to 1:8.

9.1.2 AVAILABLE COMBINATIONS

(a) Matching sulfonamides with potentiators

Some seven diamino-pyrimidines have been shown to potentiate the antimicrobial activity of sulfonamides; of these only three, trimethoprim, ormetoprim and baquiloprim, are currently used in veterinary medicine, and only the first two have been developed for fish.

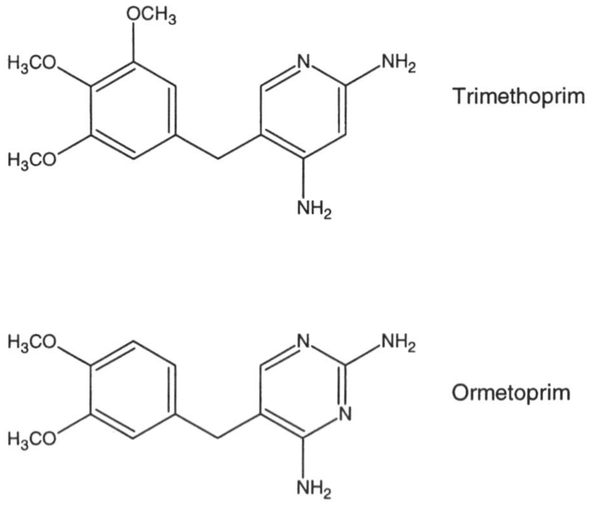

Figure 9.2.

For the synergistic antibacterial action to occur it is essential that both drugs should be at the site of infection at the same time. This means that from the large number of sulfonamides which has been developed, one with the same pharmacokinetic profile as the pyrimidine has to be selected. This is the overriding consideration in sulfonamide selection; a low *in vitro* FIC index is no guarantee of *in vivo* efficacy, although combinations with high FIC indices should not be used. In human medicine the selection of a sulfonamide is straightforward once the pharmacokinetic profile of the potentiator is known; in veterinary medicine there is a very real problem in the multiplicity of species each with its own pharmacokinetic profile for each sulfonamide and each potentiator. In fish medicine the problem is even greater than in mammal and bird medicine because of the much larger number of cultured species. It would be uneconomic to develop a separate combination for each species so a compromise is made; the sulfonamide chosen is the one with the best overall pharmacokinetic match in the important species.

(b) Trimethoprim combinations

The first commercially successful diamino-pyrimidine potentiator to be developed was trimethoprim (TMP). Trials of different ratios of TMP and sulfonamides in laboratory animals showed 1 TMP:5 sulfonamide to be optimal and this has been used in commercial products for Man and other mammals.

The first veterinary combination to be marketed was TMP and sulfadoxine, also known as co-trimoxine (trade names 'Trivetrin' and 'Borgal'). This combination was produced, as an injection, because the most important market was perceived as being for cattle, and sulfadoxine had an excellent pharmacokinetic match with TMP in that species. However while it made an adequate match in dogs it was poor for most other domesticated mammals.

Sulfadiazine (SDZ) (see Fig. 8.1) as a match for TMP in cattle is not as good as sulfadoxine but it is satisfactory. It is also satisfactory in other mammals, and has the additional advantage of being well absorbed from oral administration. It was therefore developed with TMP as an "all purpose" veterinary potentiated sulfonamide, co-trimazine, with the trade mark 'Tribrissen'. This name has frequently been used in the scientific literature as if it were a non-proprietary name for TMP-SDZ combinations; but it was actually selected by the present author from a list of trade marks registered but unused by The Wellcome Foundation Ltd., and it is not only for TMP-SDZ combinations - Tribrissen® Poultry Formula is a 1:3 combination with sulfaquinoxaline sodium, a better match for TMP in birds.

Tribrissen® 40% Powder, a premix, was developed for the medication of fish feed. The market did not justify the cost of an extensive search for the ideal sulfonamide to combine with TMP in fish, but at least TMP-SDZ was shown to be active against furunculosis, especially in rainbow trout, at dose rates which clearly indicated that synergy was occurring. Furthermore *in vitro* synergy was shown against other bacterial pathogens of fish (see Table 9.3).

McCarthy *et al.* (1974) studied the pharmaco-kinetic profiles of seven sulfonamides in combination with TMP in the blood, kidney and muscle of rainbow trout kept at 13°C. They concluded that, of those sulfonamides tested, sulfamethylphenazole was the ideal match for TMP; but they were careful to point out that this might not be the case for other species of fish, or even for rainbow trout kept at other temperatures. They noted that the synergism meant that significantly lower doses of the combination could be used than of any sulfonamide alone, and hence the known target species safety problems would be mitigated. The same workers later showed that the combination was efficacious against experimental furunculosis at 15 mg/kg/day TMP and 100 mg/kg/day sulfamethylphenazole which is half the normal dose of a sulfonamide.

In the same way, in a study of sulfisoxazole (also known as sulfafurazole) in a 5:1 ratio with TMP in yellowtail and Japanese eels, the clinically efficacious dose rates indicated that synergy was occurring. It was noted that the pharmacokinetic profiles would make the combination particularly useful for kidney infections.

Table 9.3. Synergistic bacteriostatic effects against fish pathogens demonstrated with various combinations of sulphadiazine and trimethoprim. After Kimura T. *et al*, 1983, with permission

	MIC (μg/ml)				
Organisms	SDZ alone a_0	TMP alone b_0	SDZ/TMP combination	FIC* index	Ratio of SDZ to TMP in mixture
Aeromonas hydrophila IAM 1018	3·13	0·78	0·20/0·05	0·128	4:1
A. punctata IAM 1646	3·13	0·20	0·20/0·025	0·189	8:1
A. salmonicida ATCC 14174	3·13	0·20	0·20/0·025	0·189	8:1
A. salmonicida EFDL	6·25	0·20	0·20/0·025	0·157	8:1
A. liquefaciens EFDL	50	0·78	1·56/0·05	0·095	32:1
A. punctata TUF 1	3·13	0·39	0·20/0·05	0·192	4:1
A. punctata TUF 2	6·25	0·20	0·39/0·013	0·127	32:1
A. salmonicida NCMB 833	800	0·39	200/0·10	0·500	2000:1
A. salmonicida NCMB 834	3·13	0·20	0·39/0·025	0·250	16:1
A. salmonicida NCMB 1102	50	0·20	3·13/0·025	0·188	128:1
A. salmonicida NCMB 1103	12·5	0·20	0·78/0·013	0·127	64:1
A. salmonicida Tochigi	25	0·20	1·56/0·013	0·127	128:1
A. salmonicida Tokyo	50	0·10	6·25/0·013	0·255	512:1
A. salmonicida Gifu	25	0·78	1·56/0·013	0·079	128:1
A. salmonicida Shiga 32	800	0·78	100/0·20	0·381	500:1
A. salmonicida Shiga 33	800	0·39	200/0·10	0·500	2000:1
A. salmonicida subsp. *masoucida* NCMB 2020	25	0·78	0·39/0·05	0·800	8:1
Vibrio anguillarum NCMB 6	6·25	1·56	0·78/0·05	0·157	16:1
V. anguillarum NCMB 828	3·13	1·56	0·39/0·20	0·250	2:1
V. anguillarum KAY 2	6·25	1·56	0·78/0·10	0·188	8:1
V. anguillarum VU 7601	1600	1·56	400/0·78	0·500	1024:1
V. anguillarum VCS 7601	6·25	1·56	0·78/0·20	0·250	4:1
V. anguillarum VG 7601	1600	1·56	400/0·39	0·500	1024:1
V. anguillarum VN 7601	1600	1·56	200/0·10	0·188	2048:1
V. anguillarum Y 3	6·25	1·56	0·39/0·20	0·188	2:1
V. anguillarum Y 4	6·25	1·56	0·39/0·20	0·188	2:1
V. anguillarum Y 5	6·25	1·56	0·39/0·20	0·188	2:1
V. anguillarum Y 6	6·25	1·56	0·39/0·20	0·188	2:1
V. anguillarum NS 744	1600	1·56	200/0·10	0·188	2048:1
V. anguillarum NA 7418	1·56	1·56	0·20/0·20	0·250	1:1
Escherichia coli O 26	6·25	0·05	0·10/0·013	0·276	8:1
Pseudomonas fluorescens EFDL	800	>1·56	400/0·78	1·000	512:1

* Fractional inhibitory concentration.

(c) Ormetoprim combination

Only a single ormetoprim product, a 1:5 combination with sulfadimethoxine, has been developed; this is available as a 30% premix. It was originally known in the scientific literature by its company code number Ro5-0037, and latterly by the trade mark, 'Romet-30'. It produces synergy in a wide range of mammals and fish, thus obviating the need to develop other combinations.

(d) Artemia enrichment

Verpraet *et al.* (1992) investigated the possibility of using *Artemia* enrichment (see Section 1.2.9) as a means of administering potentiated sulfonamides to larval fish. Using a TMP-sulfamethoxazole (TMP-SMX) combination they found that the concentration ratio TMP:SMX in the *Artemia* nauplii was significantly closer than the ratio in the food provided, for example a 1:5.8 ratio in the food produced a 1:2.4 ratio in the *Artemia*. In rotifers the ratio was even closer. The difference could have been due to:

1. More efficient uptake of TMP;
2. Slower metabolism and elimination of TMP; or
3. More rapid leaching of SMX (into the alkaline water used to culture *Artemia*).

However the uptake of each drug was found to be linearly related to the concentration of it in the food given to the *Artemia* or rotifers, so an appropriate ratio in them could presumably be achieved by starting with a higher proportion of sulfonamide.

9.1.3 USES

Potentiated sulfonamides are active against a wide range of bacterial infections. They have been reported to be active in infections of fish with

Aeromonas salmonicida (see especially Samuelsen *et al.* (1997)
A. hydrophila
A. liquefaciens
A. punctata
Vibrio anguillarum
V. salmonicida
Pasteurella piscicida
Yersinia ruckeri
Edwardsiella ictaluri
E. tarda

They have shown poor activity against streptococci and have none against *Pseudomonas* spp.

The normally recommended dose regimen for co-trimazine (TMP:SDZ = 1:5) is 30 mg/kg total active ingredients daily for 7-10 days. For the ormetoprim-sulfadimethoxine 1:5 combination it is 50 mg/kg/day. It may be noted that these dose rates for total active ingredients are respectively 15% and 25% of the standard rate for sulfonamides. If synergy is occurring it is easily appreciable clinically.

A dose titration conducted in channel catfish artificially infected with *Aeromonas hydrophila* showed significant although not complete control with Romet-30® at 25 mg/kg/day for 5 days. No significantly greater control was achieved with up to four times this dose rate.

9.2 Pharmacokinetics

9.2.1 TRIMETHOPRIM IN RAINBOW TROUT

(a) Radio-active tracer study

Bergsjø *et al.* (1979) studied orally administered radio-labelled TMP in rainbow trout kept in freshwater at either 7 or 15°C. At 7°C 12 hours after dosing the drug was detectable only in the stomach. By 24 hours it was distributed throughout the body with the highest concentrations in the lumen and mucosa of the gut, the kidneys, the uveal tract and the bile. At 48 hours it had accumulated in the bile but was also present in the kidneys and uveal tract; it was present in the lumen of the gut, presumably recycled in the bile. At 15°C it had a similar distribution but the movement was faster. At 3 hours there was no absorption: by 12 hours it was distributed throughout the body similarly to the 24 hour picture at 7°C. By 24 hours it had accumulated in the bile, and was also found in the skin. At 72 hours the main activity was in the skin, uveal tract, and bile, and also in the gut mucosa, presumably through recycling; there was no activity in either gills or muscle. At 144 hours the major residues were in the uveal tract and skin.

It was concluded that the bile is the main excretory route; while the kidneys may play a minor role, the gills appear not to be involved in excretion. The higher temperature halved the absorption time, and also accelerated excretion since C_{max} at 15°C was lower than at 7°C. The accumulations in the uveal tract and skin were attributed to the affinity of TMP for melanin. The persistence of TMP in the skin means that oral, as opposed to topical, administration would be an efficacious route for the treatment of skin infections, although perhaps not for ulcers where the skin has been completely eroded away.

(b) Immersion administration

It was noted in the previous chapter that sulphonamides are absorbed by fish from seawater and to a lesser extent from freshwater. In contrast to the sulfonamides, which are weak acids, TMP is a weak base, pKa 7.6. Bergsjø and Søgnen (1980) studied absorption of TMP by rainbow trout in freshwater, adapted to seawater, or in seawater but not adapted to it. In each case the water contained 75 mg/l TMP, which is close to saturation, and exposure was for 84 hours. The fish in freshwater had less than 1 ppm TMP in the plasma; the fish in seawater but not adapted to it absorbed 1 ppm in 48 hours, but the seawater adapted fish absorbed 1 ppm in 10 hours and 5 ppm over the 84 hour period. These concentrations may be compared with MICs of the order of 0.2 ppm. The livers contained 12 times the concentration in plasma; muscle concentrations rose slowly but equalled plasma by 60 hours and eventually reached about three times the plasma level. It was considered that immersion would be a useful method of administering TMP-potentiated sulfonamides to smoltifying fish since they do not eat and are subject to bacterial diseases.

(c) Plasma concentration study

Tan and Wall (1995) found that following intravenous injection the plasma kinetics of TMP in rainbow trout fit a three-compartment open model. The central compartment has a larger volume than the blood; and from it there is rapid equilibration with a peripheral compartment, and slow equilibration with a third compartment which may be muscle and bone. At equilibrium the central compartment, despite its volume, contains only 3.5% of the TMP in the body; this is presumably because the compartment is aqueous whereas TMP is lipophilic. Terminal elimination is slow, reflecting slow release from the third compartment.

As in the radio-labelled study by Bergsjø *et al.* (1979), it was concluded that renal excretion was low (estimated in this case at 14% of the injected drug). However it was further assumed that there was therefore either biotransformation or excretion through the gills. The possibility of biliary excretion appears to have been discounted.

The pharmacokinetics of TMP are independent of the dose; this contrasts with ormetoprim where they vary with the dose.

9.2.2 SULFADIAZINE IN RAINBOW TROUT

At the same time as looking at TMP in rainbow trout Bergsjø and others (1979) also studied radio-labelled sulfadiazine (SDZ) (see Fig. 8.1), since this is the sulfonamide in the main commercially available combination. This pharmacokinetic investigation was made several years after a formulation of the combination specifically for fish had been placed on the market. The marketing had been based on essentially empirical clinical evidence of efficacy and safety in rainbow trout, but in the event the laboratory study showed quite a good pharmacokinetic match in this species.

Absorption of orally administered SDZ is slow: at 7°C virtually all of a 200 mg/kg dose is still in the stomach 2 hours after dosing, and at 72 hours there is still some there. At 4 hours a little has been absorbed and is found mainly in the plasma, liver, kidney and skin. Later the accumulation is in the bile, skin and eye as is the case with TMP. By 72 hours after dosing, apart from the residue still in the stomach there is SDZ in the intestine which could be from the bile since the bile content is higher than either the blood, liver or kidney. As with TMP in rainbow trout, the main excretory pathway for SDZ is the bile; this is in contrast to the mammalian situation where SDZ is excreted in the urine.

9.2.3 ORMETOPRIM AND SULFADIMETHOXINE 1:5

There is no generally accepted non-proprietary name for this combination; it is referred to in the literature by the code number, Ro5-0037, assigned to it by the patent-holder, Roche, and later by the trade mark, 'Romet-30'. Sulfadimethoxine (SDM) is the only sulfonamide commercially available in a combination with ormetoprim, and because it is also used alone its pharmacokinetics have been reviewed in Section 8.2.3.

(a) In rainbow trout

Initial efficacy studies were conducted on rainbow trout kept at 12.5°C and artificially infected with furunculosis. Various dose rates of ormetoprim (OMP), sulfadimethoxine, sulfisoxazole or Romet-30® were surface-coated onto feed pellets using 5% gelatine as binder, and fed for 14 days. While the standard sulfonamide dose is 200 mg/kg/day, Romet-30® was shown to be efficacious at 50 mg/kg/day; furthermore a 3-day course appeared to be adequate. Ormetoprim alone at 50 mg/kg/day was not efficacious. Dose rates of Romet-30® in excess of 75 mg/kg/day were unpalatable to rainbow trout.

The pharmacokinetics of OMP have been studied at 14°C following intravenous dosage of ^{14}C-labelled drug at 8 mg/kg, which is approximately the content in 50 mg/kg Romet-30®. The plasma concentration curve is bi-exponential, with a distribution half-life of 0.54 hours and a terminal half-life after a single dose of 17.5 hours. With four daily doses there is initial cumulation with C_{max} 53 µg/ml at T_{max} after the third dose. Since distribution is rapid the distribution half-life is the same for single and multiple daily doses, but the return from the periphery to the plasma is slow, and the plasma terminal half-life for multiple doses is twice that for a single dose.

Oral bioavailability is 87% and after a single oral dose T_{max} is 8-12 hours (*cf.* 5 hours for TMP). OMP is distributed in highest concentrations to the bile, liver and kidney, but significant quantities are found also in the muscle and skin. At 38 days after dosing radio-activity was found equivalent to 0.9 ppm in skin and 0.15 ppm in muscle. Terminal half-lives calculated on the basis of total radio-activity were:

skin 9 days
fat 46 days
kidney 28 days
muscle 19 days

On the basis of the European Union MRL for TMP of 0.05 ppm, the withdrawal period for OMP in rainbow trout would be about 120 days if only muscle residues were taken into account but 160 days if skin alone were considered. In the USA the withdrawal period for Romet 30 in salmonids is 42 days; however the MRL for OMP is 0.1 ppm, and for this the muscle residue data suggest a withdrawal period of 100 days.

(b) In Atlantic salmon

The parameters in Table 9.4 were calculated from plasma concentrations observed in Atlantic salmon kept in seawater at 10°C and force fed at 1% of bodyweight a diet containing 20 g/kg 'Romet-30'.

The volumes of distribution indicate better distribution of OMP than of sulfadimethoxine into the tissues. The kidney has the highest concentration of OMP; concentration in bile, as found in rainbow trout, does not occur in Atlantic salmon.

Table 9.4 Pharmacokinetic parameters of the active ingredients of Romet 30 in Atlantic salmon

		Ormetoprim	Sulfadimethoxine
Dose	(mg/kg)	10	50
Bioavailability	(%)	85	39
C_{max} (plasma)	(µg/ml)	1.13	9.99
T_{max}	(hours)	17.6	20.3
Volume of distribution	(l/kg)	2.48	0.39
Terminal half-life	(hours)	25.6	9.9

Samuelsen *et al.* (1997) dosed Atlantic salmon in seawater by gavage at approximately 5 mg OMP/kg/day and 25 mg SDM/kg/day for 5 days, and assayed tissue concetrations by HPLC. C_{max}(plasma) was reached in 3 days and was maintained until 8 days after the beginning of medication. The accumulation of OMP in kidney was confirmed and shown to be more pronounced after multiple doses than after a single one: plasma:kidney concentrations at T_{max}(plasma) were:

after a single dose 1:1.18
after multiple doses 1:1.92

Furthermore OMP was very persistent in kidney (see Table 9.4). SDM reached a slightly higher concentration in muscle than in plasma but in the liver and kidney it was lower.

Using the EU MRLs of 0.1 mg/kg for SDM and 0.05 mg/kg for OMP, the authors conclude that OMP would be the marker residue. The withdrawal period at 10°C would be 54 days if kidney residues were ignored but 200 days if they were taken into consideration.

(c) In Chinook salmon

A residue study has been made in chinook salmon kept in seawater at 8-9°C and dosed with Romet-30® by gavage at 40 mg/kg/day for 10 days. Measurement of OMP in liver tissue proved difficult, but as SDM appeared to have the higher residues it was treated as the marker drug in liver. With a limit of detection of 0.05 ppm, both drugs were found in muscle 10 days after the end of medication, and SDM was found in the liver at the same time point; total SDM (including the N_4-acetyl metabolite) was higher in the liver than in muscle. Although the level of N_4-acetyl-SDM was significant in liver it was negligible in muscle.

(d) In channel catfish

Plakas *et al.* (1990) studied the pharmacokinetics of OMP in channel catfish using radio-labelled drug. They found a 52% bioavailability with T_{max} (plasma) at 6 hours. There was three-stage depletion from the plasma, the half-lives being 0.39, 4.9 and 49 hours. The volume of distribution was very large, indicating concentration in the tissues as is the case in salmonids. However in contrast to the position in salmonids, elimination from muscle is rapid: the musculature contains 49% of the radio-activity of an intravenous dose at 2 hours but only 0.6% at 72 hours. OMP is persistent in skin and this probably accounts for the long terminal half-life. Catfish skin is not usually eaten, so while the USA withdrawal period for 'Romet-30' in salmonids is 42 days, in channel catfish it is 3 days.

OMP is extensively metabolized in channel catfish. As proportions of the dose administered, the cumulative radio-activity excreted at 48 hours consist of 1.9% OMP, 2.5% non-polar metabolites and 20% polar metabolites.

Further reading

Bergsjø, T., Nafstad, I. and Ingebrigtsen, K. (1979) The distribution of ^{35}S-sulfadiazine and ^{14}C-trimethoprim in rainbow trout, *Salmo gairdneri*. *Acta Veterinaria Scandinavica* **20**, 25-37.

Bergsjø, T. and Søgnen, E. (1980) Plasma and tissue levels of trimethoprim in the rainbow trout, *Salmo gairdneri*, after absorption from fresh and salt water. *Acta Veterinaria Scandinavica* **21**, 18-25.

Jacobsen, M.D. (1989) Withdrawal times of freshwater rainbow trout, *Salmo gairdneri* Richardson, after treatment with oxolinic acid, oxytetracycline and trimethoprim. *Journal of Fish Diseases* **12**, 29-36.

Kimura, T., Yoshimizu, M. and Wada, M. (1983) *In vitro* antibacterial activity of the combination of sulphadiazine and trimethoprim on bacterial fish pathogens. *Journal of Fish Diseases* **6**, 525-532.

McCarthy, D.H., Stevenson, J.P. and Salsbury, A.W. (1974) A comparative pharmaco-kinetic study of seven sulphonamides and a sulphonamide potentiator, trimethoprim, in rainbow trout (*Salmo gairdneri*, Richardson). *Aquaculture* **4**, 299-303.

Oppegaard, H., Sørum, H. and Gravningen, K. (1995) *In vitro* antimicrobial activity of sulphachlorpyridazine and sulphadiazine in combination with trimethoprim against bacterial fish pathogens. *Journal of Fish Diseases* **18**, 371-376.

Plakas, S.M., Dickey, R.W., Barron, M.G. and Guarino, A.M. (1990) Tissue distribution and renal excretion of ormetoprim after intrasvascular and oral administration in the channel catfish (*Ictalurus punctatus*). *Canadian Journal of Fisheries and Aquatic Sciences* **47**, 766-779.

Samuelsen, O.B., Lunestad, B.T. and Jelmert, A. (1997) Pharmacokinetic and efficacy studies on bath-administering potentiated sulphonamides in Atlantic halibut, *Hippoglossus hippoglossus* L. *Journal of Fish Diseases* **20**, 287-296.

Samuelsen, O.B., Pursell, L., Smith, P. and Ervik, A. (1997) Multiple-dose pharmacokinetic study of Romet 30 in Atlantic salmon (*Salmo salar*) and *in vitro* antibacterial activity against *Aeromonas salmonicida*. *Aquaculture* **152**, 13-24.

Tan, W.P. and Wall, R.A. (1995) Disposition kinetics of trimethoprim in rainbow trout. *Xenobiotica* **25**, 315-329.

Verpraet, R., Chair, M., Léger, P., Nelis, H., Sorgeloos, P. and De Leenheer, A. (1992) Live-food mediated drug delivery as a tool for disease treatment in larviculture. The enrichment of therapeutics in rotifers and *Artemia* nauplii. *Aquacultural Engineering* **11**, 133-139.

10. QUINOLONES AND FLUOROQUINOLONES

10.1 Characteristics of the group

10.1.1 CHEMISTRY

(a) Structure

The quinolones, or more strictly the 4-quinolones, are a group of chemically related synthetic antibacterial agents all being carboxylic acids. Despite their name they do not all have a 4-quinolone ring molecular structure; the first to be developed, nalidixic acid, is a naphthyridine carboxylic acid. Only a limited number of the drugs have been studied for use in fish medicine, and among these oxolinic acid, which is a true 4-quinolone, is of particular interest and importance as it was originally developed in Japan specifically for fish.

It has been found that a fluorine atom at position 6 in the molecule very significantly enhances antibacterial activity and this has led to the development of a sub-group of fluoroquinolone drugs. Various substitutions at position 7 improve the pharmacokinetic penetration of the molecules. Oxygen at position 8, which destroys the true 4-quinolone ring, improves activity against Gram-positive bacteria.

(b) Solubility

Quinolones are amphoteric and most are sparingly soluble at neutral pH. Sodium salts are soluble, and as they are absorbed through fish gills they can be administered by immersion.

(c) Complexing

Quinolones form complexes with divalent cations, particularly Mg^{++}. These complexes are bacteriologically inactive so the activity of quinolones is profoundly affected by the presence of such ions. Mg^{++} ions appear to have a greater effect on the minimum bactericidal concentration (MBC) (*i.e.* the lowest concentration to kill 99.9% of bacteria) than on the minimum inhibitory concentration (MIC) (*i.e.* the lowest concentration to prevent a fourfold multiplication of c.f.u.). In the range 0.3-14.5 mMol Mg^{++} there is a linear relationship between ionic concentration and MBC. Seawater contains 54 mMol Mg^{++} and since marine fish are continually drinking water their gut contents will complex with any quinolones given in feed.

10.1.2 ANTIBACTERIAL ACTIVITY

(a) Mode of action

Quinolones inhibit the bacterial enzyme DNA-gyrase. The super-coiling of bacterial DNA requires the "nicking" of the DNA double helix, enabling it to fold over, and the enzyme heals the nicks afterwards. Inhibition of the enzyme results in breaks in the DNA.

Quinolones are usually bactericidal but a majority of them have a reduced efficacy *in vitro* at high concentrations. For example the activity of oxolinic acid against one strain of *Aeromonas salmonicida* was extremely bactericidal at 3-5 times the MIC but less so at 6-8 times the MIC; the high activity returned at 9+ times the MIC. Against another strain of the same bacterial species it was highly bactericidal at the MIC if exposure was maintained for 48 hours; at 3-5 times the MIC it was less efficacious but it was strongly bactericidal at 6+ times the MIC. In the case of flumequine this biphasic activity occurs in freshwater but not in seawater. Bacteria can develop mutation resistance but there is no plasmid-mediated transmissible resistance. Resistance to oxolinic acid normally produces cross-resistance to nalidixic acid but not to fluoroquinolones.

(b) Comparative activity

Of the large number of quinolones which have been developed only a very limited number are routinely used in fish; and of those which are used in fish nalidixic acid and piromidic acid are rarely used outside Japan. A comparative study has been made of four quinolones used or being developed for use in fish outside Japan, oxolinic acid, flumequine, enrofloxacin and sarafloxacin, together with oxytetracycline, the most extensively used antibacterial drug in fish. *In vitro* at either 4 or 15°C enrofloxacin was the most active and oxytetracycline the least; sarafloxacin was more active against most pathogens than either oxolinic acid or flumequine.

The influence of temperature was interesting: against *A. salmonicida* all four quinolones were less active (had higher MICs) at the lower temperature; this was also true for all except sarafloxacin against *Vibrio salmonicida*, but against *V. anguillarum* or *Yersinia ruckeri* temperature made no significant difference. In contrast oxytetracycline was more active against all the bacterial species at the lower temperature. It was speculated that the effect was due to lower diffusion of drug into the bacteria at the lower temperature but this does not explain the reverse effect with oxytetracycline. An alternative explanation is that the lower bacterial growth at lower temperatures would reduce susceptibility to bactericidal drugs such as quinolones but have no effect on the activity of bacteriostatic drugs such as oxytetracycline.

10.1.3 TOXICOLOGY

(a) In fish

Miyazaki *et al.* (1984) found nalidixic acid or oxolinic acid over-dosage to cause loss of appetite and dark body discolouration in yellowtail. These effects were rapidly reversible on discontinuation of the medication.

There was a macrocytic but normochromic blood picture after seven days medication which developed into a macrocytic anaemia during the recovery period. The effects were more severe with nalidixic acid than with oxolinic acid. These effects could be attributed to induced deficiency of B group vitamins but the pharmacokinetic distribution of the drugs suggested that interference with DNA synthesis was a more likely cause.

(b) In mammals
The relevance of the mammalian toxicology of quinolones to fish medicine lies in the area of consumer safety; it is of profound importance in the determination of MRLs. While other adverse effects occur at higher dose rates, at rates close to the "no effect level" the observable effect is on cartilage especially in young growing animals. This effect is in fact detectable at very low dose rates. Nevertheless EU MRLs have been determined for oxolinic acid, flumequine and sarafloxacin.

10.2 Nalidixic acid

10.2.1 USES

Nalidixic acid was the first quinolone to be developed. While it has been used to a limited extent in man and has been studied in several species of fish it has been accorded a market authorization for fish only in Japan. One reason for this is that its development was closely followed by that of other quinolones which had a greater activity (lower MIC) than nalidixic acid. Its spectrum of activity is mainly against Gram-negative bacteria but in practice this covers the majority of bacterial pathogens of fish. It has been used successfully in Japan for the control of furunculosis and vibriosis at a dose rate of 20 mg/kg/day.

10.2.2 PHARMACOKINETICS IN RAINBOW TROUT

(a) Studies reported
Two separate major studies have been reported, by Uno *et al.* (1992) in Japan and by Jarboe *et al.* (1993) in the USA. The former used a single oral dose of 40 mg/kg - twice the standard dose. The latter used single intravenous and oral doses of 5 mg/kg of the sodium salt - only a quarter of the standard dose - for their plasma concentration observations, and the same oral dose of radio-labelled sodium salt for distribution and disposition studies. The fish were kept at 15 and 14°C respectively so the observations are comparable.

(b) Absorption
Jarboe *et al.* (1993) observed complete absorption (100% bioavailability) of the, abnormally low, oral dose. They found the highest plasma concentration at 24 hours; Uno and others (1992) who made observations at more time points in the first 24 hours calculated T_{max} (plasma) to be 16.6 hours.

Nalidixic acid was detectable in all tissues examined within 30 minutes of oral dosing, and T_{max} estimates were:

- liver and kidney 12 hours
- muscle 24 hours
- bile 48 hours

Nevertheless absorption was not complete until between 36 and 48 hours. This indicates a moderate rate of absorption from the gut but a very rapid movement from the blood to the tissues.

(c) Distribution
Jarboe *et al.* (1993) calculated the steady state volume of distribution to be 965 ml/kg; other workers have reported 1010 ml/kg at 21°C. These results probably indicate at least some penetration into most tissues.

When distribution is complete, *i.e.* after 48 hours, the relative concentrations in tissues are:

bile>liver>kidney>serum>muscle>abdominal fat

Radio-label concentrations in muscle and skin are very similar. While the bile, liver and kidney have higher concentrations than the serum, the ratios between them appear to be constant. It is possible to predict concentrations in these tissues from the serum levels.

(d) Metabolism
Uno *et al.* (1992) detected a glucuronide, which must be presumed not to be antibacterial, in all tissues. At 12 hours after dosing it was found at 9% or less of the concentrations of the parent drug in serum, muscle and kidney; in the liver, presumably the site of conjugation, it was at 14% of the parent drug and in bile it was concentrated to 930% (see Table 10.3). It must be assumed that significant first pass conjugation and biliary excretion occurs.

By 7 days the levels of conjugate had equalled (in muscle) or exceeded the levels of parent drug in all tissues. The bile continued to contain vastly more than any other tissue but the kidney also contained a high level in proportion to the parent drug.

Plasma protein binding occurs, but the extent of it is unclear. Jarboe *et al.* (1993) found that *in vitro* it varied with concentration, in the range 22-28% at 14°C. Other workers have suggested about 11% at 21°C.

(e) Elimination
The bile is the predominant route of excretion, most of the drug being excreted as the glucuronide conjugate.

Jarboe *et al.* (1993) found a wide range of terminal half-lives for different tissues, ranging from 46 hours in blood to 255 hours in intestinal tissue. (The latter almost certainly reflects some reabsorption from bile). It should be borne in mind that they were reporting half-lives of radio-carbon, not necessarily of drug or glucuronide; however it seems that very little nalidixic acid is broken down.

Uno *et al.* (1992) recommended daily dosing but there are no residue data for multiple doses. Table 10.2 shows their estimates of terminal half-lives in five tissues and also the "elimination time". This latter was inadequately defined but appears to mean the time to depletion below the limit of detection of their HPLC, about 0.025 ppm. The large dose which they used was maintained above a nominal MIC of 0.2 μg/ml in serum for over 7.5 days. Since muscle and skin contain lower concentrations of drug than serum and since the half-life in serum is about 2 days, 14-16 days would appear to be an acceptable withdrawal period. There is limited evidence of much faster elimination at higher temperatures.

10.2.3 PHARMACOKINETICS IN AMAGO SALMON

Uno *et al.* (1992) made a comparative study of the position in amago salmon (*Oncorhynchus masou*) and rainbow trout (*O. mykiss*). Absorption into the blood showed a similar pattern

Table 10.1. Tissue levels of nalidixic acid in rainbow trout and amago salmon after oral administration at a dose of 40 mg/kg[1]. After Samuelsen O.B..Pursell L., Smith P. and Ervik A., 1997, with permission

Fish	Tissues	Time after administration (h)											
		0.5	1	3	6	9	12	24	48	72	120	168	240
Rainbow trout	Serum	0.32[2]	0.81	1.70	9.82	10.14	12.46	8.92	6.88	3.70	0.61	0.11	0.04
	Muscle	0.10	0.37	0.80	5.87	6.97	10.67	11.37	8.64	5.09	0.56	0.06	0.06
	Liver	2.30	2.88	3.08	14.95	13.66	15.64	11.90	10.00	7.64	1.16	0.27	0.14
	Kidney	4.35	1.62	3.69	13.25	13.09	15.35	12.43	8.09	5.04	0.93	0.26	0.09
	Bile	3.24	1.32	2.86	11.22	16.11	20.81	23.47	37.30	21.00	12.46	2.28	2.56
Amago salmon	Serum	0.78	1.18	1.88	5.61	7.90	12.40	14.96	7.76	3.49	0.66	0.13	—[3]
	Muscle	0.11	0.49	2.26	6.60	10.14	10.83	19.07	8.47	3.97	0.43	0.25	0.13
	Liver	2.68	2.30	4.02	9.13	14.11	12.40	21.12	10.89	6.59	0.94	0.87	0.43
	Kidney	1.12	1.28	2.64	10.57	13.15	15.14	24.47	12.88	6.08	0.38	0.24	0.31
	Bile	19.10	1.00	8.00	14.81	22.25	50.94	46.71	34.94	23.25	1.92	10.79	5.33

[1]The average water temperature was 15°C.
[2]μg/ml or g.
[3]Not detected.

Table 10.2. Biological half-life and elimination time of nalidixic acid in rainbow trout and amago salmon following oral administration at a dose of 40 mg/kg. After Samuelsen O.B..Pursell L., Smith P. and Ervik A., 1997, with permission

Fish	Tissues	$T_{1/2}$[1]	Et[2]
Rainbow trout	Serum	25.5	218.5
	Muscle	24.4	217.7
	Liver	30.3	238.5
	Kidney	28.8	223.9
	Bile	45.9	427.5
Amago salmon	Serum	20.8	196.4
	Muscle	28.8	251.6
	Liver	37.3	292.5
	Kidney	30.3	241.5
	Bile	63.0	556.2

[1]Biological half-life (hours).
[2]Elimination time (hours).

Table 10.3. Levels of conjugated and non-conjugated nalidixic acid in tissues after administration at a dose of 40 mg/kg in rainbow trout and amago salmon [1]. After Samuelsen O.B..Pursell L., Smith P. and Ervik A., 1997, with permission

Fish			Tissues				
			Serum	Muscle	Liver	Kidney	bile
Rainbow trout	Parent NA	12 h	13.26[2]	10.05	15.94	15.06	20.61
		7 days	0.09	0.06	0.28	0.12	2.56
	Conjugated NA	12 h	1.20	0.74	2.21	1.80	192.70
		7 days	0.25	0.07	0.39	1.60	31.95
Amago salmon	Parent NA	24 h	15.48	20.23	20.67	24.58	45.10
		7 days	0.13	0.26	0.70	0.22	11.92
	Conjugated NA	24 h	2.68	2.05	2.76	5.19	552.90
		7 days	0.39	0.14	1.53	6.56	292.90

[1]Nalidixic acid was extracted from tissue homogenates with ethyl acetate.
The residue was incubated with β-glucuronidase, and the released nalidixic acid was re-extracted with the solvent.
[2]μg/g or ml.

in the two species but distribution into the tissues showed some significant differences. In particular in amago salmon T_{max} for bile is only 12 hours, and the concentration of conjugated nalidixic acid in the bile at 24 hours was extraordinary (see Table 10.3).

Terminal half-lives for all tissnes examined except serum were longer in amago salmon than in rainbow trout so it seems almost certain that there is significant reabsorption of drug from bile in the intestine. A withdrawal perod of 21 days would appear to be appropriate.

Table 10.4. Comparative pharmacokinetics of nalidixic acid in two salmonid species

		Rainbow Trout	Amago Salmon
T_{max} (plasma)	(hours)	16.6	17.6
C_{max} (plasma)	(μg/ml)	11.4	12.3
T_{max} (bile)	(hours)	48	12
T_{max} (liver & kidney)	(hours)	12	24

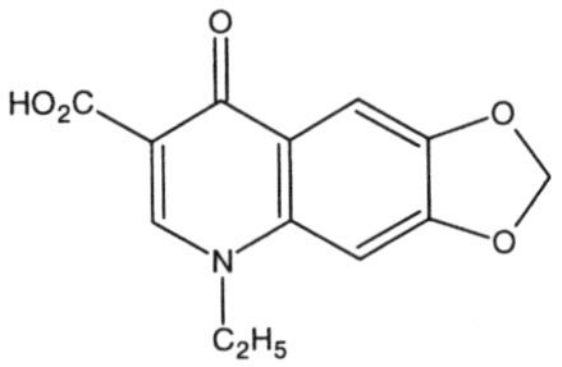

Oxolinic Acid

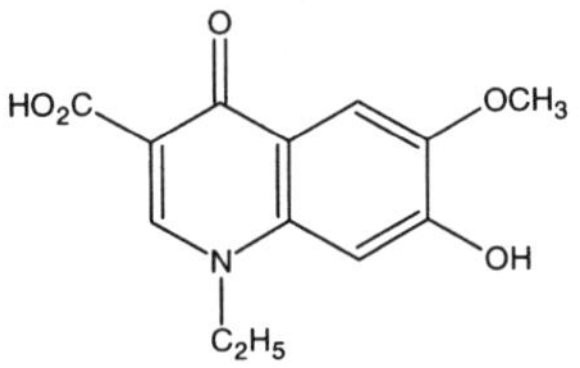

7-Hydroxy-Oxolinic Acid

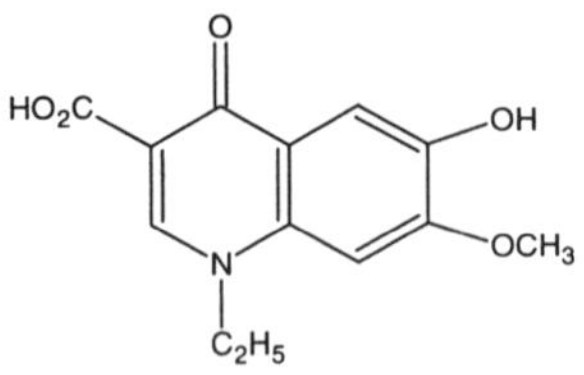

6-Hydroxy-Oxolinic Acid

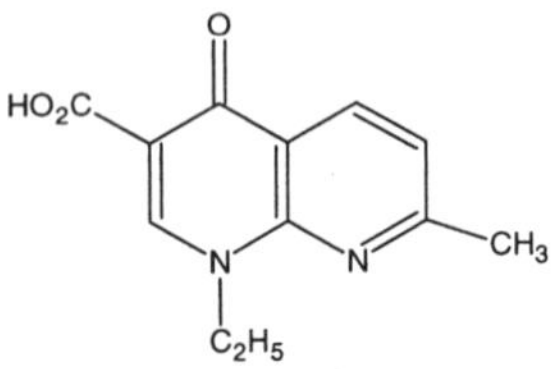

Nalidixic Acid

Figure 10.1.

10.3 Oxolinic acid

10.3.1 ORAL DOSE RATES

(a) Normal oxolinic acid

Oxolinic acid (OXA) is one of the most extensively researched of all drugs used in fish medicine. Nevertheless it would be economically impossible, and probably quite unnecessary, to conduct dose titrations in all species in which it might be used. In any case this drug was originally developed at a time when a demonstration of efficacy rather than a titration of the minimum effective dose rate was the norm. Furthermore because it was developed in Japan some of the fish species in which it was studied are normally cultured only in the Far East; and some of the species are desultory feeders making in-feed medication unsatisfactory.

The standard oral recommendation for OXA is 10 mg/kg/day in freshwater species. In marine species a higher dose rate is required because of the complexing of the drug with divalent cations; market authorizations specify different rates in different countries, usually in the range 25-50 mg/kg/day.

In field efficacy trials in the UK on a rainbow trout farm with endemic furunculosis, OXA given in feed at 5 mg/kg/day throughout the period of risk (*i.e.* water temperature above 18°C) secured a 99% survival, against a 69% survival in untreated controls. On a brown trout farm a course of 5 mg/kg/day for 10 days stopped mortalities within 24 hours of the beginning of the course.

(b) Ultrafine oxolinic acid

In the late 1980's an ultrafine formulation of OXA was developed; the mean particle diameter of this is 1 μm as compared with 6.4 μm for the "normal" drug. The use of this has been reported for yellowtail and red sea bream (both marine species), and in both there was an improved bioavailability. T_{max} (plasma) appeared to be unaffected (4 h in both species) but C_{max} was increased by factors of 1.7 in yellowtail and 2.3 in red sea bream. In a clinical study against artificial infection with *Pasteurella piscicida* in yellowtail 10 mg/kg/day of the ultrafine was found to be as efficacious as the standard 30 mg/kg/day of the normal product. This may be considered surprising as there is no reason to suppose that the ultrafine would be any less susceptible to complexing with divalent cations. 20 mg/kg/day of the ultrafine was clinically better than 10 mg/kg/day but there was no further improvement with 30 mg/kg/day. Overall these results seem to suggest that 10 mg/kg/day will absorb all available cations and leave just enough drug for efficacy equivalent to 30 mg/kg/day of normal drug.

10.3.2 PHARMACOKINETICS IN RAINBOW TROUT

(a) Bioavailability

Cravedi *et al.* (1987) studied the bioavailability (which they called "digestibility") of (presumably normal, not ultrafine) OXA in rainbow trout kept at 14°C. The fish had a mean bodyweight of 160 g and were kept in groups of 15, giving a total bodymass of 2.4 kg per group. They were fed at 20 g/day/group which is 0.83%; and the diets contained either

0.1 or 0.5% OXA, giving dose rates of 8.3 or 41.5 mg/kg/day. A significant difference in bioavailability was found, the fish medicated at the lower rate absorbing 38% and at the higher rate only 14%.

This work has been criticized for failing to take first pass metabolism into account but from the clinical viewpoint it is the drug entering the systemic circulation and hence the infected tissues which is important. The findings mean that the effective dose rates were about 3.2 and 6.0 mg/kg/day, *i.e.* differing by a factor of less than two although the feed medication rates differed by a factor of five. This obviously has considerable bearing on the efficacy of ultrafine OXA.

(b) Absorption and distribution

In addition to being affected by the drug concentration in the feed, absorption of OXA by rainbow trout is temperature-dependent. Following a single oral dose different workers have found T_{max} (plasma) to be 1 day at 16°C, 2 days at 15°C, 3 or 4 days at 10°C and 6 days at 5°C. However, subsequent distribution is very rapid, with a distribution half-life of the order of 9-30 minutes.

Jacobsen (1989) used the dose regimen approved in Denmark of 10 mg/kg/day for 10 days in fish kept at 6, 12 or 18°C. 24 hours after the end of the course the concentration in gutted fish at 6°C was 2.8 ppm and at 18°C 2.5 ppm. At 12°C the concentrations were skin 4.1 ppm, muscle 4.0 ppm, kidney 4.8 ppm and blood 1.4 ppm, the first three of these being C_{max} for the tissues. The lower concentration at 6°C was presumably due to slow absorption and at 18°C due to rapid distribution of the final dose.

The steady state volume of distribution is 1817 ml/kg but this does not necessarily mean high activity. There is a high concentration in the bile and some binding to plasma proteins. The proportion of OXA bound to plasma protein varies inversely with the concentration of OXA in the blood: 31% at 1 μg/ml, 27% at 5 μg/ml and 25% at 20 μg/ml. Affinities after distribution are:

bile>kidney>liver>muscle>blood

(c) Elimination

Whether OXA is metabolized by rainbow trout is unclear. Bjørklund *et al.* (1992) found no evidence of metabolites, but Ishida (1992) analysing bile taken 24 hours after dosing found:

29% unchanged OXA
66% glucuronide
3.4% 7-hydroxy-OXA glucuronide
1.8% 6-hydroxy-OXA glucuronide

This is the position in rainbow trout in freshwater; in seawater-adapted fish rather less OXA was found and more of the two hydroxy metabolites.

The rate of elimination of OXA from rainbow trout is also controversial: at least two authors assert that it is independent of temperature but Jacobsen (1989) and Bjørklund *et al.* (1992) both found that the terminal half-life bears an inverse relationship to temperature. It has been variously estimated by different groups to be 3.4 days at 14°C and 2.3-3.1 days at 15-17°C. A finding of 2.8 days at 8.5°C by yet another group is explicable only if the half-life is independent of temperature. Using an assay with a limit of detection of 0.01 ppm Bjørklund *et al.* (1992) found that at 16°C residues were undetectable in serum or muscle at 10 days although they persisted in liver and kidney until 25 days. At 10°C residues were undetectable at 45 days, and at 5°C they persisted beyond 55 days. Taking the limit of detection as the MRL and muscle as the marker tissue they predicted withdrawal periods with a 95% confidence limit of:

28 days at 16°C
60 days at 10°C
140 days at 5°C.

Jacobsen (1989) studied residues in whole gutted fish at 6 and 18°C, and separately in skin, muscle, kidney and blood at 12°C. At 6°C OXA was undetectable at 0.05 ppm in whole gutted fish by 22 days, and at 18°C by 8 days. At 12°C it was detectable in skin but not in muscle at 8 days. It was argued that the withdrawal period should be based on whole gutted fish because the skin may be eaten, but no observations were made on whole gutted fish at 12°C. It seems that 150 degree-days would be satisfactory for rainbow trout in freshwater given 10 mg/kg/day for 10 days of normal (not ultrafine) OXA. In this species the bioavailability and dosage of ultrafine OXA have not been determined.

(d) Rainbow trout in seawater

Ishida (1992) found that after oral administration of 40 mg/kg there was little difference in tissue concentrations at 12 hours between rainbow trout kept in freshwater and seawater. Thereafter tissue concentrations continued to rise in the freshwater fish with T_{max} at about 48 hours, and persistence as Bjørklund *et al.* (1992) found. In contrast the drug was undetectable in the seawater fish at 72 hours. A similar effect was observed after intravenous administration.

10.3.3 PHARMACOKINETICS IN CHANNEL CATFISH

(a) Intravenous injection

Kleinow *et al.* (1994) made the determinations shown in the top half of Table 10.5 from fish given a single intravenous dose of 5 mg/kg OXA. Values followed by (n) are not significantly different from others in the same row. They noted that:

1. Rainbow trout have a more rapid distribution (lower $t^{1/2}\alpha$) than channel catfish at 14°C, and a slower elimination than channel catfish at either temperature.
2. There is a clear species difference in the volume of distribution. As noted in Section 10.3.2(b) this may merely reflect the high concentration in the bile of rainbow trout.
3. In channel catfish the elimination rate is temperature-dependent.

Table 10.5. Pharmacokinetics of oxolinic acid in channel catfish and rainbow trout

	Channel catfish 24 °C	Channel catfish 14 °C	Rainbow trout 14 °C
$t^{1/2}\ \alpha$ (hours)	0.03	0.68	0.15
$t^{1/2}\ \beta$ (hours)	41	69(n)	81(n)
Vd (ml/kg)	939(n)	880(n)	1820
C_{max} (μg/ml)	3.1	3.7	2.1
T_{max} (hours)	8	24	12
Bioavailability	56%	92%	91%

(b) Oral administration

The determinations shown in the lower half of Table 10.5 were obtained from fish given a single oral dose of 5 mg/kg OXA sodium. It may be noted that:

1. C_{max} in channel catfish at either temperature is higher than in rainbow trout. This finding is unconnected with bioavailability.
2. T_{max} was shortest for channel catfish at 24°C; this may probably be attributed to the slower elimination (longer $T^{1/2}\beta$) in either species at 14°C.
3. The temperature-related differences in bioavailability in channel catfish may be attributable to the rates of intestinal evacuation. This would be particularly likely if the particle size limited the rate of absorption.

10.3.4 ORAL ADMINISTRATION IN OTHER SPECIES

(a) Atlantic salmon

Bioavailability of normal OXA in Atlantic salmon is of the order of 25%, and of ultrafine 40%.

Hustvedt *et al.* (1991) found that following intravenous injection of OXA into Atlantic salmon in seawater at 9°C the plasma concentration-time profile showed three distinct phases, the first two having half-lives of 0.3 and 5.1 hours. The second phase ended at about 24 hours, and the authors found that following oral administration the distribution phase as a whole was less easily demonstrated. In other work the same authors found that single doses of either 9 or 26 mg/kg given in feed produced only a minor difference in C_{max} (plasma) (1.7 or 2.1 μg/ml) but a considerable difference in T_{max}. The difference in T_{max} was attributed to the effect of the larger quantity of feed delaying absorption; the smallness of the difference in C_{max} could well be due to considerable distribution from blood into tissues having taken place before T_{max}. (See also Table 10.9 on page 135).

The volume of distribution was found to be 1.8 l/kg. This is the same as for rainbow trout in freshwater; but as only blood concentrations were studied the anatomical distribution is uncertain. Assays were by reversed-phase HPLC so presumably metabolites would have been excluded.

The terminal half-life following oral administration was about 42 hours. Assuming C_{max} of 2 mg/ml at T_{max} 24 hours, residues would fall below 0.01 ppm in about 14 days.

(b) Amago salmon

At 15°C T_{max} is 2 days for muscle, liver and kidney and 3 days for blood. OXA has the same order of tissue affinity as in rainbow trout but slower elimination. Whereas in rainbow trout at 15°C 10 days would be an acceptable withdrawal period, in amago salmon it takes 30 days for residues to deplete to the limit of detection after a single dose albeit at the high rate of 200 mg/kg.

(c) Yellowtail

Absorption of OXA in yellowtail has been considered in Section 10.3.1(b).

Ueno *et al.* (1988) reported that OXA was easily absorbed and very quickly distributed and eliminated. Following single oral doses of either 50 or 200 mg/kg (c/f 30 mg/kg/day recommended for marine species) T_{max} for all tissues except blood was 3 hours. The order of C_{max} was:

kidney>bile>spleen>muscle>blood>brain>liver

and the values were low compared to those for freshwater species. The blood concentration showed peaks at both 3 and 9 hours.

Following a 50 mg/kg dose the tissue levels had fallen to half the C_{max} values by 24 hours, and after a dose of 200 mg/kg concentrations in all tissues had fallen to less than 1 ppm by this time. Subsequent depletion was slow in kidney, muscle, spleen and brain. At the lower dose OXA was undetectable in blood or liver at 3 days but still detectable in the other tissues at 8 days. High ratios of drug in kidney and bile compared to blood suggest that both urinary and biliary excretion occur. Depletion of OXA following a multiple dose regimen has been reported for the ultrafine formulation: at 15°C OXA was undetectable in serum at 3 days, liver at 10 days, kidney at 16 days and muscle at 13 days.

(d) Red sea bream

In a comparative study of normal and ultrafine OXA given orally to red sea bream (*Pagrus major*), both were detectable in the blood within an hour and T_{max} (plasma) was 4 hours. C_{max} differed significantly between the formulations, being 4.24 μg/ml for the ultrafine as against only 1.53 μg/ml for the normal. Curiously a slight difference was noted in the terminal half-life, 11.5 h. for the ultrafine as against 15 h. for the normal.

10.3.5 IMMERSION TREATMENTS

(a) Goldfish

Goldfish have been used as an experimental species for the study of conditions affecting the absorption of OXA during immersion.

pH. Goldfish were immersed in 10 mg/l OXA sodium at pH 6.0, 6.9 or 7.7; the temperature in each case was 22°C.

At pH 6.0 the concentration of OXA in the fish rapidly exceeded that in the water; C_{max} was 207 μg/g at 8 hours and there was a significant decline after this to 35 μg/g at 72 hours. The values are means of three fish and at this pH the coefficients of variation were very high, possibly because the buffering of the water was poor. At pH 6.9 C_{max} was 26.0 μg/g at 12 hours and there was a slight decline afterwards. At pH 7.7 C_{max} at 12.5 μg/g was only just above the water concentration.

Temperature. Goldfish were immersed in 10 mg/l OXA sodium at pH 6.9; temperatures of 12, 17 or 22°C were used. In all cases the concentrations in the fish rose continuously throughout the 72 hours immersion. Final concentrations were 33.7, 45.4 and 48.9 μg/g at 12, 17 and 22°C respectively.

(b) Ayu

A series of experiments has been conducted on the maintenance of freshwater ("lake origin") juvenile ayu in solutions of OXA for the purpose of disease prevention during transport. After 24 hours the fish contained higher concentrations of drug than the water, and the higher the concentration in the water the greater was the excess concentration in the fish. Thus fish in water containing 1 µg/ml OXA had 2 µg/g at 24 hours and this appeared to be a steady state. Fish immersed in a 10 µg/ml solution had 11 µg/g, and in a 20 µg/ml solution had 45.8 µg/g.

In synthetic seawater uptake was lower than in either a salt solution or freshwater. Clearance followed the same pattern. The findings were attributed in part to the divalent cations in the synthetic seawater, but it was also considered that its high pH (8.5) was involved since increasing PH in freshwater reduces the uptake of OXA by goldfish.

(c) Common carp

In a dose titration against an artificial infection with *Aeromonas liquefaciens* which produced 100% mortality in untreated controls, OXA was used 4 hours after infection for single immersions for a period of 24 hours. The drug was dissolved in 0.1N NaOH and the pH adjusted to 7-8. At concentrations of 0.3 mg/l less than half the fish survived for the observation period of 12 days. At 1 mg/l more than half the fish survived while at 3 mg/l all survived.

Pharmacokinetics following immersion were not reported but a titration of single doses by gavage showed 30 mg/kg to be fully effective. At this dose the serum concentration took about 5.5 days to fall below 0.2 ppm, the limit of quantitation in the assay then used. Tissue concentrations were in the order:

liver>kidney>serum>muscle.

In an acute tolerance study no adverse reactions were noted in carp kept in 25 mg/ml OXA at 19°C.

(d) Chinese loach

A similar study to that in common carp was made on Chinese loach (*Paramisgurnus dabryanus*) artificially infected with *Chondrococcus columnaris.* All untreated control fish died within 26 hours; immersion in OXA for a period of 24 hours starting 4 hours after infection produced 100% survival. However, while fish immersed in 10 µg/ml showed no evidence of infection, one of five fish immersed in each of 3 µg/ml and 1 µg/ml developed symptoms of columnaris disease.

(e) Halibut

Samuelsen and Lunestad (1996) investigated the possibility of administering OXA by immersion to juvenile halibut (*Hippoglossus hippoglossus*) (3-5 g bodyweight), with a particular interest in using the technique to control vibriosis. Since the fish were in seawater a significantly higher concentration of drug was necessary than would be normal for freshwater on account of both pH and divalent cation complexing; a concentration of 200 mg/l (= µg/ml) was selected, and the exposure time was 72 hours. The OXA was dissolved in 0.025M sodium hydroxide.

There were no deaths. C_{max} for the abdominal organs was 73 ppm and for muscle 9.4 ppm, and the authors noted that these levels were higher than normally achieved by oral dosage. Since the elimination half-lives were 15 hours for abdominal organs and 15.6 hours for muscle, it is likely that antibacterially effective concentrations would have been maintained in the fish for several days after the end of treatment. Using an MRL of 0.05 ppm the withdrawal period was estimated to be 9 days based on abdominal organs and 8 days based on muscle.

10.3.6 ENVIRONMENTAL SAFETY

(a) Immersion administration

Where OXA is used by immersion a very high proportion of it will enter the environment unused, *i.e.* never having been absorbed by a fish. As an extreme example, if a dip contains 5% live mass of fish and if the fish actively absorb the drug to four times the concentration in the water, there will still be 80% unused. In the case of OXA this will enter the environment in solution as the sodium salt. It may be expected to remain in the aqueous compartment of the environment where it may be microbiologically active in freshwater but will be complexed and inactive in seawater. Nevertheless OXA, presumably absorbed from solution, has been found in wild pelagic species of finfish (notably coalfish and mackerel) in the neighbourhood of marine fish farms where the drug has been used.

The small proportion of OXA which is absorbed will be excreted either in urine or bile. A significant fraction of the OXA in the bile may be conjugated but some of this may be transformed back to microbiologically active compounds in the intestine of the fish. Urinary OXA will remain in solution when voided to the environment. Faecal OXA may become bound to particulate material and eventually enter sediments under the water column; however this will constitute a minute fraction of the total OXA used and in tidal or riverine sites it will be dispersed over a very wide area.

Ultimately dissolved OXA will be photo-degraded, a process affected by temperature as well as light intensity.

(b) In-feed administration

The total OXA used may be considered in three parts, that in uneaten food, that not absorbed and that absorbed. The first two of these and a proportion of the third are likely to enter the environment bound to particulate matter. In this condition the drug has been found to enter filter-feeding invertebrates such as mussels and crabs, and demersal finfish such as haddock. While it is likely to be excreted by the finfish much as it is in the farmed species, less is known about its fate in invertebrates. In practice the quantities of OXA found in sediments below marine fish cages after its use are very much lower than is the case for oxytetracycline. This may be attributed to its much higher bioavailability. In one case where it was found it had become undetectable by 6 days after the end of the course of treatment. Depuration of the sediment is by dissolution into the water column; there is no chemical or microbiological degradation.

10.4 Piromidic acid

10.4.1 USES

(a) Spectrum of activity

Like oxolinic acid, piromidic acid (PMA) was developed in Japan, but it is very little used in fish anywhere else. In Japan it has been used in amago (masou) salmon, brook trout, ayu and Japanese and European eels.

Katae *et al.* (1979) showed that *in vitro* it was active against Gram-negative bacteria and staphylococci, and had greater activity than conventional antibacterial agents for fish against *Aeromonas* spp., *Pseudomonas* spp. and *Past. piscicida*; it also showed some activity against *Flexibacter* spp. It showed slightly better activity in acid than in neutral or alkaline solutions.

(b) Dose regimen

In freshwater species 10 mg/kg/day PMA has been found to be as efficacious as 20 or 40 mg/kg/day. A 5-day course of in-feed medication was found to be effective for an outbreak of furunculosis in amago salmon. Katae *et al.* (1979) found that a single injection of 6.1 mg/kg protected goldfish against an artificial infection with *Aeromonas hydrophila*; the formulation and route of injection were not stated.

10.4.2 PHARMACOKINETICS

(a) In amago salmon

In amago salmon dosed orally at 10 mg/kg/day for 5 days the blood levels were:

0.81 μg/ml 3 hours after the 1st dose
1.50 μg/ml 3 hours after the 2nd dose
1.39 μg/ml 3 hours after the 5th dose

T_{max} (plasma) after the first dose is about 6.25 hours (and possibly as little as 3 hours in ayu). These values may be compared with MICs in the range 0.1-1.0 μg/ml. C_{max} for both liver and kidney are higher than for plasma; and excretion is in urine and bile.

Residues in amago salmon dosed orally at 40 mg/kg/day for 7 days had fallen below 0.5 ppm in 96 hours.

(b) Eels

Absorption appears to be similar to that in amago salmon but both distribution to the tissues and metabolism are very rapid. After a single oral dose of 40 mg/kg C_{max} for PMA in different tissues is in the range 12-18 μg/g at T_{max} 3-6 hours. However PMA depletes with a half-life in the range 4-7 hours being metabolised by hydroxylation in either position 2 or 3 (see Figure 10.2). C_{max} for 2-hydroxy-PMA is 4-9 μg/ml and for 3-hydroxy-PMA 1.4-4.7 μg/ml, both occurring within 24 hours; and PMA is undetectable at 0.02 ppm at 4 days. The metabolites are in turn excreted rapidly with elimination half-lives in the range 8-18 hours. Total concentrations of PMA and metabolites in kidney and liver are 1.5 times those in plasma; in muscle they are half the level in plasma.

Following a course of 40 mg/kg/day for 7 days, T_{max} is 6 hours after the final dose. C_{max} is:

PMA	8.8-16 ppm
2-hydroxy-PMA	7-12 ppm
3-hydroxy-PMA	5.5-7.6 ppm

Neither PMA nor metabolites are detectable at 11 days.

10.4.3 TARGET SPECIES TOLERANCE

The acute oral LD50 for ayu is in excess of 2000 mg/kg. In eels this dose produced an eventual 100% mortality but at 1000 mg/kg all survived.
In sub-acute tolerance tests in goldfish 50% survived 2000 mg/kg/day for 10 days; 100% survived 1000 mg/kg/day for 10 days.

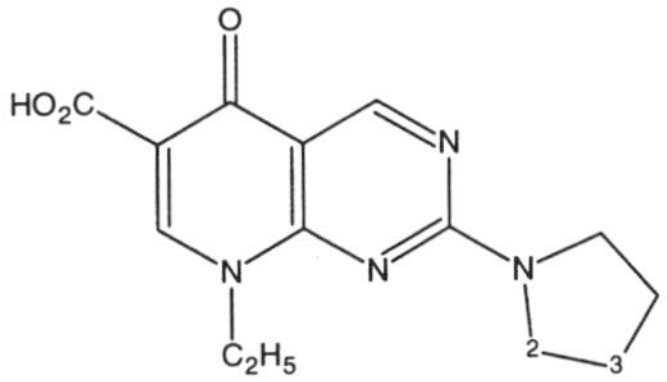

Piromidic Acid

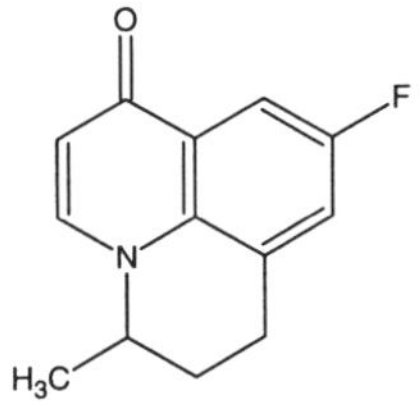

Flumequine

Figure 10.2.

10.5 Flumequine

10.5.1 USES

(a) Spectrum of activity

Flumequine (FLQ) is a fluoroquinolone or "second generation" 4-quinolone. As such its spectrum of activity covers not only Gram-negative bacteria but also fungi, protozoa and even some helminths. It is gradually replacing oxolinic acid in fish medicine because of its more appropriate pharmacokinetic profile. Furthermore its higher activity means a lower dose requirement; in marine conditions where high dose rates are necessary this makes for easier administration. In EU the MRL for FLQ in salmonids has been set at 150 μg/kg (see Section 3.6.3).

(b) Dose rates

The dose rate of FLQ is often given as a total to be divided over the period of treatment. Thus in freshwater 100 mg/kg is recommended to be given over 5-8 days. The clinician may choose between 12.5 mg/kg/day for 8 days and 20 mg/kg/day for 5 days. For seawater the total dose is in the range 125-200 mg/kg. Thus the doses in fresh and seawater have a much closer ratio than is the case for oxolinic acid.

A dose titration in yellowtail naturally infected with *Pasteurella piscicida* clearly showed 10 mg/kg/day to be the optimum rate for fish being closely monitored in tanks but 20 mg/kg/day was better under field conditions.

10.5.2 PHARMACOKINETICS IN RAINBOW TROUT

(a) Absorption and distribution

Courses of either 6 or 12 mg/kg/day for 5 days produced antibacterial concentrations in most tissues throughout the inter-dose interval but a rapid fall between 24 and 48 hours afterwards. Persistence was greatest in the liver and intestine suggesting biliary recirculation. There was also slow elimination from the gills suggesting that these organs may be a route of excretion.

(b) Elimination

Table 10.6. Elimination of flumequine from rainbow trout plasma

			3 °C	13 °C
INTRA-ARTERIAL	$t^{1/2}\ \alpha$	(hours)	6.3	2.9
	$t^{1/2}\ \beta$	(hours)	52.5	10.3
	$t^{1/2}\ \gamma$	(hours)	569	137
	Vd	(l/kg)	3.6	3.2
ORAL	C_{max}	(μg/ml)	1.07	1.91
	T_{max}	(hours)	54.5	19
	$t^{1/2}\ \alpha$	(hours)	38	16
	$t^{1/2}\ \beta$	(hours)	736	285

Sohlberg *et al.* (1994) studied the elimination of FLQ from rainbow trout plasma using single doses of 5 mg/kg intra-arterially or by oral administration at 3 and 13°C. Table 10.6 lists the salient findings, elimination being in three phases after intra-arterial injection and in two phases after oral administration; Tables 10.7 and 10.8 show the raw data.

In the oral study although T_{max} at 13° was lower than at 3°C the difference was insignificant compared to the terminal half-life; at 54.5 hours (T_{max} 3°) the residue level in the fish at the higher temperature was still higher than that in the fish at the lower temperature. Nevertheless the residues depleted faster at the higher temperature and the levels were the same for the two temperatures at 420 hours. Thereafter they were higher in the fish at the lower temperature.

Table 10.7. Plasma concentrations (μg/ml) of flumequine in cannulated rainbow trout in fresh water, after intraarterial administration of 5 mg/kg at 3°C (n = 5—6) and 13°C (n = 3—4), and intraarterial administration of 0.9% sodium chloride (control fish) at 13°C (n = 2). After Sohlberg *et al*, 1994, with permission

3°C		13°C		13°C	
Hours	Median (range) (μg/ml)	Hours	Median (range) (μg/ml)	Hours	Median (range) (μg/ml)
2	3.14 (2.82–4.15)	2	3.71 (2.49–5.71)	1	0.16 (0.13–0.22)
4	3.02 (2.15–3.41)	4	3.28 (2.54–3.50)	2	0.07 (0.00–0.13)
8	2.78 (2.26–13.4)	8	2.51 (1.84–3.05)	4	0.02 (0.00–0.04)
12	2.32 (1.90–4.94)	12	2.14 (1.90–2.37)	8	0
24	2.04 (1.70–2.87)	24	1.55 (1.26–1.94)	24	0.04 (0.00–0.08)
36	1.86 (1.51–2.57)	36	1.22 (1.09–1.36)	36	0.07 (0.05–0.09)
48	1.74 (1.28–1.97)	48	0.97 (0.81–1.28)	48	0.05 (0.00–0.10)
73	1.56 (1.10–2.00)	73	0.86 (0.70–1.05)	73	0.03 (0.00–0.05)
96	1.33 (1.00–1.50)	96	0.70 (0.53–1.08)	96	0
168	0.94 (0.64–1.23)	120	0.73 (0.36–0.98)	120	0
268	0.89 (0.61–1.04)	144	0.53 (0.27–0.98)	144	0
412	0.74 (0.37–0.98)	200	0.57 (0.40–1.10)	200	0
745	0.42 (0.17–1.02)	250	0.24 (0.00–0.85)	250	0
1070	0.28 (0.05–0.69)	350	0.16 (0.00–0.52)	350	0
1465	0.22 (0.00–0.66)	500	0.11 (0.00–0.42)	500	0

Pharmacokinetic parameters	$t/2\alpha$ (h)	$t/2\beta$ (h)	$t/2\gamma$ (h)	Vd_{SS} (l/kg)	Cl_T (l/h·kg)
3°C	6.3	52.5	569	3.6	0.005
13°C	2.9	10.3	137	3.2	0.018

Calculations were performed by estimating median values and range. The pharmacokinetic parameters were estimated from the median values by the computer program PCNONLIN, using a three-compartment open model.

Table 10.8. Plasma concentrations (μg/ml) of flumequine in cannulated rainbow trout in fresh water, after oral administration of 5 mg/kg at 3°C (n = 3—4) and 13°C (n = 3—4), and in intract fish after oral administration of 5 mg/kg at 13°C (n = 1). After Sohlberg *et al*, 1994, with permission

Cannulated fish				Intact fish	
3°C		13°C		13°C	
Hours	Median (range) (μg/ml)	Hours	Median (range) (μg/ml)	Hours	Plasma concentration (μg/ml)
2	0.28 (0.14–0.99)	2	1.39 (1.00–1.50)		
6	0.48 (0.25–2.05)	4	1.48 (1.37–1.58)		
9	0.43 (0.35–1.79)	8	1.72 (1.61–2.04)		
23	0.99 (0.55–1.53)	12	1.93 (1.61–2.00)		
35	0.94 (0.53–1.35)	24	1.85 (1.65–2.27)	24	2.4
49	1.14 (0.62–1.64)	35	1.79 (1.05–1.97)		
73	1.00 (0.66–1.98)	49	1.75 (0.82–1.87)	48	0.61
120	0.92 (0.63–1.53)	74	1.56 (0.92–2.23)	73	0.72
193	0.79 (0.68–1.16)	98	1.33 (1.09–2.22)	98	0.73
290	0.64 (0.49–0.75)	123	1.36 (0.98–2.17)		
439	0.57 (0.00–0.88)	144	1.21 (0.45–1.86)	144	0.46
771	0.43 (0.00–0.91)	194	1.21 (0.23–2.15)		
		241	1.07 (0.13–2.00)		
		337	0.90 (0.00–1.55)		
		505	0.38 (0.00–0.97)	505	0.21

Pharmacokinetic parameters	$t/2\alpha$ (h)	$t/2\beta$ (h)	C_{max} (μg/ml)	T_{max} (h)
3°C	38	736	1.07	54.5
13°C	16	285	1.91	19.1

Calculations were done by estimating median (range) values. The pharmacokinetic parameters were estimated from the median values by the computer program PCNONLIN, using a two-compartment open model.

The authors noted that temperature had a more profound effect on elimination than on distribution. In the case of the orally medicated fish this is attributed in part to the known effect of temperature on the rate of stomach emptying and hence on the arrival of the drug at the site of absorption in the intestine. However this does not adequately explain the slow elimination in the intra-arterially treated fish. The slow elimination at either temperature means that a single dose may produce clinically effective plasma concentrations for long periods.

10.5.3 PHARMACOKINETICS IN ATLANTIC SALMON

(a) Flumequine-medicated feed

Atlantic salmon kept in seawater at 8°C and offered a diet containing 5 g/kg FLQ were found to feed at only 0.45% of bodyweight, thus receiving 22.5 mg/kg/day. A steady state concentration of FLQ in plasma was reached after 3 days, and in both liver and muscle after 2 days, the levels being 14 ppm in liver, 9 ppm in muscle and 4 ppm in plasma. The drug was undetectable at a limit of detection of 0.01 ppm at 7 days after the end of treatment.

The terminal half-lives in plasma, liver and muscle were all 21 hours indicating rapid transference of drug between these tissues. In skin the half-life was 33 hours and in consequence its AUC value was twice that of muscle. If the EU MRL were to be applied to these data the musculature of 50% of fish would be expected to be clear of residues in 5 days; taking into account the persistence of FLQ in skin a withdrawal period of 7 days would probably be appropriate.

FLQ has been studied in different types of feed and the bioavailability found to be constant at about 40%. This limit may be due to malabsorption, first pass metabolism or both. Malabsorption could be related to solubility since FLQ is practically insoluble at neutral or acid pH. At pH 9 the solubility is 6 g/l.

(b) Aqualets®

Aqualets is the Alpharma trade mark for patented 2-layer co-extruded pellets consisting of an outer layer of nutrients and binding agents and a core of drug and excipients. Rogstad *et al.* (1993) studied this formulation of FLQ in Atlantic salmon in seawater using single doses by gavage of 25 and 50 mg/kg. Simultaneous observations were made on normal and Aqualets® formulations of oxolinic acid. The results are shown in Table 10.9.

It may be noted that T_{max} (plasma) increases with the dose. Whereas it is 12 hours for both plasma and liver after a dose of 25 mg/kg, it is 24 hours for muscle; this is at variance from the finding with unformulated FLQ in feed. The ratio of C_{max} (muscle) to C_{max} (plasma) is 2.9, comparable with the 2.25 for unformulated FLQ but the liver: plasma ratio is 8.6 as against 3.5.

(c) Immersion

Like other quinolones, FLQ is absorbed by to a significant extent from water, and the uptake from solution can be enhanced by the addition of surfactants to the water. Tween® 20, 40, 60 and 80, Aerosil® and Roccal® have all been shown to increase uptake by pre-smolts from a bath of 50 ppm. Roccal® was the most efficacious in this regard but at 10 μg/ml it was toxic (probably due to causing FLQ overdose). The symptoms of overdosage are similar to those of anaesthesia.

Table 10.9. Pharmacokinetic parameters for flumequine and oxolinic acid in Atlantic salmon following a sigle oral dose of Apoquin and Apoxolon Aqualets and medicated feed with oxolinic acid. Seawater temperature 5±0.2°C; fish weight 425 g

	Parameter	Flumequine (Apoquin)		Oxolinic acid (Apoxolon)	Oxolinic acid (medicated feed)	
	Dose (mg/kg b.w.)	25	50	25	25	50
Plasma	AUC_{0-t} (μg·h/ml)	110	184	46.6	47.1	58.9
	AUC_{0-t}/dose (h/l)	4.4	3.7	1.9	1.9	1.2
	C_{max} (μg/ml)	2.26	3.83	0.99	0.87	1.17
	T_{max} (h)	12	24	24	24	24
	F (%)	46	39	40	40	25
Muscle	AUC_{0-t} (μg·h/g)	402		286		
	AUC_{0-t}/dose (h/l)	16.1		11.4		
	C_{max} (μg/g)	6.58		5.72		
	T_{max} (h)	24		24		
Liver	AUC_{0-t} (μg·h/g)	729		280		
	AUC_{0-t}/dose (h/l)	29.2		11.2		
	C_{max} (μg/g)	19.5		6.83		
	T_{max} (h)	12		12		
AUC_{0-t} (muscle)/AUC_{0-t} (plasma)		3.7		6.1		
AUC_{0-t} (liver)/AUC_{0-t} (plasma)		6.6		6.0		

AUC: Area under the concentration-time curve
C_{max}: Concentration at the maximum of the absorption curve.
T_{max}: Time for concentration maximum of the absorption curve.
F: Apparent bioavailability.

Smolts with stress-inducible furunculosis (SIF) have been shown to be protected by 35 mg/ml FLQ (provided by immersion in 50 ppm for 75 minutes at 11°C). A similar treatment has protected ailing fish where untreated controls suffered 32% mortality. However immersion administration of any drug to a farmed species such as Atlantic salmon entails to a serious extent the problems of wastage (and hence cost) and environmental contamination discussed in Section 1.1.1. Hiney *et al.* (1995) investigated the use of FLQ to control SIF resulting from the transference of smolts from freshwater to seawater. To minimize the wastage they used it in the water in which the transference was actually made, with the smolts at a very high stocking density. The period of exposure was short - of the order of 15 minutes - as the transference was done by helicopter. Preliminary trials in freshwater showed that at this duration of exposure 100 μg/ml would cause the absorption of sufficient FLQ to inactivate *Aeromonas salmonicida.*

The regimen failed to control SIF. Further investigations revealed major differences in FLQ pharmacokinetics between Atlantic salmon in freshwater and in seawater. The elimination half-life was found to be significantly shorter in seawater than in freshwater for serum, muscle, kidney and liver. T_{max} for intestinal contents was 144 hours after dosing for fish in seawater. It was concluded that excretion into the intestinal lumen, presumably in the bile, occurred to a highly significant extent in fish in seawater, and that once there the drug would be complexed with divalent cations in the intestinal contents.

10.5.4 PHARMACOKINETICS IN OTHER SPECIES

(a) Brown trout

O'Grady *et al.* (1988) studied the uptake of FLQ by brown trout immersed in varying concentrations, temperatures, water hardness and pH.

- Drug concentration - At pH 7 concentrations of 50, 100 or 200 ppm all produced serum levels in excess of 30 μg/ml in an hour. T_{max} was 3 hours; and since for all concentrations C_{max} was in the range 35-50 μg/ml there was clearly no advantage (and some environmental hazard) in using 200 ppm.
- pH - As pH was increased from 6.4 to 9.0 there was a progressive decline in uptake (despite the increasing solubility of FLQ); no drug was absorbed from 50 ppm at pH 8 or from 100 ppm at pH 9.2.
- Temperature - At pH 7 C_{max} was linearly related to temperature in the range 3-15°C; 50 ppm produced C_{max} of 4 μg/ml at 3°C and 38 μg/ml at 15°C; 100 ppm produced 3 μg/ml at 3°C and 38 μg/ml at 11°C.
- Hardness - At pH 7, 11°C and 50 ppm, C_{max} was 29 μg/ml in calcium-free water. There was a steep fall in uptake to 16 μg/ml in water containing 74 mg/l calcium carbonate and a slower fall to 10 μg/ml in water containing 200 mg/l.
- Feeding - Whether the trout were feeding or not had no effect on uptake.

In fish with C_{max} (serum) of 43 μg/ml the level fell to 21 μg/ml in 24 hours and then more slowly to 11 μg/ml at 3 days. Elimination is slower following intraperitoneal injection than following immersion, presumably because absorption into the blood is slow.

Fish immersed in 50 ppm at 11°C and pH 6.5 took up 35 μg/ml in an hour and became lethargic. They recovered in an hour in unmedicated water although their drug levels would have fallen little in that time. Fish kept in the medicated water lost equilibrium at 4-5 hours, having serum levels of 69 μg/ml FLQ. Nine out of ten fish regained equilibrium after an hour in unmedicated water.

(b) European eels

Boon *et al.* (1991) found that when a single intramuscular injectionof FLQ was given to European eels at 22-25°C the plasma concentration/time curve showed 3 distinct phases. Absorption from the injection site into the blood was exceptionally slow (half-life 33 hours). This may be attributable to two separate factors: the use of a suspension for injection due to the insolubility of FLQ at neutral pH, and the poor vascularization of the musculature of eels. The distribution phase was unremarkable; but the elimination half-life was 256 hours, and this could not be attributed to the slow absorption from the injection site. A possible explanation lies in a significant route of excretion of FLQ in many species being the gills, but in the eel the gill surface is small and 60% of gaseous exchange is thought to occur through the skin.

(c) Halibut

Samuelsen and Lunestad (1996) found that higher concentrations of FLQ were absorbed by juvenile halibut, a marine species, through immersion than were normally achieved by oral dosage at 25 mg/kg/day in other species. 200 mg/l FLQ at 12°C caused 20% mortality with very high tissue concentrations; 150 mg/l caused no deaths and produced final concentrations of 85.4 μg/g in abdominal organs and 14.2 μg/g in muscle. The concentrations appeared to be still rising when the exposure was terminated after 72 hours; the rate of absorption was low compared to FLQ in freshwater species. The terminal half-life was estimated to be 10 hours for tissues, and the drug was undetectable 7 days after the end of exposure.

10.6 Other fluoroquinolones

10.6.1 ENROFLOXACIN

(a) Pharmacology in rainbow trout

Because enrofloxacin (EFX) has a higher *in vitro* activity against Gram-positive bacteria than older quinolones, it has been investigated for possible use for bacterial kidney disease (*Renibacterium salmoninarum* infection, BKD). In dose titrations 2.5 mg/kg/day for 10 days was found to be effective for BKD and 10 mg/kg/day for 10 days for furunculosis. It was also found that feed medication rates high enough to give dose rates of the order of 40 mg/kg/day were unpalatable to juvenile rainbow trout. Furthermore when the drug is given by gavage the bioavailability is significantly reduced at higher dose rates. Bioavailability is also affected by temperature, being higher at 15° than at 10°C. Overall, bioavailabilities ranging from 9% (50 mg/kg at 10°C) to 53 % (5 mg/kg at 15°C) have been recorded.

Following intravenous injection at either 5 or 10 mg/kg into fingerling rainbow trout EFX showed volumes of distribution of 3.2 and 2.6 l/kg respectively. These values are indicative of very high tissue penetration but the lower figure for the higher dose rate suggests some degree of tissue saturation. This is supported by terminal half-life results in serum at 10°C of 29.5 hours after single oral doses of either 10 or 50 mg/kg but the longer 44 hours after 5 mg/kg. The lower bioavailability at 10° than at 15°C with the same half-life means that metabolism is not extensive.

As much as 10 mg/kg/day has been recommended for the control of *Vibrio anguillarum* infection in rainbow trout; in fact 5 mg/kg/day should develop adequate serum levels for a majority of bacterial pathogens but the moderate excretion rate and high volume of distribution mean that it would take 5-9 days, depending on temperature, for a steady state to be reached. In an acute toxicity test EFX produced no gross lesions at up to 400 mg/kg. In a sub-acute test 50 mg/kg/day produced darkened skin and the fish swam lower in the water; there was eosinophilia and the spleens were congested.

(b) Pharmacology in Atlantic salmon

Stoffregen *et al.* (1993) studied the use of EFX for the control of furunculosis in Atlantic salmon as an Investigational New Animal Drug (INAD) in USA. They were faced with the problem of Romet-30® being unpalatable at the dose rate deemed necessary; the bacteria being resistant to oxytetracycline, and sulfamerazine, the only other approved antibacterial agent, not being available commercially. The EFX was mixed into the feed before pelleting, and with the objective of dosing at 10 mg/kg/day they used

1.0 g/kg to be fed at 1% in post-smolts
1.3 g/kg to be fed at 0.8% in market-weight fish
2.0 g/kg to be fed at 0.5% in broodstock

The medicated feed was palatable at all inclusion rates. In 24 hours all tissue concentrations were approximately 66% of the steady state values, these initial concentrations being a mean of 25 times the MIC. After 2 days the surviving fish were behaving and feeding normally and mortality in post-smolts was reduced from 5.8% to 0.6% per month. The tissue concentrations imply that the selected dose rate was in fact excessive despite it being only twice the normal for freshwater.

Residue levels were determined by bioassay 9 days after the end of medication and are shown together with estimated half-lives in Table 10.10. Residues were persistent in the edible tissues, and the EU MRL of 30 μg/kg (0.03 ppm) would not have been reached until about 5 months in muscle and later in skin and fat. Furthermore this study was in seawater and in salmon quinolones have longer half-lives in freshwater. EFX has in fact the longest half-life of any quinolone commonly used in fish. Following it in order of reducing persistence are FLQ, OXA and sarafloxacin.

Table 10.10. Elimination of enrofloxacin by Atlantic salmon

	Residue level after 9 days (ppm)	Estimated half-life (days)
Muscle	4.58	11
Skin+fat	7.32	22
Liver	4.58	2
Kidney	8.89	3
Gills	3.4	6

10.6.2 SARAFLOXACIN

(a) Market authorization

Sarafloxacin is now the subject of an MA in some EU member states. It is presented as a white to yellow free-flowing powder (Sarafin®) which is 100% sarafloxacin hydrochloride, equivalent to 82% sarafloxacin. It is authorized for use in the treatment of furunculosis (*Aeromonas salmonicida* infection) in Atlantic salmon in seawater at a dose regimen of 10 mg/kg/day for 5 days. The withdrawal period is 150 degree-days.

Sarafin may be incorporated in feed prior to pelleting or mixed with feed just prior to feeding. For on-farm mixing it is recommended that the product should first be suspended in edible vegetable oil. A quarter of the feed is placed in the mixer, the Sarafin-oil suspension is added and mixed, and then the remaining three-quarters of the feed is added and thoroughly mixed.

(b) Pharmacokinetics in Atlantic salmon

The pharmacokinetics of sarafloxacin (SFX) in Atlantic salmon in seawater have been reported in a series of publications by Martinsen *et al.*, reviewed in Martinsen *et al.* (1994). They found the bioavailability (F) of SFX at 8.5°C to be generally low and significantly affected by the vehicle in which it was administered. In two different feeds and in capelin oil it was in the range 3.6-7.4% whereas in corn oil it was 23.9%. The feeds delayed absorption and so T_{max} was affected both by F and the vehicles. For capelin oil T_{max} was 6 hours but for corn oil, with the substantially higher F, it was 12 hours; one of the feeds also delayed T_{max} until 12 hours, but the other delayed it to 24 hours, which was so long that significant distribution had taken place and so C_{max} was lower than with capelin oil.

Intravenous dosage showed SFX to have the very large volume of distribution (Vd) of 4.1 l/kg. A study with radio-labelled drug showed that it was rapidly distributed to all organs and tissues except the central nervous system, and, as would be expected with so large a Vd, reached a higher concentration in most tissues than in the blood. Nevertheless due to the low bioavailability the absolute values for the tissue concentrations were not high except in the intestine and bile. First pass biliary excretion is unlikely to be significant because T_{max} in bile was 24 hours, which was longer than for any other tissue except muscle.

The elimination half-life from plasma was initially thought to be 16 hours but in subsequent more detailed work it was re-assessed at 24 hours. Radio-activity became undetectable in blood and muscle at 7 and 14 days respectively, but continued to be detectable in skin and kidney at 28 days. The disposition of the drug indicated that both biliary and urinary excretion were occurring.

(c) Immersion administration to Atlantic salmon

SFX has been investigated as a possible treatment for stress-inducible furunculosis (SIF) in smolts. 10 minute immersions in 0, 50, 250 or 500 mg/l SFX with 0.01% Tween 80® added produced absorption linearly related to the concentration, and in all cases many-fold higher than the MIC.

Aerosil-OT® also enhanced absorption but to a lesser degree than Tween 80®. 5.5% sodium chloride was tried as a form of one-step hyperosmotic infiltration but proved stressful; it could only be used for 5 minutes as 10 minutes produced some fatalities.

(d) In other species

In addition to Atlantic salmon, Martinsen *et al.* (1994) conducted their radio-labelled study in rainbow trout, cod and turbot, all in seawater. F was much higher in cod than the other species, resulting in C_{max} (plasma) an order of magnitude higher at T_{max} only 4 hours after oral administration. The authors speculated that this could have been due to the specialized intestinal epithelium and branched pyloric caeca characteristic of the cod family (*Gadidae*).

Also in cod the distribution of SFX was different from the other species, the liver concentration being low and the muscle concentration being high. This was attributed to SFX being poorly accumulated in lipids; cod is a lean species with fat reserves in the liver, whereas salmonids have adipose tissue in the musculature.

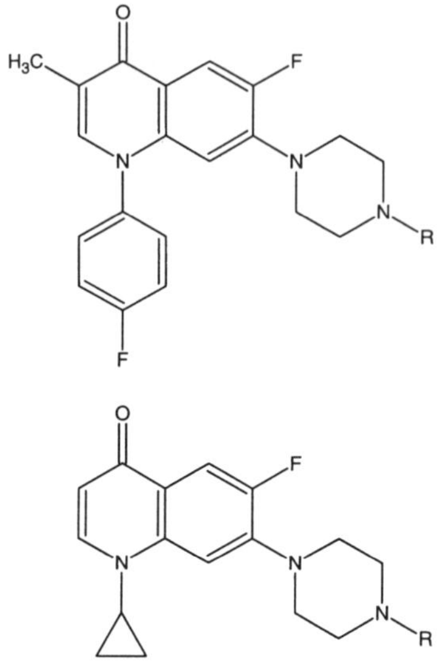

R=H Sarafloxacin
R=CH_3 Difloxacin

R=H Ciprofloxacin
R=C_2H_5 Enrofloxacin

Figure 10.3.

10.6.3 DIFLOXACIN

A dose titration has been conducted on difloxacin in the treatment of Atlantic salmon artificially infected with furunculosis. Initially it was found that there was no significant difference in efficacy between 2.5, 5 and 10 mg/kg/day for 5 days. Subsequently 1.25, 2.5 and 5 mg/kg/day were tried for 5 or 10 days and again there was no significant difference in the results.

10.6.4 CIPROFLOXACIN

Nouws *et al.* (1988) studied the pharmacokinetics of ciprofloxacin in the freshwater species, common carp, African catfish and rainbow trout. Each species was kept at an appropriate temperature, *i.e.* 20, 25 and 12°C respectively, and the determinations in Table 10.11 were made.

Table 10.11. Major pharmacokinetic parameters of ciprofloxacin in three fish species

		Common carp	African catfish	Rainbow trout
Plasma protein binding		22%	20%	23%
Volume of distribution	(l/kg)	3.1	5.6	4.7
AUC in plasma	(μg.h/ml)	104	55	53
Terminal half-life	(hours)	14.5	14.2	11.2

Of the species studied, African catfish had the greatest volume of distribution, although in all three it was very large, as is the case with SFX in Atlantic salmon in seawater. Additionally, in African catfish distribution of the drug was very rapid, equilibrium being reached only 7 hours after an intravenous injection of 15 mg/kg, probably an unnecessarily large dose. In the other two species intramuscular injections were also investigated and these produced elimination rates which were slow initially, probably due to slow absorption from the injection site. The elimination half-life was of the same order of magnitude in all three species. There was no evidence of catabolism or conjugation of the drug in any species.

Further reading

Bjørklund, H., Eriksson, A. and Bylund, G. (1992) Temperature-related absorption and excretion of oxolinic acid in rainbow trout (*Oncorhynchus mykiss*). *Aquaculture* **102**, 17-27.

Boon, J.H., Nouws, J.M.F., van der Heijden, M.H.T., Booms, G.H.R. and Degen, M. (1991) Deposition of flumequine in plasma of European eel (*Anguilla anguilla*) after a single intramuscular injection. *Aquaculture* **99**, 213-223.

Bowser, P.R., Wooster, G.A., St. Leger, J. and Babish, J.G. (1992) Pharmacokinetics of enrofloxacin in fingerling rainbow trout (*Oncorhynchus mykiss*). *Journal of Veterinary Pharmacology and Therapeutics* **15**, 62-71.

Cravedi, J.-P., Choubert, G. and Delous, G. (1987) Digestibility of chloramphenicol, oxolinic acid and oxytetracycline in rainbow trout and influence of these antibiotics on lipid digestibility. *Aquaculture* **60**, 133-141.

Hiney, M.P., Coyne, R., Kerry, J., Pursell, L., Samuelsen, O.B. and Smith, P. (1995) Failure of Flumisol bath treatments during commercial transport of Atlantic salmon smolts to prevent the activation of stress inducible furunculosis. *Aquaculture* **136**, 31-42.

Hustvedt, S.O., Salte, R. and Vassvik, V. (1991a) Absorption, distribution and elimination of oxolinic acid in Atlantic salmon (*Salmo salar* L.) after various routes of administration. *Aquaculture* **95**, 193-199.

Hustvedt, S.O., Salte, R., Kvendset, O. and Vassvik, V. (1991b) Bioavailability of oxolinic acid in Atlantic salmon (*Salmo salar* L.) from medicated feed. *Aquaculture* **97**, 305-310.

Ishida, N. (1992) Tissue levels of oxolinic acid after oral or intravascular administration to freshwater and seawater rainbow trout. *Aquaculture* **102**, 9-15.

Jacobsen, M.D. (1989) Withdrawal times of freshwater rainbow trout, *Salmo gairdneri* Richardson, after treatment with oxolinic acid, oxytetracycline and trimethoprim. *Journal of Fish Diseases* **12**, 29-36.

Jarboe, H., Toth, B.R., Shoemaker, K.E., Greenlees, K.J. and Kleinow, K.M. (1993) Pharmacokinetics, bioavailability, plasma protein binding and disposition of nalidixic acid in rainbow trout (*Oncorhynchus mykiss*). *Xenobiotica* **23**, 961-972.

Katae, H., Kouno, K., Takase, Y., Miyazaki, H., Hashimoto, M. and Shimizu, M. (1979) The evaluation of piromidic acid as an antibiotic in fish: an *in vitro* and *in vivo* study. *Journal of Fish Diseases* **2**, 321-335.

Kleinow, K.M., Jarboe, H.M., Shoemaker, K.E. and Greenlees, K.J. (1994) Comparative pharmacokinetics and bioavailability of oxolinic acid in channel catfish (*Ictalurus punctatus*) and rainbow trout (*Oncorhynchus mykiss*). *Canadian Journal of Fisheries and Aquatic Sciences* **51**, 1205-1211.

Martinsen, B., Horsberg, T.E., Ingebrigtsen, K. and Gross, I.L. (1994) Disposition of ^{14}C-sarafloxacin in Atlantic salmon *Salmo salar*, rainbow trout *Oncorhynchus mykiss*, cod *Gadus morhua* and turbot *Scophthalmus maximus*, as demonstrated by means of whole-body autoradiography and liquid scintillation counting. *Diseases of Aquatic Organisms* **18**, 37-44.

Miyazaki, T., Nakauchi, R. and Kubota, S.S. (1984) Toxicological examinations of oxolinic acid and nalidixic acid in yellowtail. *Bulletin of the Faculty of Fisheries, Mie University* **11**, 15-26.

O'Grady, P., Moloney, M. and Smith, P.R. (1988) Bath administration of the quinolone antibiotic flumequine to brown trout *Salmo trutta* L. and Atlantic salmon *Salmo salar* L. *Diseases of Aquatic Organisms* **4**, 27-33.

Oida, T., Kuono, K., Katae, H., Nakamura, S., Sekine, Y. and Hashimoto, M. (1982) Levels of piromidic acid and its metabolites in eel tissues. *Bulletin of the Japanese Society for Scientific Fisheries* **48**, 1599-1608.

Rogstad, A., Ellingsen, O.F. and Syvertsen, C. (1993) Pharmacokinetics and bioavailability of flumequine and oxolinic acid after various routes of administration to Atlantic salmon in seawater. *Aquaculture* **110**, 207-220.

Samuelsen, O.B. (1997) Efficacy of bath-administered flumequine and oxolinic acid in the treatment of vibriosis in small Atlantic halibut. *Journal of Aquatic Animal Health* **9**, 127-131.

Samuelsen, O.B. and Lunestad, B.T. (1996) Bath treatment, an alternative method for the administration of the quinolones flumequine and oxolinic acid to halibut *Hippoglossus hippoglossus* and *in vitro* antibacterial activity of the drugs against some *Vibrio* sp. *Diseases of Aquatic Organisms* **27**, 13-18.

Stoffregen, D.A., Chako, A.J., Backman, S. and Babish, J.G. (1993) Successful therapy of furunculosis in Atlantic salmon *Salmo salar* L. using the fluoroquinolone antimicrobial agent enrofloxacin. *Journal of Fish Diseases* **16**, 219-227

Ueno, R., Horiguchi, Y. and Kubota, S.S. (1988a) Levels of oxolinic acid in cultured yellowtail after oral administration. *Nippon Suisan Gakkaishi* **54**, 479-484.

Ueno, R., Okumura, M., Horiguchi, Y. and Kubota, S.S. (1988b) Levels of oxolinic acid in cultured rainbow trout and amago salmon after oral administration. *Nippon Suisan Gakkaishi* **54**, 485-489.

Uno, K., Kato, M., Aoki, T., Kubota, S.S. and Ueno, R. (1992) Pharmacokinetics of nalidixic acid in cultured rainbow trout and amago salmon. *Aquaculture* **102**, 297-307.

11. OTHER SYSTEMIC ANTIBACTERIAL AGENTS

11.1 Nitrofurans

11.1.1 GROUP CHARACTERISTICS

(a) Activity

The nitrofurans are synthetic antimicrobial agents with a broad spectrum covering not only Gram-positive and Gram-negative bacteria but also several types of protozoan parasites. They are normally bacteriostatic but can be bactericidal at high doses. They are more active in acid conditions but have low potency compared to other antibacterial drugs and high doses are necessary.

Bacterial resistance develops only slowly; where it occurs there is complete cross-resistance to all other drugs in the group but no cross-resistance to other drug groups.

Most nitrofurans are very poorly absorbed from the gastro-intestinal tract in mammals and birds and are useful only against pathogens which are external or in the gut lumen. Nitrofurantoin is an exception; it is absorbed but is immediately excreted so is useful for urinary tract infections. In fish also absorption of most nitrofurans from the gastro-intestinal tract is poor, but furazolidone is well absorbed by fish and where it is used it is normally administered in feed. Some nitrofurans not absorbed from the gastro-intestinal tract are absorbed from water and can be administered by immersion. This again does not apply to all; one of the best known nitrofurans in mammalian medicine, nitrofurazone, has been found not to be absorbed by gilt-head sea bream (*Sparus aurata*) or Mozambique tilapia (*Sarotherodon mossambicus*) in seawater nor by common carp in freshwater.

Nitrofurans give positive results to some *in vitro* tests for carcinogenicity and are banned, at least for food-producing species, in a number of countries including the whole of the EU and North America. For ornamental fish, where much medication is actually carried out by amateurs, it needs to be recognized that nitrofurans do constitute a hazard to operators.

(b) Nitrofurans used in fish medicine

The two nitrofurans most commonly used in fish medicine are furazolidone and nifurpirinol. In addition, nitrofurantoin has been investigated for possible use. Another compound, sodium nifurstyrenate, which is chemically related although not strictly a nitrofuran, has been used in Japan in a variety of fish species.

11.1.2 FURAZOLIDONE

(a) Salmonids

Heaton and Post (1968) studied the use of furazolidone (FZD) in four species of salmonid, representing three genera, *viz.* brook trout (*Salvelinus fontinalis*), brown trout (*Salmo trutta*),

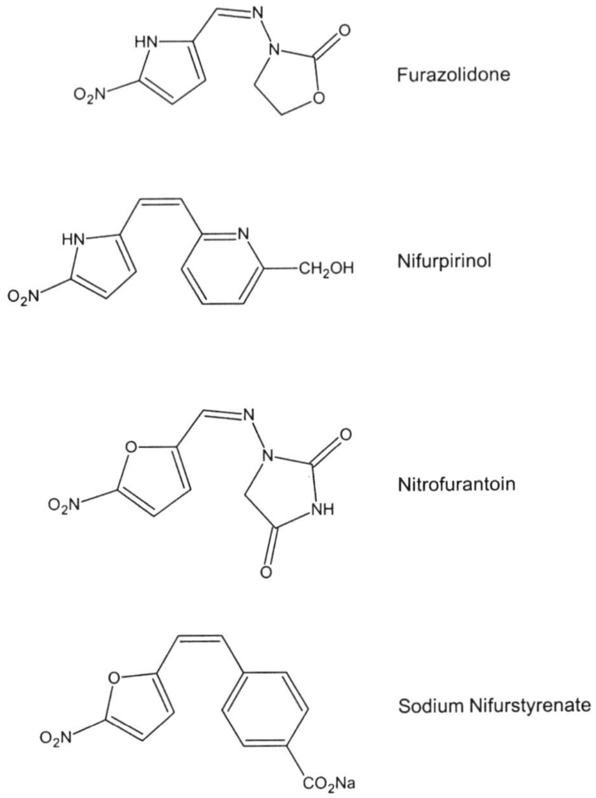

Furazolidone

Nifurpirinol

Nitrofurantoin

Sodium Nifurstyrenate

Figure 11.1.

rainbow trout (*Oncorhynchus mykiss*) and cutthroat trout (*O. clarki*). An in-feed dose regimen of 25-35 mg/kg/day for 20 days proved reliably efficacious in all species although the cutthroat trout appeared to need less than the others. There were limits to the medication rates for feeds to be acceptable to the fish; these were estimated to be:

for brown trout 2.0 g/kg
for rainbow trout 4.2 g/kg
for brook trout 3.8 g/kg

In the case of cutthroat trout, fingerlings would accept only 3.3 g/kg (reported as "150 g/100 lb") whereas "catchables" would take 3.75 g/kg ("170 g/100 lb"). In another trial hybrid rainbow x cutthroat trout medicated in feed at 152-194 mg/kg/day (an inexplicably high dose rate) for a year consumed less than they were offered and grew at only half the rate of unmedicated controls. However the FZD medication completely prevented whirling disease which had an 18% incidence in the controls.

In the four species study by Heaton and Post (1968) only musculature was assayed for residues, the limit of detection being 0.1 ppm. T_{max} (muscle) was 10 days and 15 days for brown and cutthroat trout respectively, although the medication was continued for 20 days. The production of inducible enzymes by the fish was suggested as a possible reason for the fall in concentration of drug after T_{max} and before the end of medication. Residues were below the level of detection by 10 days in all species, and in rather less time in brook trout.

In target species toxicity trials brook trout were found to tolerate single doses of up to 2200 mg/kg FZD and brown and rainbow trout tolerated 500 mg/kg/day for 14 days or 50 mg/kg/day for 40 days.

(b) Channel catfish

The oral bioavailability of FZD in channel catfish has been estimated to be 28% from feed and 58% from solution.

Plakas *et al.* (1994) found that following intravenous injection the parent drug was eliminated extremely rapidly from plasma; the half-life is about 16 minutes at 24°C. Following oral administration T_{max} for plasma is 1.2 hours and the drug is undetectable (<0.02 ppm) at 5 hours. The half-life of the parent drug is 0.6 hour, but the half-life of radio-carbon label is 35 hours, indicating significant metabolism of FZD. The parent compound is 40% bound to plasma protein; but the percentage binding increases with time, suggesting that the metabolites bind to a greater extent than the parent drug.

Renal excretion accounts for about 55% of the ^{14}C in an intravenous dose. Residual radio-carbon in muscle at 8 hours is equivalent to 0.274 ppm FZD but only about 10% of this is actually parent drug, and at 24 hours parent drug is undetectable (<0.001 ppm). The radio-carbon is however persistent in bound metabolites and at 168 hours is equivalent to 0.059 ppm furazolidone (half-life 97 hours). The bioavailability and toxicity to consumers of these bound metabolites is unknown.

(c) Environmental safety

In a tank trial using an artificial sediment into which FZD had been homogenized, the drug had a low persistence at 4°C (half-life 18 hours). The parent drug was undetectable at 10 days as was antibacterial activity. Metabolites persisted for much longer. The FZD initially killed half the bacteria in the sediment, and the flora took several months to recover even though there was no detectable antibacterial activity for most of the time.

11.1.3 NITROFURANTOIN

Stehly and Plakas (1993) studied the pharmacokinetics of nitrofurantoin (NFT) in channel catfish using intravenous radio-labelled drug and oral unlabelled drug. As for FZD in this species, NFT in plasma is eliminated very rapidly: an intravenous dose of 1 mg/kg was undetectable (<0.005 ppm) at 4 hours, and the mean residue of a dose of 10 mg/kg was just detectable after 8 hours although the regression line calculated from the data went below the level of detection. This is illustrated in Figure 11.2 where the different terminal elimination slopes confirm an active elimination process for the radio-label. Despite this rapid depletion there was limited tissue distribution; clearance was rapid with half-lives of 23 minutes for the 1 mg/kg dose and 46 minutes for the 10 mg/kg dose.

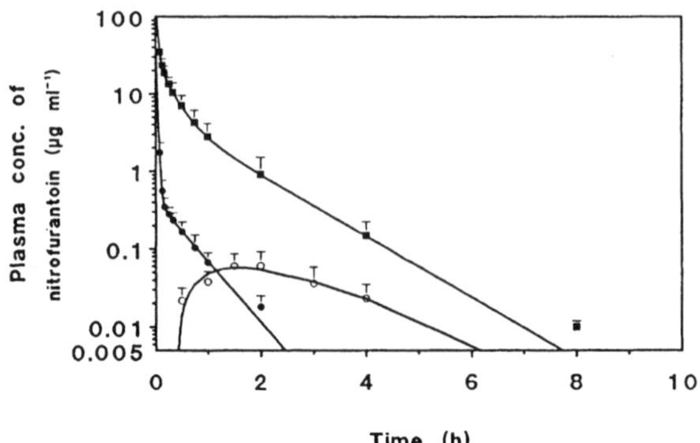

Figure 11.2. Mean plasma concentrations (µg ml^{-1}) of parent nitrofurantoin in channel catfish after intravascular dosing at 1 mg kg^{-1} (●) (n = 5) or 10 mg kg^{-1} (■) (n = 4) and after oral dosing at 1 mg kg^{-1} (○) (n = 5). The intravascular curves represent the best fit for the means (±s.d.). The oral curve represents the fit obtained by curve stripping. After Stehly G.R. and Plakas S.M., 1993, with permission.

Some NFT is metabolised before elimination: it was calculated that as little as 5 minutes after the intravenous dose only 80% of the radio-label in the plasma was parent NFT.

Following oral administration the bioavailability was 17% and T_{max} (plasma) 2 hours. The limited tissue distribution meant that muscle residues were very low; the main residues were in the excretory tissues, *i.e.* liver, kidney and gills. The urine contained predominantly unchanged NFT but the proportion of metabolites increased with time.

11.1.4 NIFURPIRINOL

Nifurpirinol (NPL) was originally developed by the Japanese company Dainippon specifically for aquacultural use. It is sold under the trade mark 'Furanace' and this name is frequently used in the scientific literature about it. Although it is active when given orally it is usually administered by immersion and can be used in both fresh and seawater.

(a) Coho salmon

Amend and Ross (1970) used coho salmon in a dose titration of NPL for columnaris disease. The drug was administered by flush medication (see Section 1.1.4) in a trough at 21°C. Fish were exposed to infection 24 hours previously and given 1 hour treatments at 0, 0.25, 0.5, 1.0, 2.0 ppm. All medications except the 0.25 ppm level prevented the development of the disease and that level reduced mortality. A second test was then conducted on fish in which symptoms were already established. Following exposure to infection mortalities began on the third day and so treatment was started on the fourth day. Concentrations of 0.5 and 1.0 ppm were each used for two flushes of 1 hour at an interval of 24 hours and both regimens controlled the disease.

A further trial investigated the ability of NPL to prevent the recurrence of the disease after the end of medication. It was found that even where the disease was well established (cumulative mortality over 17%) a single flush at 1.0 ppm was effective in controlling mortality with no recurrence.

With a single treatment at 1.0 ppm a bioassay showed T_{max} for all tissues to be less than 1 hour. After the end of the treatment residue levels fell rapidly; they were below the limit of detection (0.02 ppm) in all tissues except kidney by 24 hours, and in the kidney by 48 hours. At no time, even immediately at the end of treatment, was NPL detected in muscle tissue.

In an acute toxicity trial there was no mortality after exposure for 1 hour to 10 ppm NPL (which is a virtually saturated solution). In a chronic toxicity trial fish were treated at 1.0 ppm 5 days a week for 8 weeks. No gross or histological changes were noted except an increase in immature erythrocytes and a decrease in the leucocyte count.

(b) Rainbow trout

Egidius and Andersen ((1979) investigated the use of NPL in rainbow trout kept in seawater, stressed by an abrupt temperature change from 9° to 12°C, and artificially infected with *Vibrio anguillarum*. The drug prevented any mortality if used up to 4 days after infection; and at 6 or 8 days afterwards it held mortality down to 40-50% as against 80% in unmedicated controls.

The interest of this work lies largely in the formulation used. The authors reported using 0.2 g per 100 ml seawater but their discussion makes it clear that it was in fact 0.2 mg per 100 ml (2 μg/ml). This is stated to be a virtually saturated solution at the temperature of the water used, and hence it would have been impractical to produce a large volume of solution at such a concentration without the use of a solubilizer. They used dimethylformamide at 20 ml per 0.2 mg NPL.

(c) Channel catfish

Mitchell and Plumb (1980) conduced a dose titration of NPL in channel catfish artificially infected with *Aeromonas hydrophila*. They also accepted the need for a solubilizer, and prepared a 0.4% w/v solution in isopropyl alcohol. They concluded that the optimum concentration for efficacy was 2 mg/l (= 2 ppm) and concentrations of less then 1.5 mg/l were unsatisfactory. Absorption is obviously slower in channel catfish than in salmonids because the optimum immersion time was 6.5 hours. Recrudescence of the disease 4 days after treatment led to the use of a repeat immersion on day 5 and the conclusion that 3 days would be an ideal interval between treatments.

The 96 hour “median tolerance limit” (the concentration at which 50% of the fish population show any adverse reaction) was 0.96 mg/l at pH 7.6 and 21°C. The observed adverse reactions were lesions initially in the skin of the body and tail and progressing to the underlying musculature. The lesions started as areas of loss of sheen which soon became petechiated. As damage progressed the skin split or eroded exposing the underlying musculature which was haemorrhagic, and in severe cases the erosions extended as far as the vertebrae.

(d) Atlantic salmon

NPL is active *in vitro* against *Paramoeba* spp., the causative organisms of amoebic gill disease which is an important infection of Atlantic salmon in seawater in Tasmania. The cheapest treatment of this disease in terms of material costs is a 4-hour bath in freshwater, but this procedure is of course extremely labour-intensive. NPL given in feed has been proposed as an economic alternative.

11.1.5 SODIUM NIFURSTYRENATE

This drug is used almost exclusively in Japan, and although it is said to be of value in a number of fish species there appears to be relatively little published about it in the English language.

In yellowtail injected intravenously with sodium nifurstyrenate in phosphate-buffered saline at a dose of 10 mg/kg, the drug had a distribution half-life of 36 minutes and a volume of distribution of 2.99 l/kg. This suggests that it would probably be of value for systemic infection in this fish species. The terminal half-life in plasma is 7.66 hours.

11.2 Chloramphenicol

11.2.1 USES

Chloramphenicol is a broad-spectrum antibiotic now rarely used systemically in veterinary medicine for two reasons. The first is that it causes aplastic anaemia, and as there appears to be no 'no effect level' for this action there are concerns about the action of residues in foods of animal origin. The second reason is that it is one of the very few drugs useful for typhoid fever, and there is concern that any veterinary use might cause the development of transmissible resistance in bacteria which then might transmit it to *Salmonella typhi*.

Chloramphenicol has been found useful for the prophylaxis of carp dropsy (caused by *A. liquefaciens*); single doses by intraperitoneal injection varying from 12 to as much as 80 mg/kg have been recommended. It has also been used in the treatment of trout ulcer disease (caused by *Haemophilus piscium*) and furunculosis (caused by *A. salmonicida*) at 50-75 mg/kg/day. In species such as salmonids which consume feed quickly chloramphenicol may be given in feed, but otherwise it is given by injection.

11.2.2 TARGET SPECIES SAFETY

Chloramphenicol does not cause any identifiable lesions in salmonids but prolonged administration of it has been known to cause retarded growth in some species.

In eels, chloramphenicol injected intraperitoneally at therapeutic doses has an identifiable effect on the blood cells. Of cells in the erythrocyte series there is a significant fall in the count of circulating erythroblasts by 48 hours after dosing; a few erythrocytes are vacuolated with enlarged nuclei and lipid droplets in the vacuoles. By 72 hours 40% of all cells in the series are vacuolated; they are of variable diameter and there is evidence of fragmentation. After a second dose the blood can take 14 days to return to normal. In the leucocyte series the monocytes, especially in the spleen and liver, are vacuolated and have enlarged nuclei; eventually they become phagocytosed. By 72 hours the counts of platelets and heterophils are down and those of monocytes, basophils and lymphocytes are up. Normality returns in about 7 days.

11.2.3 PHARMACOKINETICS IN RAINBOW TROUT

Chloramphenicol is almost completely absorbed (bioavailability >99%) in rainbow trout.

In a study with radio-labelled chloramphenicol 67% of the radio-activity was excreted in 48 hours; and the cumulative distribution of the radio-activity over 5 days was:

64% lost in faeces (presumably after biliary excretion)
16% excreted in urine
12% in aquarium water (presumably excreted through the gills)
7.5% residue in the carcass

The highest faecal excretion in any one 8-hour period occurred at 32-40 hours whereas in urine the highest was 8-16 hours. In addition to chloramphenicol the urine contained a glucuronide conjugate, chloramphenicol base and chloramphenicol alcohol (see Figure 11.3). There appeared to be no metabolism of the nitro group as does occur in mammals.

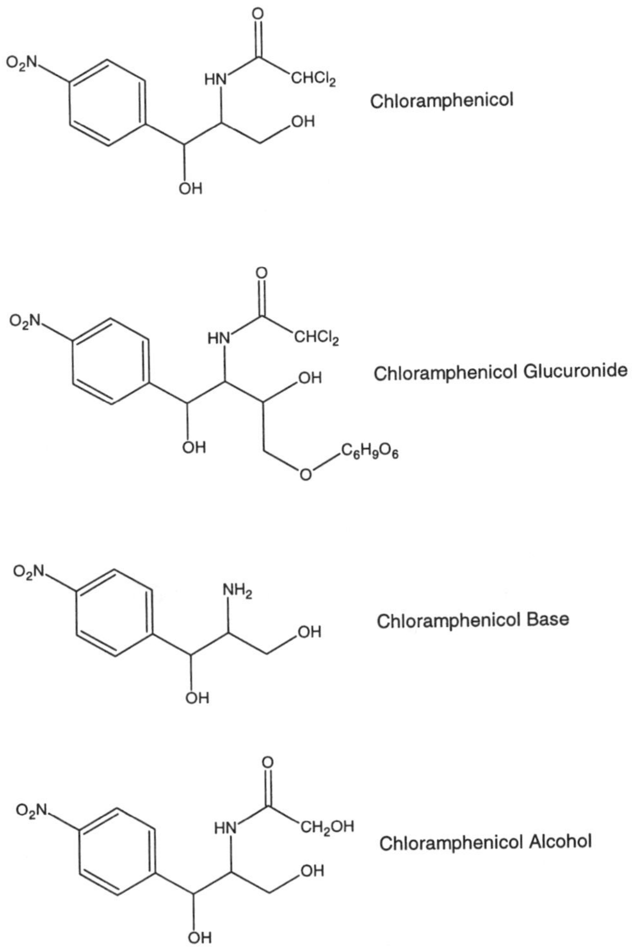

Chloramphenicol

Chloramphenicol Glucuronide

Chloramphenicol Base

Chloramphenicol Alcohol

Figure 11.3.

There were high concentrations of radio-activity in the bile and intestine and a lower concentration in the muscle; however because of the relative mass of the musculature this tissue contained 2.5% of the administered dose.

11.2.4 PHARMACOKINETICS IN SOCKEYE AND COHO SALMON

(a) Sockeye salmon in seawater

Chloramphenicol given as a single dose of 50 mg/kg in feed to immature sockeye salmon in seawater was rapidly absorbed; high tissue levels were reached by 3 hours and maintained until 12 hours. During this period levels in liver, muscle and kidney were respectively 37, 18 and 14 ppm in one study and 45, 22 and 19 ppm in a replicate. After 12 hours there was rapid depletion and by 24 hours residues were quantifiable only in the liver. At 7 days no chloramphenicol was detectable.

(b) Coho salmon in seawater

In coho salmon in seawater a similar profile with respect to time was found but the concentrations were much lower. In the liver the mean level during the 3-12 hour period was 4.4 ppm in one study and 8.6 ppm in a replicate; by 24 hours the liver residue was below the level of quantitation. The fish ate readily and were presumed to have taken as large a dose as the sockeye salmon; the different residue values were attributed to very rapid excretion in coho in seawater. Chloramphenicol may be of little use in coho salmon in seawater; if it were to be used they would require larger doses than sockeye. In either species twice daily dosage is necessary.

(c) Sockeye and coho salmon in freshwater

In freshwater the values for both species were the same as for sockeye salmon in seawater. Coho salmon given the higher dose of 75 mg/kg had tissue levels proportional to the dose.

11.2.5 PHARMACOKINETICS IN COMMON CARP

Kozlowski (1964), using a colorimetric assay, found that in year-old carp kept at 12-14°C and injected intraperitoneally with chloramphenicol at 40 mg/kg, C_{max} (blood) was 36 µg/ml at T_{max} 16 hours. Doses of 40, 80 and 160 mg/kg given at 7-10°C also resulted in T_{max} at 16 hours; C_{max} was not proportional to the doses. Furthermore the blood level following the 160 mg/kg dose fell by about 45% between 16 and 24 hours after dosing and thereafter was little higher than that resulting from the 80 mg/kg dose. The author noted that increases in efficacy could not necessarily be expected from increased dosage.

In 2-year-old carp the levels in muscle and spleen by 24 hours after injection were slightly higher than in blood, and in kidney they were nearly three times as high, suggesting that excretion is primarily in the urine.

Elimination of chloramphenicol is dependent on both temperature and the age of the fish. For a 40 mg/kg dose in 1 year-olds at 12-14°C the results show an exponential decline from 33 µg/ml at 20 hours with a half-life of approximately 15 hours; at 7-10°C the half-life is of the order of 40 hours. 2-year-olds injected at half the dose rate of 1-year-olds nevertheless had higher blood levels at 48 hours.

11.3 Florfenicol

11.3.1 USES

(a) Advantages over chloramphenicol

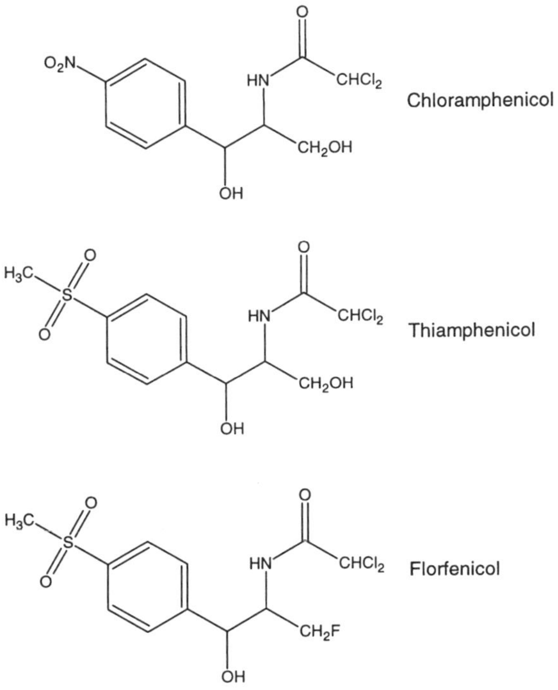

Chloramphenicol

Thiamphenicol

Florfenicol

Figure 11.4.

Thiamphenicol and florfenicol differ from chloramphenicol in having a methyl-sulfonate group in place of a nitro group in the molecular structure, and this is claimed to avoid the causation of aplastic anaemia, making the compounds acceptable for use in food-producing species. Florfenicol further differs from thiamphenicol in having fluorine atom in the molecular structure where thiamphenicol has a hydroxyl group. Bacteria having the usual form of resistance to chloramphenicol and thiamphenicol, which is the ability to acetylate this hydroxyl group, remain fully susceptible to florfenicol. However a transmissible plasmid with resistance to florfenicol has been found.

Fukui *et al.* (1987) found that for six common bacterial pathogens of fish florfenicol had *in vitro* MICs as as low as, or lower than, chloramphenicol or thiamphenicol. While this was true for *Streptococcus* spp., the only Gram-positive bacteria tested, for all three drugs their MICs were four to eight times as high as those of the Gram-negative bacteria.

(b) Use in Japanese aquaculture

Fukui *et al.* (1987) also studied the *in vivo* activity of different doses of florfenicol in artificially infected fish:

1. In goldfish infected with *Vibrio anguillarum* by injection and immediately treated by gavage with a drug suspended in gelatinized starch, 2.5 mg/kg florfenicol was comparable with 10 mg/kg of either chloramphenicol or thiamphenicol or 50 mg/kg oxytetracycline.
2. In Japanese eels infected with *Edwardsiella tarda* 5 mg/kg given in a gelatine capsule was protective irrespective of the resistance status of the bacteria to chloramphenicol.
3. In eels dosed 18 hours after infection florfenicol had a similar activity to oxolinic acid.
4. In yellowtail infected with *Pasteurella piscicida* florfenicol at 6.3-25 mg/kg in food was superior to other antibacterial agents commonly used in fish.

Other workers have found that in yellowtail artificially infected with *Past. pseudotuberculosis* florfenicol gave full protection in a single dose at 50 mg/kg; results were dose-dependent at rates in the range 5-10 mg/kg/day for 5 days. A more critical study showed that florfenicol gave better results in pelleted feed than in a minced fish diet; where it is surface-coated onto pellets a binder should be used.

(c) Furunculosis in Atlantic salmon

Inglis *et al.* (1991) conducted a dose titration of florfenicol in the control of furunculosis in Atlantic salmon parr. The fish were artificially infected while being kept at 19°C and hence under considerable heat stress. Food crumbs were surface-coated by mixing them with pure drug and then spraying them with 0.5% gelatin. Medication rates were designed to give 20, 10 and 5 mg/kg/day when fed at 0.75% per day (*i.e.* 8, 4 and 2 g florfenicol in 3 kg food). Medicated diets were started 24 hours after an infection which produced 75% mortality in unmedicated controls. The results showed dose-related prevention of mortality with all rates significantly better than unmedicated controls and 20 significantly better than 5 mg/kg/day; 10 mg/kg/day was not significantly different from either 5 or 20 mg/kg/day.

While this dose titration was conducted on parr in freshwater, Nordmo *et al.* (1994) reported a series of field trials on naturally infected Atlantic salmon at seawater sites. Fish treated with florfenicol at 10 mg/kg/day for 10 days were compared with positive controls given the drug and regimen normally used on the site. At two sites the positive control drug was oxolinic acid, at two flumequine and at one site trimethoprim-sulfadiazine. In all cases florfenicol gave significantly better results. In all cases the drugs were surface coated onto feed pellets using fish (capelin) oil as adhesive.

(d) Rainbow trout fry syndrome

In a summary of current knowledge about rainbow trout fry syndrome (RTFS), Branson (1998) mentioned studies of the sensitivity of 47 different isolates of *Flavobacterium psychrophilum* to ten different antibiotics. Enrofloxacin, sarafloxacin and florfenicol showed *in vitro* activity, but enrofloxacin had no therapeutic effect in subsequent *in vivo* trials. Florfenicol at 10 mg/kg/day for 10 days was highly effective.

11.3.2 PHARMACOKINETICS IN ATLANTIC SALMON

(a) Plasma analysis

Martinsen *et al.* (1993) studied Atlantic salmon in seawater at 11°C using single doses of 10 mg/kg florfenicol given either intravenously or orally in heat-extruded feed pellets. Plasma analysis was by HPLC. They found the bioavailability to be very high (96.5%) as is the case with chloramphenicol. Following the oral dosing C_{max} was estimated to be 4.0 µg/ml at T_{max} 10.3 hours, but they noted that these parameters could be profoundly affected by the method of incorporation in the feed and the quantity of feed given. In fact Horsberg *et al.* (1996) giving surface-coated pellets to fish at 10°C (a negligibly different temperature) obtained values of C_{max} (plasma) 9.1 mg/ml at T_{max} 6 hours.

The volumes of distribution found in the two studies were 1.1 and 1.3 l/kg respectively; they suggest a satisfactory distribution for clinical purposes with tissue levels similar to those in the plasma. With the florfenicol given in the heat-extruded pellets C_{max} was five times the highest *in vitro* MIC found by Fukui *et al.* (1987), but the elimination half-life was 12.2 hours suggesting that daily dosage would be necessary. Multiple doses given by surface-coating pellets led to an increase in feed consumption, from 0.51% on the first day to 0.74% on the 10th day, with the plasma florfenicol concentration correlating with the consumption of medicated feed. The plasma concentration had a mean of 8.4 mg/ml and never fell below 6 mg/ml during medication; on day 8 it reached 12 µg/ml.

(b) Radio-labelled drug study

Following an intraperitoneal injection of radio-labelled florfenicol as a 5% solution in capelin oil, radio-activity was found in all tissues by 3 hours; C_{max} in most was 12 hours but in the kidney and urine it was 3 days. Radio-activity in the brain was low, indicating a limited movement across the blood-brain barrier. By 2 days after dosing radio-activity in most tissues had fallen significantly, but there were still high levels in the eye, meninges, skin, intestinal mucosa and the liver and bile; levels in the kidneys and urine were still rising. By 7 days radio-activity was undetectable except in the kidney, skin, choroidea, liver and intestinal mucosa; it was still detectable in the kidney and choroidea at 56 days.

(c) Florfenicol amine

In muscle unmodified florfenicol constituted 90% of the radio-activity in Atlantic salmon at 6 hours after dosing, but by 3 days it was only 20%. In the same period florfenicol amine, the de-acetylation metabolite, rose from 7% to 70%. Both florfenicol and the amine were detectable in the blood on the day after the end of a 10-day oral course, but thereafter only the amine was found. The amine would additionally have accounted for virtually all of the radio-activity found in the liver at 7 days after injection.

The amine is microbiologically inert but it is the marker residue, *i.e.* the compound to which the MRL is assigned. In muscle it has a half-life of 49 hours and in liver 56 hours. In a residue depletion study the upper 95% confidence band intercepted the limit of quantitation (0.05 ppm) at 27 days for liver and 18 days for muscle.

Further reading

Amend, D.F. and Ross, A.J. (1970) Experimental control of columnaris disease with a new nitrofuran drug, P-7138. *Progressive Fish-Culturist* **32**, 19-25.

Branson, E.J. (1998) Rainbow trout fry syndrome: an update. *Fish Veterinary Journal* **2**, 63-66.

Egidius, E. and Andersen, K. (1979) The use of Furanace against vibriosis in rainbow trout *Salmo gairdneri* Richardson in salt water. *Journal of Fish Diseases* **2**, 79-80.

Fukui, H., Fujihara, Y. and Kano, T. (1987) *In vitro* and *in vivo* antibacterial activities of florfenicol, a new fluorinated analog of thiamphenicol, against fish pathogens. *Fish Pathology* **22**, 201-207.

Heaton, L.H. and Post, G. (1968) Tissue residues and oral safety of furazolidone in four species of trout. *Progressive Fish-Culturist* **30**, 208-215.

Horsberg, T.E., Hoff, K.A. and Nordmo, R. (1996) Pharmacokinetics of florfenicol and its metabolite florfenicol amine in Atlantic salmon. *Journal of Aquatic Animal Health* **8**, 292-301.

Horsberg, T.E., Martinsen, B. and Varma, K.J. (1994) The disposition of ^{14}C-florfenicol in Atlantic salmon (*Salmo salar*). *Aquaculture* **122**, 97-106.

Inglis, V., Richards, R.H., Varma, K., Sutherland, I.H. and Brokken, E.S. (1991) Florfenicol in Atlantic salmon, *Salmo salar* L., parr: tolerance and assessment of efficacy against furunculosis. *Journal of Fish Diseases* **14**, 343-351.

Kozlowski, F. (1964) Chloromycetin levels in the blood and some tissues of carps in the prophylactic treatment of dropsy. *Bulletin of the Veterinary Institute in Pulawy* **8**, 188-195.

Marking, L.L. and Piper, R.G. (1976) Carbon filter for removing therapeutants from hatchery water. *Progressive Fish-Culturist* **38**, 69-72.

Martinsen, B., Horsberg, T.E., Varma, K.J. and Sams, R. (1993) Single dose pharmacokinetic study of florfenicol in Atlantic salmon (*Salmo salar*) in seawater at 11°C. *Aquaculture* **112**, 1-11.

Mitchell, A.J. and Plumb, J.A. (1980) Toxicity and efficacy of Furanace on channel catfish *Ictalurus punctatus* (Rafinesque) infected experimentally with *Aeromonas hydrophila*. *Journal of Fish Diseases* **3**, 93-99.

Nordmo, R., Varma, K., Sutherland, I.H. and Brokken, E.S. (1994) Florfenicol in Atlantic salmon, *Salmo salar* L.: field evaluation of efficacy against furunculosis in Norway. *Journal of Fish Diseases* **17**, 239-244.

Plakas, S.M., el Said, K.R. and Stehly, G.R. (1994) Furazolidone disposition after intravascular and oral dosing in the channel catfish. *Xenobiotica* **24**, 1095-1105.

Stehly, G.R. and Plakas, S.M. (1993) Pharmacokinetics, tissue distribution and metabolism of nitrofurantoin in the channel catfish (*Ictalurus punctatus*). *Aquaculture* **113**, 1-10.

Yasunaga, N. and Tsukahara, I. (1988) Dose titration study of florfenicol as a therapeutic agent in naturally occurring pseudotuberculosis. *Fish Pathology* **23**, 7-12.

PART THREE

OTHER CHEMOTHERAPEUTIC AGENTS

12. SYSTEMIC ANTI-PROTOZOAL AGENTS

12.1 Fumagillin

12.1.1 PRESENTATION

Fumagillin is an antibiotic produced by the parasitic fungus, *Aspergillus fumigatus*, and acts by inhibiting RNA synthesis. It is acidic and is normally presented as the dicyclohexylamine (DCH) ($(C_6H_{11})_2NH$) salt. This salt is sparingly soluble in water but moderately soluble in ethanol.

Fumagillin DCH is heat-labile so it cannot be incorporated in feed before pelleting. The pure drug may be surface-coated onto pellets by dissolving it in 95% ethanol and spraying the solution onto the pellets. Palatablility is improved by top-coating the medicated pellets with 1:1 mixture of cod liver oil and ethanol. A 2% premix (Fumidil B®) is available commercially, and this may be surface-coated onto pellets by conventional methods.

12.1.2 USES

(a) Indications

Fumagillin has low antibacterial and anti-fungal activity and is primarily active against protozoa. It was formerly used for amoebic dysentery in Man, and Fumidil B is used against *Nosema apis* in bees by dissolving it in sugar solution.

In fish medicine fumagillin DCH has been used for:

Microsporea: *Enterocytozoon salmonis* in chinook salmon
Loma salmonae in Pacific salmon
Pleistophora anguillarum in eels
Sphaerospora testicularis in sea bass
Sphaerospora renicola in common carp
Myxosporea: *Myxosoma (Myxobolus) cerebralis* in rainbow trout.

It has also been shown to be active in the control of proliferative kidney disease (PKD), a condition of freshwater salmonids for which a putative protozoan pathogen has been seen histologically but never isolated and cultured.

Fumagillin has been tested but found inactive against infections of *Myxobolus cyprini* and *Thellohanellus nikolskii*.

(b) Swimbladder inflammation of common carp
This disease has a complex aetiology but it appears to be initiated by *Sphaerospora renicola*. Molnár *et al.* (1987) found that in carp infected intraperitoneally, a diet containing 0.1% fumagillin DCH effectively prevented the development of disease.

In a natural infection of juvenile carp in a pond the regimen gave significant but not complete protection. Symptoms developed but disappeared in a short time and it was concluded that the fumagillin was more efficacious against the later stages of the parasite life cycle. A difficulty was that little food was eaten until the fifth week of medication and it was suggested that a higher inclusion rate in the diet might have given better results.

No observations were made on residues but the authors commented that if the drug were only to be used in juvenile carp there would be little concern about 2-3 year-old table fish.

(c) *Enterocytozoon salmonis infection*
A trial of fumagillin for artificial *Enterocytozoon salmonis* infection in chinook salmon has been reported. The drug was found to prevent proliferation of the parasites rather than killing them. The regimen used was 0.1 g/kg fumagillin DCH in a diet fed at 1.5% for 3 weeks starting immediately after the infection. This equates to 1.5 mg/kg/day fumagillin DCH or about 1.0 mg/kg/day fumagillin; in trials against other parasites 3 mg/kg/day has been found to be a minimum for clinical efficacy.

(d) Myxosoma cerebralis infection
This parasite is the cause of whirling disease, an economically important condition of freshwater salmonids, especially rainbow trout kept in earth ponds. The nervous symptoms result from lesions produced in the head cartilages before ossification, and the infection is therefore of importance only in juvenile fish.

In small scale trials using artificial infection of rainbow trout, El-Matbouli and Hoffman (1991) used a diet containing 0.1% fumagillin DCH fed at 1% for 4 months starting 1 month after infection; this is equivalent to 10 mg/kg/day. They found that of 30 medicated fish none developed symptoms but in six the infection did become established; in unmedicated controls 18 out of 30 developed symptoms of whirling disease and the infection became established in a further four. The same medication rate was used in a second trial but starting only 14 days after infection and continuing for only 50 days. One out of 30 fish developed symptoms and the infection became established in a further two, while in unmedicated controls 22 out of 30 developed symtoms and the infection became established in all the rest. In both trials where parasites were seen histologically in treated fish they were deformed; it was postulated that they were non-viable but had not been phagocytosed because they were in cartilage and therefore remote from circulating phagocytes. It was concluded that fumagillin could both prevent clinical outbreaks and reduce infection rates.

The same consumer safety considerations would apply here as mentioned above for the common carp.

(e) Sphaerospora testicularis infection

This infection destroys testicular tissue in sea bass and hence reduces the fecundity of males. Control has been attempted using Fumidil B® at a rate to give 1 g/kg fumagillin in the feed and feeding at 1.5%, which is 15 mg/kg/day, for 8 weeks. It was found that even at this (relatively high) dose rate the medication did not affect well-established infections, but it did advance spermiation by 2 months. As had been found with *Enterocytozoon salmonis* infection in chinook salmon at a much lower dose rate, fumagillin was more parasitistatic then parasiticidal.

(f) Loma salmonae infection

This disease affects the gills of Pacific salmon in both fresh and seawater. Chinook salmon have been treated with a course of fumagillin for 30 days starting 10 days after artificial infection. The feed medication rate used was 1 g/kg fumagillin DCH, and feeding was at 1% giving a dose rate of 10 mg/kg/day. Subsequent histology of the gills of treated fish showed no evidence of infection, whereas parasites were found in 8 of 11 and 5 of 11 in tanks of unmedicated fish. The infection had not been severe enough to cause any mortality and medication had been started soon after it; so while control was achieved, the efficacy of fumagillin in advanced cases was untested.

(g) Pleistophora anguillarum infection

Kano *et al.* (1982) studied the use of fumagillin DCH in Japanese eels artificially infected with *P. anguillarum*. Infection can be achieved by either oral or immersion administration of this pathogen, and preliminary work showed that immersion treatment of oral infection did not work. This was attributed to the spread of infection from the fish gut to the musculature *via* the gut wall and peritoneal fluid, not *via* the blood. Following immersion infection, symptoms could be prevented by immersion in 60.5 ppm fumagillin for 120 hours starting 2 days after infection.

The authors were able to prevent symptoms following oral infection by using fumagillin orally at 250 mg/kg/day for 30 days from the time of infection. An interesting feature of this regimen is that there was no mention of toxicity; it would produce 100% mortality in salmonids. The authors then studied various oral dose regimens and concluded that 5 mg/kg/day for 60 days was the most economic. Nevertheless relapses occurred after the end of medication. An intermittent regimen of four periods of medication at 50 mg/kg/day for 5 days with intervals of 5 days between them was also efficacious if started within 6 days after infection. However this uses more drug than a steady 5 mg/kg/day and relapses still occurred after the end of treatment.

(h) Proliferative kidney disease

Hedrick *et al.* (1988) found fumagillin to be active against artificial infection with PKD in chinook salmon. 10 mg/kg/day for 7 weeks starting immediately after the artificial infection was protective but the fish became inappetent after 6 weeks. 5 mg/kg/day for 6 weeks starting a week after the artificial infection was also protective; there was no inappetence but there was histological evidence of loss of haemopoietic tissue. The authors commented that although there are some parts of California where there is a natural disease challenge for as much as 7 months in the year, fumagillin is suited to chinook salmon because this species has only a short rearing period in freshwater before migrating to the sea. Furthermore a minimum of 3 years at sea should constitute a very adequate withdrawal period.

Higgins and Kent (1996) used fumagillin DCH at 3.0 mg/kg/day for 14 days in a field trial in coho salmon. They found that although untreated control fish recovered, the treated fish had a lower prevalence of disease and more rapid recovery. At this dose rate there were no toxic side effects or kidney histopathology.

Problems have been encountered in using fumagillin in natural infections of rainbow trout in Europe; these are related to both timing of medication and tolerance of the drug by fish already debilitated by the disease. Infection heavy enough to produce symptoms of disease will also produce good protective immunity; premature antiparasitic medication leaves the fish susceptible to reinfection. Treatment of severely affected fish requires high dose rates, such as 10 mg/kg/day, and these may be toxic to some individuals. Since where PKD occurs at all the prevalence is virtually 100%, a possible solution to the timing problem is to take samples of fish at intervals during the period of risk (normally when the water temperature rises above 12°C), and examine Giemsa-stained kidney smears for parasites. The tolerance problem is usually addressed by using an intermittent dose regimen - three periods of medication lasting 10-15 days each, separated by equal periods without medication.

12.1.3 TARGET SPECIES SAFETY

The safety margin of fumagillin is narrow, at least in salmonids. In most species studied its main effects are on the erythropoietic system, including, in some cases, a reduction in the size of the spleen and kidneys after prolonged treatment. In some cases it has also been observed to cause vacuolation of the renal epithelial cells and, again after prolonged treatment, loss of appetite and consequent reduced growth rates. Wishkovski *et al.* (1990) studied 40 g rainbow trout treated for 4 weeks at either 1.5-2, 3.25-5, 7.5-10 or 15-20 mg/kg/day. Anorexia occurred at the two highest dose rates but mortality only at the highest. All the mortalities occurred in the final week suggesting a cumulative toxic effect.

Depletion of haematopoietic cells occurs in chinook salmon treated with fumagillin at 10 mg/kg/day for 10 weeks. Similar effects are seen in rainbow trout with doses in excess of 7.5 mg/kg/day for 4 weeks or 3.75 mg/kg/day for 8 weeks.

In sea bass dosed at 15 mg/kg/day for 8 weeks significant decreases in haemoglobin, haematocrit (packed cell volume) and erythrocyte count have been observed together with an increase in mean corpuscular haemoglobin content. In this species there was no visible change in the haematopoietic tissue so the effect of fumagillin appears to be simply destruction of erythrocytes. Recovery of haemoglobin concentration, erythrocyte count and mean corpuscular haemogobin content takes about 2 months after the cessation of treatment; later still (about 5 months) there is a rebound in haemoglobin concentration and erythrocyte count.

12.2 Nitroimidazoles

12.2.1 USES IN MAMMALS AND BIRDS

The nitroimidazoles are a group of synthetic antimicrobial agents active against protozoa and obligate anaerobic bacteria. Few drugs of this group are now in use because they give positive responses to some *in vitro* mutagenicity tests. Metronidazole is one which remains in use in human medicine; and dimetridazole (DMZ) has been used in veterinary medicine particularly for swine dysentery and for histomoniasis (blackhead) and other protozoal infections in poultry and game birds. In the EU, DMZ has the anomalous legal status of being in Annex IV (therapeutic use prohibited) of Regulation 2377/90 but permitted for the prevention of histomoniasis when used as a feed additive under Directive 70/524/EEC. For the latter purpose it is available as a premix.

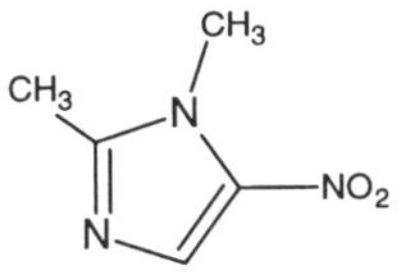

Figure 1.12.

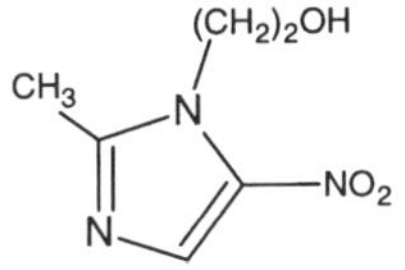

12.2.2 USE IN FISH

(a) Food-producing fish

The use of DMZ for the treatment of 'white spot' (*Ichthyophthirius multifiliis* infection or "Ick") in rainbow trout was investigated by Rapp (1995). The rationale for the work was that in Germany malachite green is banned for food-producing fish species, and the belief that immersion in potassium permanganate was the only other possible treatment but had serious target species safety problems. (In fingerlings at 16-17°C it was said to have "had a catastrophic effect"). Combinations of potassium permanganate immersion with in-feed DMZ at various different temperatures were tested.

At 10-13°C in juvenile rainbow trout (mean weight 20 g) the disease was controlled by in-feed DMZ at 28 mg/kg/day for 10 days. This regimen was tried with and without simultaneous immersion treatments with 2 ppm potassium permanganate, and the DMZ alone was equally efficacious and better tolerated. Other workers have found a 10 day course of in-feed DMZ to be effective for white spot in European catfish.

In rainbow trout at 6-8°C, DMZ enters the musculature quickly and is eliminated quickly. At 18° higher levels accumulate in the musculature; this may be due to increased bioavailability rather than any slower elimination.

(b) Ornamental fish

Nitroimidazoles have been recommended for bath treatment of some ornamental species, especially for 'hole-in-the-head' disease, a necrotic fistulous condition thought to be initiated by the protozoan, *Hexamita* (syn. *Octomita*). Dipping in 7 ppm metronidazole or 5 ppm DMZ is usual. Where malachite green is not available nitroimidazoles may also be used for 'velvet disease' caused by the protozoan, *Oodinium*, but twice the concentration used for 'hole-in-the-head' disease is recommended. Where a concentrated solution of DMZ is not commercially available it is possible to prepare a metronidazole dip solution by grinding tablets intended for human medicine with a pestle and mortar.

Further reading

El-Matbouli, M. and Hoffmann, R.W. (1991) Prevention of experimentally induced whirling disease in rainbow trout *Oncorhynchus mykiss* by fumagillin. *Diseases of Aquatic Organisms* **10**, 109-113.

Hedrick, R.P., Groff, J.M., Foley, P. and McDowell, T. (1988) Oral administration of fumagillin DCH protects chinook salmon *Oncorhynchus tshawytscha* from experimentally-induced proliferative kidney disease. *Diseases of Aquatic Organisms* **4**, 165-168.

Higgins, M.J. and Kent, M.L. (1996) Field trials with fumagillin for the control of proliferative kidney disease in coho salmon. *Progressive Fish-Culturist* **58**, 268-272.

Kano, T., Okauchi, T. and Fukui, H. (1982) Studies on *Pleistophora* infection in eel, *Anguilla japonica* - II Preliminary tests for application of fumagillin. *Fish Pathology* **17**, 107-114.

Molnár, K., Baska, F. and Szekely, C. (1987) Fumagillin, an efficacious drug against renal sphaerosporosis of the common carp *Cyprinus carpio. Diseases of Aquatic Organisms* **2**, 187-190.

Rapp, J. (1995) Treatment of rainbow trout (*Oncorhynchus mykiss* Walb.) fry infected with Ichthyophthirius (*Ichthyophthirius multifiliis*) by oral administration of dimetridazole. *Bulletin of the European Association of Fish Pathologists* **15**, 67-69.

Wishkovski, A., Groff, J.M., Lauren, D.J., Toth, R.J. and Hedrick, R.P. (1990) Efficacy of fumagillin against proliferative kidney disease and its toxic side effects in rainbow trout (*Oncorhynchus mykiss*) fingerlings. *Fish Pathology* **25**, 141-146.

13. EXTERNALLY APPLIED ANTIMICROBIAL AGENTS

13.1 Formalin

13.1.1 USES

(a) Spectrum of activity

Formalin is a very widely used anti-parasitic agent. It is used routinely against many protozoans parasitic on the skin or gills of fish, including *Chilodonella* spp., *Epistylis* spp., *Ichthyobodo (Costia) necator, Ichthyophthirius multifiliis, Scyphidia* spp. and *Trichodina* spp. It has little activity against a majority of bacteria but has been used for bacterial gill disease which is a mixed infection caused by *Flavobacterium branchiophilum* together with protozoans. It is also useful for the monogenetic fluke parasites of fish gills and skin, respectively *Dactylogyrus* and *Gyrodactylus* spp., and also *Cleidodiscus* spp.

Quite apart from its therapeutic use formalin has a place in prophylaxis. Some of the above parasites stimulate protective immunity but this is temporary and temperature- and nutrition-related; it is low in early spring when over-wintered parasites become active. Additionally fish are less able to withstand formalin treatment at this time. In consequence prophylactic medication with formalin in early autumn is advisable for all fish normally at risk from these parasites.

In the case of white spot, *Ichthyophthirius multifiliis* infection, formalin is thought to have little or no activity against the parasitic trophont stage of the life-cycle; it is however active against the free-living tomite stage through which infection is transmitted between hosts. This means that effective use of formalin in an infected fish population involves repeated treatment of the water. Since the parasite's life-cycle is temperature-dependent so must be the frequency of repeat treatments: a common recommendation is three treatments at 5 day intervals at 16°C, and three treatments at 14 day intervals at 10°C. Wahli *et al.* (1993) found formalin to be unsatisfactory for the control of white spot in fingerling rainbow trout. They tried both 25 and 100 ppm and in both cases they described the *in vivo* effects on the parasites as "none" and on the fish as "negative". The low concentrations they used might account for the lack of efficacy, but they examined only one fish per time point and appear to have been looking for a reduction in trophont colony count which is not the main action of formalin. The negative effect of formalin on the fish may have been due to its use where there was already stress due to overt parasitic infestation.

Formalin is a standard general disinfectant used in hatcheries for the prevention of infections of eggs, the most important being fungi of the genus *Saprolegnia*.

(b) Presentation

Formalin is a virtually saturated solution of formaldehyde in water. At normal water temperatures it contains 34-38% w/w formaldehyde. For aquacultural purposes it is usually supplied containing 12% methanol to slow down the polymerization of the formaldehyde. Polymerization occurs faster in sunlight and at low temperatures; formalin should therefore be stored in the dark and at temperatures above 10°C. The polymer, paraldehyde, is extremely toxic to fish and any white precipitate should be filtered out of formalin before it is diluted for use.

Diluted solutions are normally designated by their content of formalin, not of formaldehyde; for example a use concentration obtained by diluting formalin 1:6000 will be designated "0.017% formalin" or "167 ppm formalin" but actually contains about 60 μg/ml (ppm) formaldehyde.

(c) Dosage

The normal use dilution of formalin for fish is 1:6000 or 167 ppm, but 1:4000 or 250 ppm is necessary for some parasites, notably *Chilodonella* and *Epistylis*. Exposure is normally for 30-60 minutes; it should not exceed 60 minutes on any one occasion but treatments may be repeated daily.

For eggs the customary dilution is 1:600 (1667 ppm), *i.e.* 10 times the normal concentration for fish, although in a dose titration Marking *et al.* (1994) found 1000 ppm to be the optimum. Intermittent exposures of 15 minutes each are used to prevent fungal infection.

(d) Administration

When formalin is used on fish attention must always be given to the oxygenation of the water. Formalin is a reducing agent which will absorb oxygen from water; in addition it is toxic to fish gills and will lower the efficiency of gaseous exchange. The major indications for its use are bacterial or parasitic gill diseases which means that gaseous exchange is usually impaired before treatment starts. For this reason treatment should preferably be done at the time of day when the water temperature is lowest, and aeration should always be provided. If the fish show symptoms of respiratory distress - accelerated or exaggerated opercular movements, or gasping at the water surface - exposure should be terminated by flushing with fresh, well-aerated water. The oxidation product of formaldehyde is formic acid and it is common practice to add limestone chips to water as a buffering agent.

Pure formalin should never be added to water containing fish; a preliminary dilution of the order of 1:100 should be made, and this should be added to the water and immediately mixed thoroughly. In ponds the preliminary dilution should be added at several different places to avoid creating areas of high concentration.

Formalin should not be used at all if:

- The oxygen level in the water is below 5 ppm;
- there is a heavy phytoplankton bloom (which would be killed by the formalin and then instead of contributing photosynthetic oxygen would decay and increase biological oxygen demand); or
- the water temperature is above 27°C (see also Section 13.1.2(b)).

Where formalin is used in culture systems with recirculated water it may kill the nitrifying bacteria in the biofilter. Filters should be isolated from the system during exposure and not reconnected until after the formalin has been flushed away. It should be remembered that the filter may be a reservoir for free-living stages of some species of parasites.

For further practical details of the procedure for administering formalin to freshwater farmed fish see Klontz (1994).

In seawater the toxicity of formalin to gill epithelia affects osmoregulation as well as gaseous exchange. Where formalin is used in seawater, changes in salinity during treatment may increase its toxicity to fish, and steps should be taken to avoid this. In particular the preliminary dilution and the terminal flushing should be done with water of the salinity to which the fish are acclimated.

13.1.2 SAFETY ASPECTS

(a) Effects on target species

In all species the adverse effect of formalin is on the gill epithelium and the symptoms of toxicity are those of hypoxia. In one trial rainbow trout immersed in 200 ppm for 1 hour showed a 30% increase in ventilation rate, but recovered to control values in a few hours. Subsequent daily treatments produced less intense responses and quicker recoveries.

(b) Tolerance by salmonids

Species vary in their susceptibility to formalin. Salmonids and centrarchids (sunfishes) are particularly susceptible and should not be exposed to concentrations higher than 167 ppm at temperatures above 10°C or to higher than 250 ppm at below 10°C. LC_{50} values for various periods of exposure have been determined but they vary not only with species of fish but also with temperature, pH and hardness of freshwater and salinity of seawater.

Speare and McNair (1996) conducted a chronic toxicity test on juvenile rainbow trout exposing them to 200 mg/l (= 1:5000 dilution) for 1 hour twice a week for 12 weeks. They found no effects on appetite, growth rate, food conversion ratio or body condition; the treated fish had better fin condition and fewer corneal opacities than the controls. Powell *et al.* (1996) conducted a similar procedure with juvenile Atlantic salmon. They divided their fish into "large" (>16 cm) and "small" (<16 cm) groups and treated some of each with 167 ppm and 250 ppm formalin. All groups were immersed for 1.5 hours on each of five occasions at 2 week intervals. All fish gained in weight but four of the six groups had reduced condition.

In addition to water chemistry genetics has a significant effect. The rainbow trout is considerably more susceptible than other salmonids and within rainbow trout there are heritable strain differences.

(c) Tolerance by common carp and African catfish eggs

Theron *et al.* (1991a) have investigated the tolerance of the eggs of carp and African catfish (*Clarias gariepinus*) to formalin. Only a single treatment was given because at a water temperature of 25°C the incubation period for carp is 60 hours and for African catfish only 48 hours. They tested various exposure times, using a constant concentration of 2000 mg/l. Carp eggs proved to be considerably more tolerant of formalin than catfish eggs: thus carp eggs tolerated 30 minutes exposure, whereas catfish eggs had slightly higher mortality than controls after as little as 5 minutes exposure, and an economically significant mortality after 15 minutes.

(d) Tolerance by African catfish larvae and juveniles

In a parallel study of newly hatched fish Theron *et al.* (1991b) used 4, 12 and 20 day-old African catfish and a formalin concentration of 200 mg/l. The 4 and 12 day-olds had negligible losses after 30 or 60 minutes exposure; both ages of fish also tolerated 90 minutes initially but suffered considerable mortality over the ensuing days, so that by 12 days after exposure the survivals were only 23% for 4 day-olds and 44% for 12 day-olds. The 20 day-olds were the most sensitive of the catfish groups having immediate losses correlating with exposure time and rising to 61% after 90 minutes. However the losses were only 2% after 30 minutes and this would presumably be long enough for efficacy.

(e) Tolerance by common carp

Theron *et al.* (1991b) were unable to test 4 day-old carp because of their sensitivity to handling stress. For the 12 and 20 day-old fish they found, in contrast to the African catfish, that it was the younger ones which were the more sensitive. After 60 minutes exposure the mortality in the 12 day-old carp was 33% but in the catfish it was negligible; 20 day-old carp nearly all survived 90 minutes exposure whereas this was the most sensitive age in catfish.

(f) Tolerance by bass

Striped bass (*Morone saxatilis*) fingerlings are particularly susceptible to formalin toxicity and formalin should not be used with this species.

(g) Operator safety

Formalin must be handled with care. Formaldehyde is harmful if inhaled or absorbed through contact of formalin with the skin. Formalin is seriously irritant if swallowed; the patient should immediately drink plenty of water, vomiting should be induced and then demulcents may be taken. In the eyes formalin, even in considerable dilution, is very irritant; first aid treatment is flushing with water for a considerable time, *e.g.* 15 minutes.

13.2 Malachite green

13.2.1 CHEMISTRY

Malachite green (MG) is a synthetic organic dye. It has the same colour as the copper-containing mineral, malachite, but has no chemical relationship to it at all. In the presence of water it exists as an equilibrium between two forms, the dye ion which is soluble in water, and the (colourless) non-ionic pseudo-base (actually a carbinol) which is virtually insoluble in water but soluble in lipids. The predominating form depends on pH: at pH 4 it is wholly ionized; at higher pHs it combines covalently with hydroxyl anions, and at pH 10 it is wholly in the carbinol form. The pKa is 6.9. On a change in pH the equilibrium

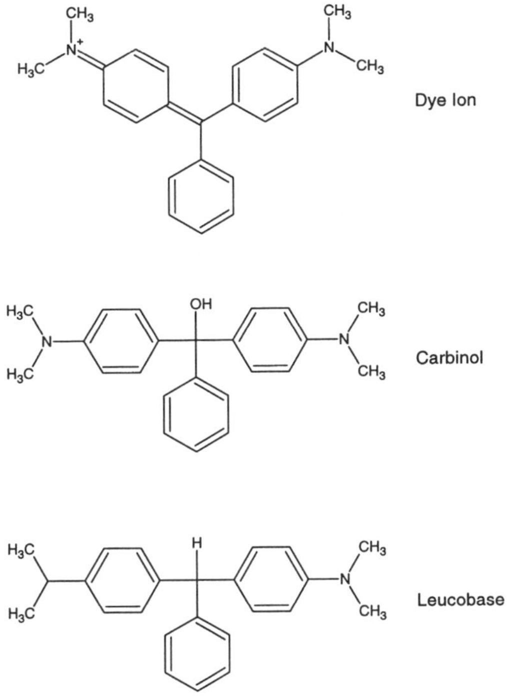

Figure 13.1. Forms of malachite green.

is re-established only slowly: 50% of the change occurs within about 3 minutes but full equilibrium may not be reached for 10 hours. (See also Alderman, 1991).

The dye form of MG, being ionic, exists only in a salt, and is commonly supplied as the hydrochloride, oxalate, acetate or, less commonly, formate. Most of the MG that is manufactured is used in dyeing, and satisfactory results are obtained with material of variable purity with the nature of the impurities unknown. Concentrations tend to be expressed in terms of dyeing power rather than chemistry. A hydrochloride precipitated with zinc chloride is commonly used. This product may be toxic to fish, and for safety aquaculturists are advised to use "zinc-free" MG although the toxicity actually resides in the impurities rather than the zinc. An alternative is to use MG oxalate which is usually free of toxic impurities.

MG is also available in solution, often as a 50% mixture of hydrochloride and acetate. The oxalate salt tends to come out of solution with time.

13.2.2 USES

(a) Spectrum of activity

Of the two chemical forms of MG only the coloured cations have any anti-microbial activity; but it is the colourless carbinol which, being soluble in lipids, will pass through biological membranes.

MG has a similar spectrum of activity to formalin. Additionally it shows *in vitro* activity against Gram-positive bacteria but this is rarely exploited. Its efficacy against fungi is greater than that of formalin and this is its major use. It is effective against many species of protozoan ectoparasites but only at concentrations which may be toxic to the fish.

Clifton-Hadley and Alderman (1987) showed that MG has activity against proliferative kidney disease (PKD) in salmonids. It is in fact one of the very few drugs with this activity (see also fumagillin, Section 12.1), and this has led to political pressure for its continued availability on animal welfare grounds when its use might otherwise be banned on toxicological grounds.

(b) Dosage and administration

MG is always administered by immersion. It is absorbed through the gills and the eventual equilibrium concentration is higher in the fish than in the water.

For PKD Clifton-Hadley and Alderman (1987) found both the efficacy and the toxicity to be dependent on concentration and time - a function of the "area under the curve" (AUC) in a concentration/time graph. They used flush treatments (see Section 1.1.4) where the inflow of unmedicated water was uninterrupted; the depletion of the drug therefore began immediately it was added to the fish tanks. They found a single treatment of initially 0.15 ppm to be ineffective, and a single treatment initially at 1.5 ppm to be partially effective but with evidence of MG toxicity. The ideal appeared to be 3 treatments at initially 1.0 ppm; the intervals between treatments were either 4 and 4 days or 4 and 7 days. In subsequent field trials they found (Alderman and Clifton-Hadley, 1988) 1.6 ppm in a bath for 40 mins, or a flush with a median concentration of 3.2 ppm (108 ppm-mins over 4 hours) to be satisfactory. 7, 14 and 21 day intervals were tried and while 7 days gave the best results 21 days was adequate.

Scott (1993) recommended 100 ppm-minutes over a 4 hour period and explained by an example how this was to be achieved. It would appear to be a similar regimen to the one which Clifton-Hadley and Alderman (1987) found toxic; its safety to the fish could well depend on the pH of the water.

In the UK the normal dose rate for the control of external parasitisms is 1 ppm for up to 1 hour. However Machova *et al.* (1996) noted that previous studies on elimination had used short term baths but that, "A long-term (6 days) therapeutic bath in malachite green is used very frequently in rainbow trout culture." In their study they used 0.2 mg/litre (= 0.2 ppm) for 6 days at 12-14°C with the implication that this was a typical regimen in the Czech Republic.

MG has been used for its anti-fungal activity in adult male trout in the breeding season. Hormonal effects on the skin of such animals and their tendency to fight means they are liable to injury and secondary invasion by fungi. Such use carries no risk to the consumer because sexually mature fish are unmarketable for food.

(c) Dosage and administration to eggs

It has been alleged that in the treatment of eggs the use of zinc-free MG is particularly important because MG increases the permeability of the vitelline membrane to zinc.

The usual fungus control regimen using MG is 2 ppm for 15 minutes every 24 hours. However Willoughby and Roberts (1992) have presented *in vitro* evidence that this is excessive. They found that 0.25 ppm MG oxalate used continuously would kill the infective stages of *Saprolegnia parasitica* but that as little as 0.05 ppm would inhibit fungal growth even though allowing spore germination. They found evidence that young mycelial growths were more susceptible to MG than older ones and postulated that pulses of 0.25 ppm for 15 minutes every 24 hours would keep growth under control. If this is true the regimen should be adopted because MG is an undesirable environmental contaminant.

For warm water species with short incubation periods such as channel catfish, African catfish and carp the usual procedure is to dip the eggs rather than flush them. A standard regimen is 15 ppm for 10 secs.

(d) Legal aspects

Meyer and Jorgenson (1983) found evidence that MG is mutagenic in both trout eggs and rabbits. As a result a number of countries have imposed bans on its use in food-fish production. However in aquaculture it has some unique merits, in particular its activity against PKD, and complete bans would create serious animal welfare problems; in at least two EU member states it is banned but well known to be used. In other countries there are restrictions on its use: for example Machova *et al.* (1996) note an MRL of 10 μg/kg in the Federal Republic of Germany in 1989, and a 6 months withdrawal period in the Czech Republic. In the UK the British Trout Association has published formal advice to its members not to use MG in fish weighing more than 5 g. The EU residue surveillance programme (see Section 3.6.4) is known to include MG in the substances tested for in fish; but this will not necessarily improve control over UK fish farmers because the fish tested are bought at the point of sale to the consumer and their provenance (even the country!) is not always known. In the USA MG was used in federal facilities (mainly hatcheries) until 1993 and effluent water was passed through activated charcoal (see Marking *et al.*, 1990). Since that time it has been permitted for use only for winter chinook salmon, an endangered species.

Alderman (1991) has pointed out the intractable problem that for a “mature” medicinal product like MG with no patent protection no company can afford to conduct the toxicological studies to obtain a market authorization in the EU. This would still be true even if, as is most certainly not the case, there was assurance of a market authorization once the work had been done. This lack of assurance is making fish farming organizations hesitant to sponsor such studies; an unequivocal demonstration of toxicity would put them in a more difficult position than the current one of not knowing.

(e) Operator safety

Apart from the chronic toxicological effects described in the next section, MG has produced acute fatalities in rabbits which accidentally aspirated a solution of it. Operators working with MG on a fish farming scale should wear protective (waterproof) bib-aprons, gloves and boots and, particularly when using crystalline MG, face masks.

13.2.3 TOXICOLOGY

(a) Mutagenicity

Meyer and Jorgenson (1983) using doses of MG which they acknowledged to be much higher than normal for aquaculture showed a variety of mutational effects in rainbow trout eggs. 500 ppm for 1 minute daily for 5 days produced chromosomal aberrations. 1 ppm for 1 hour daily produced mitotic defects and chromosome fractures. 0.1 ppm for 3-5 hours blocked cell division, and in fish it inhibited the repair of injured fins (but also prevented necrosis). Immersion regimens sufficiently mild to avoid acute toxicity nevertheless caused tumours in the gastro-intestinal tracts of juvenile trout.

The level of abnormalities in eggs was dose-related. There were delays in hatching which were also dose-related, but the percentage hatch was improved by treatment presumably because of the control of fungi. Growth was depressed in those apparently normal fish which hatched after treatment.

Meyer and Jorgenson (1983) reported earlier evidence that tumours induced in rats persisted in subsequent generations; and mutagenesis has also been shown in small ornamental species of finfish used as laboratory animals. In the latter study the concentrations used had to be low enough to obviate acute toxicity but the exposure was continuous for 7 days. So while the findings of mutagenicity are incontrovertible they have only been shown with extreme overdosage and a question remains concerning their cause. The provenance and purity of the MG used by Meyer and Jorgenson (1983) and in several of the earlier studies were inadequately defined and the toxicity could have been due to impurities.

(b) Target species safety

Maximum no-effect levels of MG vary widely between species; and within species they vary widely according to water conditions, especially pH. In general it is more toxic, and less efficacious for external parasites and fungi, in alkaline conditions because it is the carbinol form which is absorbed.

Machova *et al.* (1996) surveyed earlier reports that MG, like formalin, damages gill epithelium and so depresses respiratory gaseous exchange. It also acts through enzyme inhibition at the sub-cellular level. The toxicity can be reversed by cytochrome C but in normal circumstances recovery is slow and the toxicity is cumulative. Frequent treatments should therefore be avoided. It is also advisable not to feed, handle or otherwise stress fish for 24 hours before and after treatment. This is to reduce oxygen demand and hence gill ventilation and uptake of MG, and also to reduce the oxygen demand of tissues affected by the compound. Particular care is needed where there is any gill disease.

(c) Tolerance by common carp and African catfish eggs

In parallel with their studies on formalin Theron *et al.* (1991a) used single doses of MG on carp and African catfish eggs. They tested various exposure times with two different concentrations, said to be 750 and 1500 mg/l but presumably μg/l was intended. The exposure times were very much shorter than for formalin - a matter of seconds rather than minutes.

With carp eggs	satisfactory results were achieved with 750 µg/l for 10, 20 or 40 sec; with 1500 µg/l a 5 sec exposure was insufficient for anti-fungal action; 10-30 sec was satisfactory, and 60 sec was toxic.
With catfish eggs	a 10 sec exposure at either concentration produced higher mortality than in control eggs; 5 sec exposure to 1500 µg/l was satisfactory.

(d) Tolerance by catfish and carp larvae and juveniles

Theron *et al.* (1991b) also studied immersion of newly hatched African catfish and carp in MG, the water being at pH 7.4-7.8. They state that they used 100 mg/l; this is a very high concentration and not surprisingly they concluded that it should not be used for longer than 10 sec in African catfish. It was anti-fungal at this exposure.

Juvenile carp were more sensitive to MG than African catfish of the same age, and it was concluded that MG should not be used at all in juvenile carp. This may be true for concentrations of the order of 100 ppm, but in fact it has been recommended for use by prolonged exposure at 0.1 ppm. However a problem with long-term exposures is that MG is adsorbed onto organic material and becomes unavailable unless continually renewed. In their 6-day study using 0.2 ppm Machova *et al.* (1996) transferred their rainbow trout to freshly prepared baths each day.

(e) Toxic effects

Rainbow trout given 7 treatments at 7-day intervals showed a slight loss of appetite after each treatment but they recovered well before the next treatment. Liver histopathology was cumulative showing degeneration of hepatocytes and eventual disorientation of the hepatic laminae; vascular stasis was followed by focal perivascular and peribiliary coagulative necrosis. In the gills the secondary lamellae showed apical clubbing and separation of the epithelia; later the epithelia became hypertrophic with deformed cells with large nuclei.

Treatment with MG has been shown to have significant effects on the blood cell picture of several salmonid species. It produces a profound lymphopenia in Atlantic salmon; this may well be a stress response mediated by cortisol, but it is almost certainly immunosuppressive and is therefore of importance when the MG is being used to treat an immunogenic infection. Daily exposure of rainbow trout to 2.27 ppm MG had no significant effect on most blood parameters but it did depress the neutrophil count to 50% of control values. After the first few treatments it did not cause any rise in blood cortisol whereas previously untreated fish did produce a cortisol response to MG. It was concluded that cortisol depresses lymphocytes; MG depresses neutrophils.

13.2.4 PHARMACOKINETICS

(a) In eggs

Meinertz (1995) studied the absorption of radio-labelled MG by rainbow trout eggs. Flushes of 1.0 ppm were given on day 0 and then every 3 days until day 24 when hatching began, and again on day 31 when hatching ended. The eggs were kept in a flow of well water at pH 8, so the majority of the MG would have been in the carbinol form. MG was absorbed cumulatively by the eggs and a plateau was not reached.

Most of the drug was in the leucobase form, a reduction metabolite of the carbinol, in the eggs and fry, and it was shown that this compound depurated at the same rate as the radio-activity. The half-life of the absolute weight of drug was 13.3 days, but due to the growth of the fry its concentration fell faster with a half-life of 9.7 days.

(b) In rainbow trout

MG is not absorbed when given orally and in any case fish refused to eat pellets soaked in 10 ppm MG. Nevertheless the efficacy of MG in the control of proliferative kidney disease is in itself evidence of significant absorption from immersion. It is in fact actively absorbed, presumably through the gills, reaching a higher concentration in the fish than in the water. Alderman and Clifton-Hadley (1993) studied the distribution and elimination of MG in rainbow trout at 8 and 16°C following immersion in 1.6 ppm for 40 minutes; with the water at pH 7.6 (95% of MG as carbinol) this was an acceptable treatment for PKD. Table 13.1 shows their major findings.

Table 13.1. Major pharmacokinetic parameters for malachite green in rainbow trout*. After Alderman D.J. and Clifton-Hadley R.S., 1987, with permission

				Half-lives		
Tissue	Temperature (°C)	C_{max}	T_{max}	$t½\pi$	$t½\alpha$	$t½\beta$
Serum	8	13·0 mg ml^{-1}	0 h	5·37 min	2·15 h	0·62 d
	16	13·5 mg ml^{-1}	0 h	17·07 min	7·33 h	14·5 d
Liver	8	9·0 ppm	0 h		2·10 h	13·18 d
	16	16·5 ppm	0 h		1·98 h	5·68 d
Kidney	8	8·0 ppm	0 h		11·78 h	30·03 d
	16	34·0 ppm	0 h		6·46 h	5·81 d
Muscle	8	7·8 ppm	90–120 min		5·43 h	3·86 d
	16	10·8 ppm	90–120 min		8·35 h	2·89 d

* C_{max} = maximum concentration detected in tissue.
T_{max} = time at which maximum concentration occurred.
$t½\pi$ = distribution half-life in first of three compartments in serum three compartment model.
$t½\alpha$ = distribution half-life in second of three compartments in serum three compartment model, or initial elimination half life in tissues.
$t½\beta$ = elimination half life in serum and second exponent of elimination in tissues.

In serum MG appeared rapidly, rose throughout the exposure and fell rapidly afterwards. C_{max} and the distribution half-life were little affected by temperature. In the liver C_{max} was temperature-dependent but high at both temperatures, with T_{max} at the end of exposure or within the next 20 minutes; concentrations remained near C_{max} for 24 hours. In the kidney the peaks shown in Table 13.1 were transient and further peaks developed later: at 16°C there were peaks of 25 ppm at 8 hours and 30 ppm at 12 hours, and at 8°C of 8.8 ppm at 5 hours.

Both uptake and elimination by muscle were temperature-dependent, and the authors point out that since PKD is a summer disease MG will be used to control it at high water temperatures and will persist as the temperature falls. The authors conclude that a withdrawal period would have to be at least 600 degree-days. Observance of this would preclude slaughter of the fish for the table in the same growing season, and treated fish might well have lost their immunity to PKD by the following season. It is also important to recognize that if MG is used in a hatchery or farm upstream of a second farm the probability is that the downstream fish will absorb detectable quantities of MG.

Machova *et al.* (1996) using their 6-day exposure to 0.2 ppm found that the coloured ionic form of MG was eliminated fairly rapidly. In contrast the colourless carbinol was still detectable in muscle and skin (but not liver) 8 months after the end of treatment, and in muscle at 10 months but not 12 months.

(c) In channel catfish

Poe and Wilson (1983) reported a somewhat bizarre case of channel catfish flesh turning green on storage. The colouration was traced to the treatment of the fish pre-slaughter by immersion in MG. Absorption (presumably as the carbinol) had taken place and the drug would have been metabolized to the colourless leucobase. On prolonged storage the drug on the surface of the fish would have oxidized back to the carbinol and the fish would have become acidic converting the carbinol to the dye. Subsequent studies of various pre-slaughter immersion regimens showed that the colour nearly always developed in visceral fat before muscle, presumably because the carbinol would be preferentially absorbed into this tissue.

13.2.5 ENVIRONMENTAL FATE

MG will be converted in the environment to the microbiologically inactive carbinol and absorbed onto organic particles. It will eventually be destroyed by oxidation.

13.3 Leteux-Meyer mixture

13.3.1 THE PROBLEM

Leteux and Meyer (1972) noted the limited efficacy of either formalin or malachite green for the treatment of white spot (ichthyophthiriasis) at concentrations which were less than toxic for a majority of American species of cultured fish. Concentrations of formalin safe for the target species normally need to be used repeatedly, leading to unacceptable materials and labour costs on commercial farms. A similar problem arises with MG although there is considerable variation in tolerance between species.

In rainbow trout, a relatively tolerant species, immersion in 7 ppm MG for 1 hour produced 85% mortality; and in channel catfish, a susceptible species, exposure to 0.19 ppm produced 50% mortality at 24 hours and 100% within 96 hours. Goldfish, the standard laboratory species, were tested in water of pH 7.1 and total alkalinity 150 ppm. When immersed in 0.6 ppm they were severely stressed after 4 hours exposure and all were dead after 12 hours exposure. At 0.4 ppm they were stressed after 12 hours exposure and all dead after 96 hours exposure.

13.3.2 THE SOLUTION

It was shown initially in goldfish and then in the less drug-tolerant channel catfish that mixtures of malachite green and formalin could be formulated which were both efficacious for ichthyophthiriasis and safe for the target species. In the treatment of channel catfish infected with ichthyophthiriasis there was evidence of synergistic, not simply additive, anti-parasitic activity, although in the absence of a suitable *in vitro* test the synergism has not been quantified.

Leteux-Meyer mixture, which is now widely used, is formulated as a concentrate by dissolving 3.3 g malachite green in 1 litre of formalin. It is used diluted 1:40,000 (2.5 ml in 100 litres, designated 25 ppm) or 1:67,000 (1.5 ml in 100 litres designated 15 ppm). These dilutions are designated according to their formalin content; they also contain 0.083 ppm and 0.05 ppm MG respectively. The higher concentration is used for exposures of up to 1 hour, the lower for more prolonged treatment. A single treatment is usually efficacious for infections with *Costia (Ichthyobodo), Trichodina, Chilodonella, Scyphidia, and Trichophyra*; three applications at intervals of 2 days are normally used for *Ichthyophthirius*. The mixture was found not to be effective for the fluke, *Cleidodiscus*.

13.4 Chloramine-T

13.4.1 USES

(a) Indications

Chloramine-T is used as a disinfectant, and as a treatment for bacterial gill disease and occasionally fin-rot. Bacterial gill disease is a multi-factorial infection in which the bacterium, *Flexibacter branchiophilum*, is usually present, and the protozoa, *Ichthyophthirius, Ichthyobodo, Chilodonella* and *Trichodina* may all be involved. It is traditionally treated with either quaternary amines or chloramine-T but the former have narrow therapeutic margins. Fin-rot is a mixed infection of bacteria and fungi, usually secondary to trauma.

Chloramine-T has been shown to prevent water-borne *Aeromonas salmonicida* infection in brook trout if the fish are treated immediately after exposure to infection. It has no effect on established disease. The organism is known to colonize gill mucus and this is assumed to be the site of action of the chloramine-T.

In comparison with formalin, chloramine-T has greater activity against bacteria but less activity against protozoa.

(b) Chemistry

Chloramine-T is a sodium salt whose anion slowly decomposes in water to the hypochlorite anion and hence to the weak hypochlorous acid (HClO). This in turn decomposes to chlorine and oxygen. The degradation product of chloramine-T is p-toluene-sulfonamide (pTSA). Hypochlorous acid is a strong disinfectant, the hypochlorite anion less so; in consequence the activity of chloramine-T is greater in acid conditions. The total of the hypochlorous acid, hypochlorite and chlorine in a solution is known as the free available chlorine.

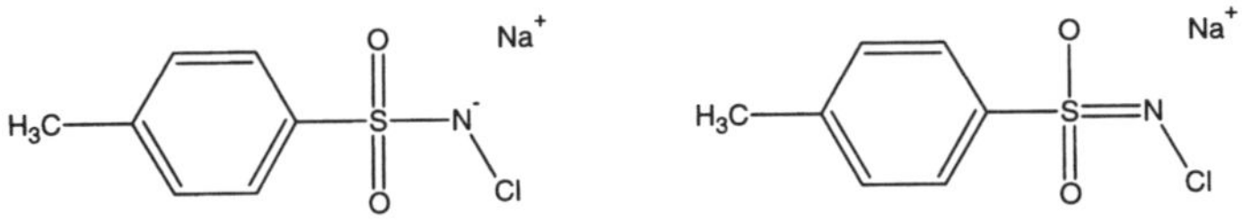

Chloramine-T

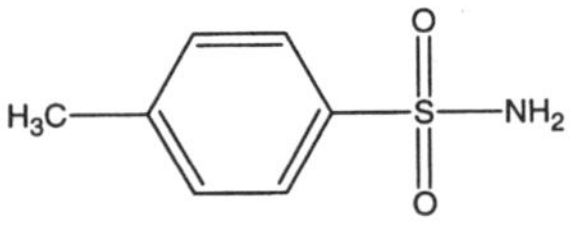

para-Toluene-Sulphonamide

Figure 13.2.

Chloramine-T is regarded as a safer disinfectant than chlorine because the latter combines with organic matter to form carcinogenic trichloro-methanes. However, because it is unstable in aqueous solution chloramine-T is supplied as a powder. This can cause burns or sensitization on skin contact and sensitization on inhalation; it is injurious on contact with the eyes and harmful if swallowed. Protective clothing must be worn when handling chloramine-T.

Chloramine-T is a strong oxidizing agent in both acid and alkaline solutions and should not be used with other chemicals, particularly reducing agents such as formalin. It should not be dispensed in metal containers.

(c) Dosage

Table 13.2 shows initial concentrations recommended by Cross and Hursey (1973) for flow-through systems, *i.e.* flush treatment. It should be noted that because the activity of chloramine-T resides in the degradation product hypochlorous acid, both its antimicrobial activity and its toxicity to fish (see Section 13.4.2(a) below) fall with increasing pH. A curvilinear relationship between pH and the 2 h LC_{95} of *I. multifiliis* was shown, with rapidly increasing activity at higher pH. The non-linearity would have been due to a substantial extent to the direct effect of pH on the parasites - in water at pH 8 all parasites die in 2.5 hours with no chloramine-T.

Table 13.2. Initial concentrations of chloramine-T for flush treatments

pH of water	Concentration (ppm)	
	Soft water	Hard water
6.0	2.5	7.0
6.5	5.0	10
7.0	10	15
7.5	18	18
8.0	20	20

For dip treatments concentrations of 8.5-10 ppm are recommended, with exposure for 1 hour daily for 3 days. Alternatively lower concentrations in the range 5-6.5 ppm may be used for 3 hours daily. As noted above, bacterial gill disease is usually a mixed infection of bacteria and protozoa and a comparative trial of exposures to 10 ppm chloramine-T and 167 ppm formalin showed both to have a significant therapeutic action but neither to entirely sterilize the infection. It was suggested that the ideal regimen might be an initial dip in formalin followed by two in chloramine-T.

15 ppm chloramine-T for 1 hour was found to control an artificial infection of brook trout with *Aeromonas salmonicida* while formalin was inactive. *In vitro* the MIC of *A. salmonicida* was in the range 5-12.5 ppm chloramine-T for 60 minutes or 1.5-3 ppm for 24 hours.

In their investigations of the treatment of *I. multifiliis* infestations Cross and Hursey (1973) showed that for each of the fish species they studied it was possible to plot on a concentration/pH graph:

1. A linear 1 wk LC_{50} for the fish
2. A linear 1 wk LC_5 for the fish (at a lower concentration)
3. A curvilinear LC_{95} for the parasite.

They could therefore construct a curvilinear line mid-way between 2 and 3 above which would give the optimum therapeutic concentration at any pH. They found that such concentrations could be safely maintained for several weeks.

(d) Administration in tanks

The procedure recommended by Cross and Hursey (1973) is as follows:

(i) Tanks with running water

1. The tanks are carefully siphoned to remove as much detritus as possible;
2. The flow of water through the tank is adjusted to give a replacement time of approximately 4 hours;
3. The water is gently aerated;
4. The recommended dose of chloramine-T is added to the water. If a stock solution is used it should be freshly prepared.

Stages 1, 2 and 4 are repeated thrice daily, say at 0900, 1300 and 1700 h. (At 1700 h the concentration is approximately twice the therapeutic dose but over a 24 h period the average concentration approximates to the therapeutic).

(ii) Tanks with static water

Stages 1, 3 and 4 of the above treatment are retained. In addition any filtration equipment is turned off. Such a system operating in the tank would rapidly deactivate the chloramine-T and, in case of a biological filter, which depends for its action on a well-developed micro-fauna and flora, the future functioning of the filter would be affected by the disinfectant action of the chloramine-T. The dose is applied once daily until all signs of the infection have disappeared. Once a week half the water is siphoned from the tank and replaced by new water.

13.4.2 TOXICOLOGY

(a) Factors affecting toxicity

The toxicity of chloramine-T to the target species is increased by soft water and low pH. Increases in toxicity are indicated by reductions in LC_{50} and LC_{95}, and Cross and Hursey (1973) showed that in soft water within the pH range of normal supplies there are linear relationships between pH and both LC_{50} and LC_{95}. The general principle of greater toxicity in more acid water applies also to hard water but the linear relationship has not been shown. Toxicity is reduced by a high organic loading of the water. Cross and Hursey (1973) also reported that chloramine-T was less toxic at low temperatures, but other work suggests that high temperature accelerates the onset of symptoms of toxicity rather than altering the eventual effect.

Larson et al. (1977) showed that, at least in some species of salmonids, tolerance was profoundly affected by age during the first few weeks of life. Tolerance was high at hatching, but fell rapidly during the alevin and fry stages. It rose again during the juvenile stage, but reached a plateau below that at hatching.

(b) Symptoms of toxicity

Powell *et al.* (1994) conducted a series of trials on rainbow trout exposed to chloramine-T for 1 hour on two consecutive days per week for 4 weeks; the water conditions were pH 7.4, temperature 11°C and hardness 209 mg/l. They found that at 5 ppm chloramine-T there was even distribution of fish and only a slight hyperventilation; the fish became darker after the 3rd week. At 10-20 ppm the fish crowded at the top of the tank, had increased amplitude and frequency of ventilation and became darker after the first treatment; however they recovered in 24 hours and fed normally. At 20-30 ppm there were erosions of fins and the caudal peduncle and severely affected fish swam erratically and lost equilibrium.

At all concentrations the treated fish grew more slowly than untreated controls. Where chloramine-T has been used weekly for 11 weeks the adverse effect on the specific growth rate (*i.e.* the weight gain over a period as a proportion of the initial weight) is normally confined to the first 3 weeks. Feeding during this time is unaffected so the food conversion ratio is adversely affected. After 3 weeks the fish appear to become acclimated to the treatment.

(c) Toxic effects

Chloramine-T, like formalin, affects the gill epithelia causing the symptoms of hypoxia. In rainbow trout there is hyperplastic thickening of the lamellae and intercellular oedema, and eventually epithelial erosions. Mucus production is variously reported as increased and reduced. There is a dose-dependent fall in plasma calcium, sodium and chloride ions together with a reduced haematocrit, indicating a loss of osmoregulation and influx of water.

Despite the hyperventilation [CO_2] is elevated by treatment. This implies a reduced conductance by the gills which could be due to increased mucus production or the epithelial thickening. If the mucus production is increased this is probably a response to the irritant hypochlorite ions produced by degradation of the chloramine-T.

13.5 Copper sulfate

13.5.1 USES

(a) Indications

Copper sulfate has been used in the control of bacterial gill disease but cannot be recommended; its mode of action is astringent rather than antibacterial. It is active against protozoa and has a particular indication for *Amyloodinium* and *Cryptocaryon* infections in marine aquaria where the target species tolerance is much better than in freshwater.

Copper sulfate is lethal to marine invertebrates and elasmobranchs so it cannot be used in community aquaria containing these organisms. It is also toxic to fish which have been exposed to a reduced osmotic pressure without time to acclimate; the loss of sodium ions exacerbates copper toxicity.

(b) Dosage

Copper sulfate is administered only by immersion, and unlike the other compounds covered in this chapter the exposure is usually prolonged, for example for 10 days. Dipping is therefore not appropriate and the compound is added to the pond or aquarium water. The normal procedure is to prepare a stock solution of 400 mg/l copper sulfate pentahydrate. Concentrations are normally considered in terms of copper ions rather than the salt, and use of this stock solution at 1 ml per litre gives a copper concentration of 0.1 ppm.

Use concentrations are in the range 0.1-0.2 ppm, the higher concentration being used in hard water where copper carbonate is precipitated. It is essential to test the copper content of the water daily and adjust it by adding stock solution or making water changes.

13.5.2 TOXICOLOGY

(a) LC_{50}

Ling *et al.* (1993) conducted a series of studies on the use of copper sulfate for ichthyophthiriasis in goldfish. Table 13.3 shows their findings of LC_{50} of copper ions at various times after the end of exposure.

Table 13.3. Average water quality in aquaria in which LC_{50} values for copper sulphate in goldfish at different time intervals were determined. After Ling K.H. *et al*, 1993, with permission

Time (h)	LC_{50}[1] at 95% confidence limits (μg/l)	pH	Dissolved[2] oxygen (mg/l)	Ammonia (total nitrogen content) (mg/l)	Total hardness ($CaCO_3$ mg/l)
24	365	7.2±0.2	7.6±0.2	2.3±1.1	77.5±2
48	355	7.3±0.1	7.2±0.8	2.4±1.0	78.5±1
72	301	7.2±0.2	7.5±0.2	2.3±1.2	77.2±2
96	288	7.1±0.3	7.3±0.3	2.5±1.3	76.4±3
144	276	7.1±0.2	7.3±0.3	2.7±1.0	78.3±2
192	268	7.1±0.2	7.3±0.5	2.6±1.1	79.3±1

[1] LC_{50} at each time interval was computed by moving average (Chapman and Stevens, 1978).
[2] Average temperature for tank water is 28.5±2°C.

These results may be compared with reports of 96 h LC_{50} for other species of:

740 µg/l in bluegill sunfish
600 µg/l in fathead minnows
408 µg/l in rainbow trout
46 µg/l in adult coho salmon
32 µg/l in juvenile chinook salmon
0.15 µg/l in striped bass in freshwater
8 µg/l in striped bass in brackish water

(b) Sub-lethal toxicity

Although copper sulfate has an antimicrobial action, Baker *et al.* (1983) showed that it also has an adverse effect on salmonids: sub-lethal concentrations render them more susceptible to infections. Thus various salmonid species were shown to be made more susceptible to Infectious Haemopoietic Necrosis, *Vibrio anguillarum* and *Yersinia ruckeri.*

Water hardness affects the toxicity of copper sulfate and Baker *et al.* (1983) referred to concentrations in terms of toxic units (TU), 1 TU being the 96 hour LC_{50}. Using an injected *V. anguillarum* challenge after 96 hours exposure they found that maximum susceptibility was after 0.08-0.20 TU in chinook salmon and after 0.18-0.31 TU in rainbow trout. The reason why higher sub-lethal concentrations cause less susceptibility is not known.

13.5.3 ABSORPTION

Ling *et al.* (1993) exposed four groups of goldfish to 288 µg/l copper sulfate for periods of up to 120 minutes and immediately afterwards challenged them with *I. multifiliis*, with the results shown in Table 13.4. The correlation of exposure time with protection shows that copper ions were being absorbed over the whole period of exposure. Copper levels were high in the liver and kidney and low in the fins, skin and muscle.

Table 13.4. Effect of different prior exposure times to copper sulphate (288 µg Cu^{2+}/1) on the protection of goldfish from *I. multifiliis*. After Ling K.H. *et al*, 1993, with permission

Group	$CuSO_4$ exposure (min)	No. uninfected / Total no. fish	Percentage of survival[1]
I (control)	-	2/36	5.6
II test	15	0/9	0.0
III test	30	3/9	33.3
IV test	60	5/9	55.6
V test	120	8/9	88.9

[1]Calculated at 14 days after exposure to free-swimming tomites (2000 tomites/fish).

Three groups of fish were then exposed to a lower concentration for 2 weeks and challenged at intervals after the end of exposure. Copper ions returned from the fish to the water but Table 13.5 shows that a significant level remained in the fish for at least 3 weeks.

Table 13.5. Fish exposed to 2 weeks of 255 μg Cu^{2+} /l copper sulphate solution and transferred to copper-free water for different durations before challenge with *I. multifiliis*. After Ling K.H. *et al*, 1993, with permission

Groups (weeks after the last $CuSO_4$ treatment)	No. uninfected / Total no. fish	Percentage of survival[1]
I (control)	1/27[a]	3.7
II (1 week)	8/9	88.9
III (2 weeks)	6/9	66.7
IV (3 weeks)	5/9	55.6

[1]All fish exposed to live tomites at the concentration of 2000 tomites/fish for *I. multifiliis* infection. Percentage of survival was calculated at 14 days after initial exposure to free-swimming tomites.
[a]Pooling of the 3 controls used for different weeks.

Further reading

Alderman, D.J. (1985) Malachite green: a review. *Journal of Fish Diseases* **8**, 289-298.

Alderman, D.J. (1991) Malachite green and alternatives as therapeutic agents. In *Aquaculture and the environment* (eds. de Pauw, N. and Joyce, J.) European Aquaculture Society Special Publication No. 16, Gent, Belgium, pp. 235-244

Alderman, D.J. and Clifton-Hadley, R.S. (1988) Malachite green therapy of proliferative kidney disease in rainbow trout: field trials. *Veterinary Record* **122**, 103-106

Alderman, D.J. and Clifton-Hadley, R.S. (1993) Malachite green; a pharmacokinetic study in rainbow trout, *Oncorhynchus mykiss* (Walbaum). *Journal of Fish Diseases* **16**, 297-303.

Clifton-Hadley, R.S. and Alderman, D.J. (1987) The effects of malachite green upon proliferative kidney disease. *Journal of Fish Diseases* **10**, 101-107.

Cross, D.G. and Hursey, P.A. (1973) Chloramine-T for the control of *Ichthyophthirius multifiliis* (Fouquet). *Journal of Fish Biology* **5**, 789-798.

Klontz, G.W. (1994) Autumnal fish health management. *Pacific Coast Aquaculture* **November/December 1994**, 8-11 & 22.

Larson, G.L., Hutchins, F.E. and Schlesinger, D.A. (1977) Acute toxicity of inorganic chloramines to early life stages of brook trout (*Salvelinus fontinalis*). *Journal of Fish Biology* **11**, 595-598.

Leteux, F. and Meyer, F.P. (1972) Mixtures of malachite green and formalin for controlling *Ichthyophthirius* and other protozoan parasites of fish. *Progressive Fish-Culturist* **34**, 21-26.

Ling, K.H., Sin, Y.M. and Lam, T.J. (1993) Effect of copper sulfate on ichthyophthiriasis (white spot disease) in goldfish (*Carassius auratus*). *Aquaculture* **118**, 23-35.

Machova, J., Svobodova, Z., Svobodnik, J., Piacka, V., Vykusova, B. and Kocova, A. (1996) Persistence of malachite green in tissues of rainbow trout after a long-term therapeutic bath. *Acta veterinaria Brno* **65**, 151-159.

Marking, L.L., Leith, D. and Davis, J. (1990) Development of a carbon filter system for removing malachite green from hatchery effluents. *Progressive Fish-Culturist* **52**, 92-99.

Marking, L.L., Rach, J.J. and Schreier, T.M. (1994) Evaluation of antifungal agents for fish culture. *Progressive Fish-Culturist* **56**, 225-231.

Meinertz, J.R. (1995) Residues of [^{14}C]-malachite green in eggs and fry of rainbow trout after treatment of eggs. *Journal of Fish Diseases* **18**, 239-247.

Meyer, F.P. and Jorgenson, T.A. (1983) Teratological and other effects of malachite green on development of rainbow trout and rabbits. *Transactions of the American Fisheries Society* **112**, 818-824.

Poe, W.E. and Wilson, R.P. (1983) Absorption of malachite green by channel catfish. *Progressive Fish-Culturist* **45**, 228-229.

Powell, M.D. and Speare, D.J. (1996) Effects of intermittent formalin treatment of Atlantic salmon juveniles on growth, condition factor, plasma electrolytes and haematocrit in freshwater and after transfer to seawater. *Journal of Aquatic Animal Health* **8**, 64-69.

Scott, P. (1993) Therapy in aquaculture. In *Aquaculture for veterinarians* (ed. Brown, Lydia) Pergamon Press, Oxford, pp. 131-152

Speare, D.J. and MacNair, N. (1996) Effects of intermittent exposure to therapeutic levels of formalin on growth characteristics and body condition of juvenile rainbow trout. *Journal of Aquatic Animal Health* **8**, 58-63.

Theron, J., Prinsloo, J.F. and Schoonbee, H.J. (1991a) Treatment of *Cyprinus carpio* L. and *Clarias gariepinus* (Burchell) embryos with formalin and malachite green: effect of concentration and length of treatment on their survival. *Onderstepoort Journal of Veterinary Research* **58**, 239-243.

Theron, J., Prinsloo, J.F. and Schoonbee, H.J. (1991b) Investigations into the effects of concentration and duration of exposure to formalin and malachite green on the survival of the larvae and juveniles of the common carp *Cyprinus carpio* L. and the sharptooth catfish *Clarias gariepinus*. *Onderstepoort Journal of Veterinary Research* **58**, 245-251.

Willoughby, L.G. and Roberts, R.J. (1992) Towards strategic use of fungicides against *Saprolegnia parasitica* in salmonid fish hatcheries. *Journal of Fish Diseases* **15**, 1-13.

14. ECTOPARASITICIDES

14.1 Metazoan ectoparasites

A useful summary of the range and life-cycles of the important metazoan ectoparasites of finfish was provided by Southgate (1993). A majority of them are treated with formalin or sodium chloride and virtually the only ones for which specific drug regimens have been developed are the sea-lice, which despite their name are not insects but copepods. The important species are *Lepeophtheirus salmonis*, the salmon louse, and the related *Caligus* spp. Another group of copepods, the freshwater fish lice, *Argulus* spp., which includes the carp louse, can be controlled by the same range of parasiticides with the exception of hydrogen peroxide. The active ingredients of these drugs are for the most part insecticides.

Several genera of monogenean (*i.e.* having only a single host species in the life cycle) flukes are parasitic on fish. *Gyrodactylus* spp. are skin parasites and hence obviously ectoparasites, but there are several genera parasitic on the gills, including *Dactylogyrus* in cold waters and *Diplectanum, Microcotyle, Epibdella* and *Polylabris* in the Mediterranean Sea. Drugs developed primarily for sea-lice, notably trichlorfon and ivermectin, have shown varying degrees of activity against these flukes.

The anchor worm (*Lernaea* spp.) is a crustacean parasite of worldwide importance in freshwater. Organo-phosphorus compounds are effective against the juvenile stages but not the adults. There is a report of the successful treatment of goldfish by intramuscular injection of ivermectin at 16 μg/kg.

14.2 Organo-phosphorus compounds

14.2.1 GROUP CHARACTERISTICS

(a) Mode of action

Organo-phosphorus compounds are anti-cholinesterases, that is, they inhibit the enzyme, acetyl-cholinesterase (AChE), which catalyses the hydrolysis of acetyl-choline, the neuro-muscular transmitter in both vertebrates and arthropods. This inhibition occurs after ingestion of organo-phosphorus compounds by fish as well as mammals, and in arthropods they are contact poisons because they readily penetrate the cuticle. In vertebrates organo-phosphorus compounds are hydrolysed rapidly after which the AChE regenerates and hydrolyses the accumulated acetyl-choline; arthropods do not have the enzymes to hydrolyse organo-phosphorus compounds quickly, so their neuro-muscular transmissions and consequent muscular contractions are prolonged.

The main aquacultural use of organo-phosphorus compounds is for the control of sea-lice, although they have also been used for other external metazoan parasites. Of the 10 stages in the life-cycle of the sea-louse, the compounds are effective against only the last three or four, known as pre-adults and adults. The immediately preceding stages, known as chalimi, do contain AChE but for reasons unknown the compounds have no activity against them.

(b) Target species tolerance
Lethal doses of organo-phosphorus compounds vary with species. At 20°C tilapia will survive 1 mg/l (= 1 ppm) dichlorvos for 15 days; carp will survive 30 mg/l for 30 minutes, either of which regimens would be totally lethal to Atlantic salmon. Mexican mojarra (*Cichlasoma urophthalmus*) will survive 20 ppm trichlorfon at 28°C. In carp trichlorfon at normal use rates causes higher mortality than dichlorvos. Dichlorvos is absorbed by carp in sufficient quantities to depress AChE activity in brain, liver and spleen for 3 days. It persists longer in the kidney because there is enzymatic degradation in the liver and final elimination is through the kidneys. Exposed carp are thought to have lowered detoxification capacity and to be more sensitive to other pollutants.

(c) Timing of treatment
Timing of the use of organo-phosphorus compounds is critical because of the limits on the stages of sea-lice for which they are active and because they have no residual or prophylactic action. Premature use will be ineffective, and delayed use will result in both fish mortality and reproduction by the parasites. All infestations consist of lice at a variety of stages of the life-cycle and this further complicates timing decisions. Treatment will eliminate adults and hence prevent the reproduction of new larvae; but it will not eliminate any chalimi which may be on a fish, and these will continue development to become adults. Examination of fish 10-20 days after treatment is therefore recommended to determine whether a second treatment is necessary. A third treatment may be necessary on rare occasions but if all the pens at a site are treated at the same time no further treatment should be necessary for a considerable time.

14.2.2 TRICHLORFON (METRIPHONATE)

(a) Presentation and administration
Trichlorfon was the first organo-phosphorus compound to be developed for the control of sea-lice. The insecticide Neguvon® consisting of 97% trichlorfon and 3% stabilizers was used. It was initially administered orally but this resulted in low target species tolerance, attributed to a large variation between individual fish in their feed intake. Now it is always administered as a bath (see Section 1.1.5). To do this a preliminary dilution must be made and added at several points around the bath. Oxygen must be sparged into the bath throughout the exposure time.

(b) Use in Atlantic salmon
The concentration of trichlorfon which used to be recommended was 300 g/m^3 (the units giving expression to the normal use of the compound in baths typically of 500 m^3).

The duration of exposure ranged from 15 to 60 minutes according to temperature, 15 minutes being recommended for 12°C and 60 minutes for 3°C. The recommendation is no longer followed, however, because investigation of sporadic cases of mass mortality in Atlantic salmon showed that they all occurred at 12°C or above and brought to light the importance of the conversion of trichlorfon to dichlorvos.

A more recent recommendation has been to use 300 g/m^3 at temperatures of 6°C or lower, ranging down to 15 g/m^3 at temperatures of 14°C or above.

Residues of trichlorfon in Atlantic salmon are not persistent. They fall to below 1 µg/kg (<0.001 ppm) within 12 days. Withdrawal periods of 21 days and 30 days have been applied in Norway and Germany respectively.

(c) Conversion to dichlorvos

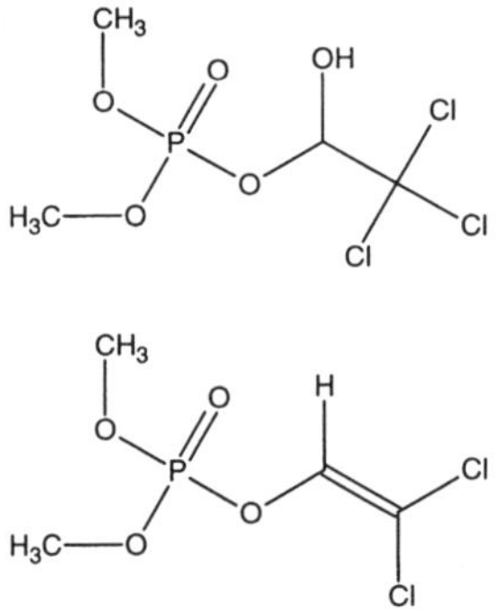

Figure 14.1.

Trichlorfon is gradually converted to dichlorvos in water. The latter compound is about 100 times more potent an AChE inhibitor than trichlorfon, and it is more lipid-soluble and hence more rapidly absorbed (by both vertebrates and arthropods). Thus although trichlorfon is active against arthropods, in water the effective compound is dichlorvos. The rate of action of trichlorfon, and hence both the dose and target species toxicity, depends on the rate at which it is converted to dichlorvos. This is significantly affected by temperature, pH, sunlight and aeration of the water.

Aeration of the water during exposure used to be recommended. Whatever its value to the fish may have been, Samuelsen (1987) showed that the procedure had a profound chemical effect on the drug. It could reduce the half-life of trichlorfon by up to 67% and that of dichlorvos by up to 70%. Under certain conditions the recommended concentration of trichlorfon could degrade to more than the recommended concentration of dichlorvos in less than 2 hours. Other workers have shown more recently that Neguvon may contain up to 0.3% dichlorvos. If this were to be used at 300 g/m^3 the bath would contain 0.9 g/m^3 dichlorvos at the beginning of exposure when the recommended use concentration of dichlorvos is 1.0 g/m^3!

(d) Use in Masou salmon

Japanese workers have reported the use of trichlorfon for the control of *Argulus coregoni* in Masou (amago) salmon. This is a warm, freshwater species and they were working with water at 20°C and pH 7.0. They obtained 100% kill of the parasites at:

6.25 ppm for 60 minutes
12.5-50 ppm for 30 minutes
100 ppm for 10 minutes

Curiously, at higher concentrations of trichlorfon (*e.g.* 12.5 ppm for 60 min) or longer exposures (*e.g.* 100 ppm for 20 min) they found slightly reduced efficacy. Lower concentrations gave very poor results.

These dose rates are an order of magnitude lower than the recommendation for sea-lice on Atlantic salmon. Quite apart from the possibly greater sensitivity of the parasites, it must be recognized from the temperature that they must have been working with a high concentration of dichlorvos. A possibly remarkable feature of the results was the survival of the fish, which may have been an effect of species, pH or salinity. In Masou salmon, and brook, rainbow and brown trout at 16-17°C they found no mortality in 800 ppm for 60 minutes, but high mortality after 18 hours in as little as 6.25 ppm.

14.2.3 DICHLORVOS (DDVP)

(a) Presentation

The insecticide formulation of dichlorvos, Nuvan®, was the first to be tried in aquaculture and following its success a specific formulation, Aquagard® SLT, was developed. ("SLT" stands for sea lice treatment). This is an emulsifiable liquid containing 50% w/v dichlorvos with excipients consisting of 8% emulsifiers and 42% di-*n*-butyl phthalate (DBP).

(b) Dosage and administration

The recommended dose regimen is 1 ppm dichlorvos (2 ppm Aquagard SLT) for 30 to 60 minutes. In fact this is rather arbitrary because in normal use the drug becomes stratified with poor mixing between layers in the enclosed netpens; concentrations may vary between 0.5 and 3.5 ppm between layers. Furthermore efficacy rises with temperature; 0.5 ppm is fully efficacious at 16°C or above. The bath must be sparged with oxygen at several points during exposure, and Aquagard, pre-diluted with water from the body of water containing the fish, should be added at these points to ensure rapid dispersion. Fish should be monitored during exposure, especially during the second 30 minutes, and the tarpaulins enclosing the pen should be dropped immediately if there are any symptoms of distress.

There is a serious operator hazard in the use of dichlorvos. It is a neuro-toxin which can be absorbed by inhalation, ingestion and through the skin. It should be used only by trained operators wearing protective overalls, rubber gloves and faceshields.

(c) Target species toxicity

The symptoms of dichlorvos toxicity are muscular rigidity, the fish congregating at the bottom of the netpen and falling on their sides. The smaller the fish the more rapidly the symptoms develop.

At a farm site where fatalities occurred in some but not all netpens it was found that the dead fish had less thn 20% of the normal AChE activity; survivors in affected groups had 45-48%, and unaffected groups had levels over the wide range 23-62%.

Even though dichlorvos has ceased to be detectable in surviving fish inhibition of AChE persists. It may continue for as long as 14 days after a treatment; and a repeat treatment during this period would entail a higher risk of mortality. Correlation between muscle residues of dichlorvos and AChE inhibition is low. Similarly only small amounts of dichlorvos have been detected in the brain of dead fish despite substantial depression of brain AChE activity.

Oxygen deficiency enhances the anti-AChE activity of dichlorvos. It is not clear whether this effect is due to increasing the activity of dichlorvos on the enzyme, delaying the metabolism of dichlorvos by the fish, or delaying the regeneration of AChE.

(d) Residues

Only a few studies have been conducted on residues of dichlorvos in Atlantic salmon not least because the drug becomes undetectable well within two weeks after death, and there would be no point in treating fish which were to be harvested so soon afterwards. The few studies that have been published have generally examined fish which died as a result of treatment, and there is some evidence of profound changes in the disposition of residues within the first 24 hours after death.

Some published results are *prima facie* conflicting, for example that residues in liver are higher than those in muscle, and that at death residues were 0.08 ppm in muscle and 0.07 ppm in liver. However one study, inevitably using different fish at the two time points, found residues declining from detectable to undetectable in muscle during the first 24 hours after death, but doubling in level in the liver during the same period. If the latter finding was correct, reconstitution of dichlorvos from liver metabolites seems inherently more probable than transference from muscle in a dead fish!

Withdrawal periods which have been applied are 14 days in Norway and 4 days in the UK; in Ireland they are 7 days at temperatures above 10°C and 10 days below 10°.

(e) Environmental safety

The usefulness of dichlorvos (or of any other sea-lice treatment) depends on its toxicity to arthropods, and since it is released into the environment after use it is a potential hazard to invertebrate wildlife, and indeed farmed aquatic invertebrates. On release it remains in the water compartment of the environment and is subject to both dispersion (and hence dilution) and hydrolysis. Dispersion depends on water movement, which in a sea-loch or fjord includes both flushing due to entry of freshwater and tidal flows, and it will vary according to the point in the water-body at which the release occurs. Hydrolysis is to some extent dependent on temperature. The widespread use of dichlorvos in Scotland has occasioned several studies on the rates of dispersion in different types of sea-loch, and on the basis of these the Scottish Environmental Protection Agency controls the quantities and frequency of releases in any one sea-loch. McHenery *et al.* (1991) reviewed the assembly and application of the data.

Molluscs are relatively resistant to dichlorvos: mussels and periwinkles will survive 1.0 ppm for 60 min, the normal therapeutic regimen for Atlantic salmon. Farmed Manila clams and Japanese oysters survived this concentration for 6 hours but the valves opened because of relaxation of the adductor muscles. The therapeutic regimen is lethal to test crustacea (lobster larvae and adults) and zooplankton and phytoplankton. Crustacea appear to be the most susceptible of the aquatic taxa: the 96 h LC_{50} for lobster larvae (5.7 μg/l) is less than 5% of that for herring larvae (122 μg/l). For lobster larvae the 96 h LC_{50} may in fact be little less than the 12 h LC_{50}: lobsters can recover from exposures which are not immediately lethal.

The formulation of dichlorvos first used in sea-lice control was Nuvan, and it was shown that its toxicity to some wildlife taxa, especially phytoplankton, was due more to the excipients than to the dichlorvos. Aquagard SLT contains only 50% dichlorvos and is more toxic than pure dichlorvos to phytoplankton and to juvenile Atlantic salmon, so the environmental fate of its 42% content of DBP is of concern. Hydrolysis to phthalic acid and butanol is the main chemical degradation of DBP, but it is slow. Dichlorvos remains in the water column; but the solubility of DBP in water decreases as salinity increases, and as much as 20% of the DBP in Aquagard SLT may initially be deposited in sediment. It depurates more rapidly in active than in sterilized sediment so microbial degradation may be important (in seawater as well as in sediment).

Some of the data on toxicity to lobsters (and other wildlife species) may in fact be irrelevant to the field situation. In field trials of dichlorvos no effect has been found in lobsters adjacent to treated cages. This may be due to the dichlorvos remaining in the upper layers of the water column until dispersed by water movements, with only insignificant concentrations diffusing down to the seabed. Furthermore it must be remembered that any part of the environment which is denuded of wildlife by a non-persistent chemical will very rapidly be recolonized from the surrounding areas.

(f) Use in France

Messager and Esnault (1991) developed a dichlorvos regimen adapted to French aquacultural requirements. French conditions differ from Norwegian, Scottish and Irish ones in that the main species farmed at marine sites is the rainbow trout; water temperatures are higher; the farm sites are smaller, and in many cases there are shellfish (especially oyster) farms close by.

They showed efficacy and target species safety with two regimens, a short duration one of 15 ppm for 1 minute and a long duration one of 1 ppm for 30 minutes, the latter being comparable to that used for Atlantic salmon. They considered the short duration regimen to be better suited to French conditions because it could be adapted to virtually eliminate environmental pollution with dichlorvos. Their procedure was to prepare a tank containing 3 g dichlorvos in 200 l seawater (= 15 ppm); the fish were netted 10 at a time, dipped for 1 minute, and put into another tank of seawater. In this way they were able to treat 680 fish in 1 h 50 min using only 3 g dichlorvos. Furthermore, because the dichlorvos was in a tank it could be chemically degraded before discharge. The rate of hydrolysis is pH-dependent: it has a half-life of about 6 days in seawater but is rapidly hydrolysed in more alkaline conditions. It can be completely degraded by adding sodium hydroxide to pH 10 for 18 hours and then neutralizing with hydrochloric acid before discharge. The innocuity of the solution to be discharged was shown by its safety to red shrimp (*Palaemon serratus*), a species exquisitely sensitive to dichlorvos.

14.2.4 AZAMETHIPHOS

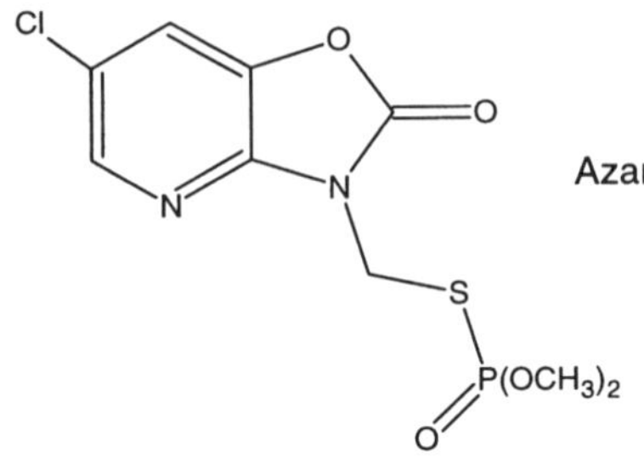

Figure 14.2.

(a) Advantages

Azamethiphos is the latest organo-phosphorus compound to be developed for the control of sea-lice by bath administration. The spectrum of sea-lice stages for which it is active is the same as for dichlorvos. It is safer to the operator having a tenth of the toxicity of dichlorvos to mammals; furthermore the commercial formulation, Salmosan®, is a wettable powder and so can be supplied in water-soluble sachets. Nevertheless it does still present a significant hazard to the operator: waterproof coveralls, heavy duty nitrile gauntlets and a faceshield should be worn when working with Salmosan. A further advantage of azamethiphos is that although it is less toxic than dichlorvos to mammals it is much more active against sea-lice, meaning that lower weights need to be transported and handled.

Salmosan is extensively used in Norway; it has replaced Aquagard SLT in the UK, and at the time of writing it is the only medicinal product authorized for sea-lice control in Canada.

(b) Target species tolerance

Initial laboratory trials of efficacy at 1, 0.1 or 0.01 ppm for 1 hour showed the highest concentration to be acutely toxic to Atlantic salmon and the lowest to be fully efficacious against adult sea-lice. Three treatments at weekly intervals using 0.1, 0.3 or 0.5 ppm for 1 hour resulted in no mortalities at 0.1 or 0.3 ppm; at 0.5 ppm there were mortalities after the second treatment. At 0.3 ppm although there were no mortalities the fish did become lethargic and took a few hours to recover. At concentrations below 0.1 ppm the fish may show a hyperactive response.

Studies of the brains of fish given repeat treatments showed that 0.3 ppm caused significant AChE depression on each occasion but that there was substantial recovery within the week before the next treatment. This is a further advantage of azamethiphos over dichlorvos.

Wrasse tolerate concentrations of azamethiphos which are lethal to Atlantic salmon, so they can safely be used as cleaner fish to supplement the action of the drug.

(c) Efficacy

Roth *et al.* (1996) studied the effect of azamethiphos on sea-lice. They showed that *in vitro* the action was protracted giving substantially higher cumulative mortalities at 24 hours than immediately after the end of treatment. Louse counts on fish immediately after treatment showed azamethiphos to be 2.4-4.5 times as active as dichlorvos but by 24 hours the difference had increased to 7.6-10-fold. Field trials at three sites gave unsatisfactory results at two and very good results against pre-adult and adult lice at the third. The *in vitro* studies had shown varying sensitivities between lice from different locations. It was suggested that previous use of dichlorvos at some sites may have caused some degree of resistance and hence cross-resistance to other organophosphorus compounds.

O'Halloran and Hogans (1996) used Salmosan at a farm on the Atlantic seaboard of Canada. They used 60 g (30 g azamethiphos) in 300 m^3 seawater (0.1 ppm azamethiphos) for 30 minutes at 11.6°C. They had no mortalities and the fish ate voraciously 20 minutes after the end of exposure; efficacy was 100% against gravid female lice, 98.3% against pre-adults and 68% against chalimus larvae.

(d) Administration

Salmosan is a 50% wettable powder available in 20 g and 100 g water-soluble sachets, so the recommended concentration of 0.1 ppm azamethiphos is obtained by one 20 g sachet in 100 m^3 or one 100 g sachet in 500 m^3. A preliminary concentrated solution should be made and in the interests of operator safety it is recommended that this is done on land wearing protective clothing. The required number of sachets for any given netpen are put into a screw-topped polythene jar and shaken with not less than 100 ml water per 20 g Salmosan; it is recommended that distilled water is used. Immediately before use this preliminary concentrated solution is diluted in 200 litres of seawater and stirred together with rinsings from the polythene jar. The latter dilution is added to the bath at oxygen diffuser points.

The water should be oxygenated during treatment. The period of exposure before the tarpaulins are removed should be not more than 60 minutes at temperatures below 10°C and not more than 30 minutes above 10°C. If the fish show signs of distress, for example falling on their sides, the tarpaulins should be dropped immediately and the water oxygenated vigorously.

(e) Environmental safety

A series of government-sponsored trials have been conducted in the Canadian province of New Brunswick on the toxicity of azamethiphos to invertebrate wildlife. It was shown that, like dichlorvos, azamethiphos at normal use concentrations is significantly toxic only to crustacea; although anomalously it is less toxic to green crabs than to shrimps, and American lobsters are exquisitely sensitive. The mean 96 h LC_{50} for American lobsters was determined to be 0.84 µg/l, as compared to a use concentration of 100 µg/l.

Field studies were then made of the concentration of azamethiphos in water and its effect on sentinel invertebrates close to cages where it was used. Concentrations in water were monitored at a farm site with a depth of 22 m at mid-tide. Samples were taken at the edge of the treatment cage immediately after the tarpaulin was dropped and in the effluent drift at up to 150 m from the treatment cage. The assay had a limit of detection of 0.1-0.2 μg/l and a limit of quantitation of 0.5 μg/l (compared to a recommended use concentration of 100 μg/l). In two trials no azamethiphos was detected at any point or depth. This was attributed to rapid mixing and dilution after the tarpaulin was dropped. In one of the trials water samples were taken from inside the tarpaulin during treatment and showed concentrations of 60 μg/l at the surface, 6 μg/l at mid-depth and 59 μg/l at the bottom of the bath. This indicates poor mixing but also suggests that the volume of the bath may have been underestimated.

American lobsters, shrimps, scallops and soft-shell clams were used as test organisms in cages within the bath, at varying distances from the bath and attached to a free-floating drogue which remained in the treatment effluent as it dispersed in the current. In the cage within the bath all the lobsters died within an hour, but there was no mortality in any of the other species. In the other cages there was no mortality in any species up to 24 hours after the tarpaulin was dropped apart from a 2% mortality in the shrimps; even the shrimp mortality was probably unconnected with the azamethiphos because it occurred up-current from the treatment cage. It was noted that the experiment was designed to study acute toxicity and had no relevance to repeated exposures at short intervals. It should also be noted that on each occasion only a single cage was treated. The most economic way to use a sea-lice parasiticide is to treat all the cages at a site in rapid succession, and preferably to treat all the sites in a body of water with in a day or two of each other.

(f) Consumer safety

The MRL for azamethiphos approved in the EU is 100 μg/kg. Owing to the low concentration of azamethiphos used and its rapid metabolism in fish the withdrawal period approved for Salmosan in UK is 24 hours.

In the Canadian trials mentioned above invertebrates were collected by a SCUBA diver from sites below and near the treatment cages. No azamethiphos residues were found in any of them.

14.3 Hydrogen peroxide

14.3.1 MODE OF ACTION

The mode of action of hydrogen peroxide on sea-lice is not definitely established. Its effect on bacteria is believed to be due to its decomposition to hydroxyl radicals and on protozoa, and possibly on flukes, due to intracellular oxidation. Gas, presumably oxygen, emboli have been seen in affected sea-lice and they float up to the surface of the water. This flotation may be the only way in which hydrogen peroxide acts on adult sea-lice because Hodneland *et al.* (1993) showed that all affected lice in a tank trial started to recover after about 1 hour. This was in marked contrast to the effects of organo-phosphorus compounds from which there was no recovery.

Hodneland *et al.* (1993) pointed out that the recovered lice might be capable of reinfesting the fish from which they had been detached, other fish on the farm or wild fish. Other commentators have suggested that the likelihood of reinfestation would vary between sea-lice species. *L. salmonis* when removed from salmon show little swimming activity and are thought not to transfer between hosts in normal circumstances. In contrast *Caligus* spp. pre-adults and adults are active swimmers and normally do transfer between hosts.

McAndrew *et al.* (1998) studied the activity of hydrogen peroxide against several of the life-cycle stages of *L. salmonis*. They found that reinfestation by adult females occurred to a significantly greater extent in unmedicated controls than in treated fish. This seems to imply a residual activity of the drug in the fish, presumably in its skin mucus. They also found that hydrogen peroxide killed copepodids and nauplii, and significantly reduced the hatchability of the egg-strings of females from treated fish. These actions could not have been due to the formation of oxygen emboli and flotation.

14.3.2 ADMINISTRATION AND DOSAGE

Hydrogen peroxide is used as a bath, as are organo-phosphorus compounds, but there are two important differences in the method of administration. It is supplied as 35% or 50% aqueous solutions, both of which are extremely corrosive to all tissues, so it has to be mixed into the water through equipment which prevents any contact with it by the operator; and because it decomposes releasing oxygen on contact with organic material no oxygenation of the water is necessary.

Thomassen (1993) showed that at either 1.0 or 1.5 g/l maximum removal of sea-lice was reached by about 20 minutes; there was therefore no advantage in continuing exposure for more than 30 minutes at the most. He also showed that efficacy was related to concentration, but only over a narrow range of concentrations. Efficacy rises with temperature but toxicity to the target species rises even more so that the compound is not recommended for use in water above 14°C, and in Norway it is not used above 10°C. Below those temperatures the normal use concentration is 1.25-1.5 g/l.

The safe and efficacious range of concentrations is narrow and it is difficult to estimate the volume of water enclosed in the tarpaulin round a netpen; there are, therefore, problems in the calculation of the correct volume of concentrated hydrogen peroxide to be added. It is normal to add about 80% of the estimated requirement, allow 2 or 3 minutes for mixing and then conduct a rapid assay of the hydrogen peroxide present in a sample of the water. Any additional requirement can be read from tables.

14.3.3 TARGET SPECIES TOXICITY

(a) Factors affecting toxicity

Hydrogen peroxide is toxic to Atlantic salmon and there are a number of reports of levels of mortality under various conditions. The factors increasing toxicity are:

- higher temperature
- longer exposure
- higher concentration
- smaller fish

In general it may be said that at 10°C 1.25 g/l is safe but at higher temperatures some mortality in the fish may be inevitable. At 6°C there is a wide margin of safety. The temperature effect on toxicity is probably due to accelerated degradation of the hydrogen peroxide in the fish. It leads to problems in clinical use because the life-cycle of the sea-louse is shorter and hence the concentration of infective stages is greater at higher temperatures. The best way to use hydrogen peroxide having regard to target species safety is to treat all the farms in a body of water in winter.

The increased toxicity in smaller fish applies to Atlantic salmon stages in seawater. Where hydrogen peroxide is used as an anti-fungal drug in freshwater, eggs, alevins and fry are relatively tolerant of it but older fish are more sensitive. There are also species differences: Atlantic salmon are more tolerant then either chinook salmon or rainbow trout. Tilapia have been found unable to tolerate 0.24 g/l for 1 hour, but this trial was conducted in water at 30°C.

(b) Symptoms of toxicity

Diametrically opposed symptoms of toxicity have been reported by different groups of observers. Both noted an initial escape response; but one found this to be followed by increased ventilation with the fish congregating at the surface of the water, and the other said there was reduced ventilation with the fish settling at the bottom of the cage. With mild toxicity there is loss of balance and on transference to unmedicated water the fish may be unable to swim at first. This is followed by a period of short bursts of random swimming interspersed with drifting. Full recovery may take 2 hours. Severe overdosage is fatal but may take 24 hours to produce 100% mortality. If severely overdosed fish are transferred to unmedicated water there may be an initial recovery of balance but this is followed by lethargy; after 2 hours there are short bursts of random swimming, as in the mildly affected fish, but with loss of equilibrium in the drifting phases. These times of lost equilibrium become extended until eventual death.

Histologically, lesions are seen in the gills. They consist initially of epithelial hypertrophy and hyperplasia resulting in some fusion of filaments. In more severe cases there is bleeding and necrosis of the epithelia and finally "lifting" of epithelium over considerable areas.

14.3.4 CONSUMER AND ENVIRONMENTAL SAFETY

Hydrogen peroxide degrades only to oxygen and water and so is considered essentially safe both to consumer and the environment. In the EU it has been assigned to Annex II of Regulation 2377/90 - see Section 3.6.2.

In the USA hydrogen peroxide has been classified as "low regulatory priority", but this applies to its use to control fungal infections of fish and fish eggs. This is a freshwater use and no consideration appears to have been given to its anti-parasitic use in seawater. As an anti-fungal treatment it appears to be comparable to formalin for eggs and better for fish.

14.4 Ivermectin

14.4.1 MODE OF ACTION

Ivermectin increases the production of the inhibitory neuro-transmitter gamma-amino-butyric acid (GABA) at nerve endings, and enhances the binding of GABA to receptors which, in invertebrates, occur in muscles as well as synapses. Ivermectin thus causes paralysis. In vertebrates GABA is confined to the central nervous system and ivermectin does not easily cross the blood-brain barrier.

14.4.2 MERITS AND DEMERITS

(a) Merits

The merit of ivermectin is that it is active against ectoparasites when administered orally. Its possible use for sea-lice was first studied at the University College of Galway in Ireland: an alternative to dichlorvos was needed because:

1. Bath treatment is laborious;
2. Bath treatment is virtually impossible in very large netpens;
3. Bath treatment is stressful to fish, increasing susceptibility to infections including pancreas disease; and in salmon newly put into seawater it increases grilse formation;
4. Dichlorvos is a neurotoxin which is absorbed through the skin; the recommended protective clothing is uncomfortable, especially in summer, and often rejected by operators, and hence there have been cases of operator poisoning;
5. The safety margins of both dichlorvos and hydrogen peroxide to fish are narrow, especially in summer;
6. Hydrogen peroxide is very expensive.

The ivermectin dose rate used in mammals, 0.2 mg/kg, was shown to be effective when given in feed to Atlantic salmon. A commercially available 1% injectable solution was used surface-coated onto pellets with 5% gelatin as binder.

(b) Demerits

Høy *et al.* (1990) concluded from a pharmacokinetic study of ivermectin in Atlantic salmon that it was unsuitable for use in that target species, mainly on grounds of consumer safety. Others have found no problems of target species safety with repeated use in practice.

The main demerit of ivermectin is its questionable environmental safety. It is toxic to a wide range of invertebrate species, but Canadian work has shown that for shrimps, a sensitive taxon, it is toxic when ingested in food but not when it is in the water. It will of course be present in uneaten food beneath cages, and is likely to persist there because it is sparingly soluble in water.

Burridge and Haya (1993) found the no-observed effect concentration (NOEC) in feed to be 2.6 µg/g (= ppm) for shrimps. They pointed out that for fish fed at 1% per day the normal dosage of 0.05 mg/kg body weight requires a feed medication rate of 5 ppm, so any uneaten food falling to the seabed would contain twice the NOEC for shrimps.

14.4.3 DOSE REGIMEN

No complete dose titration has ever been conducted on ivermectin, and there is no market authorization for its use in fish anywhere in the world. This being so, there can be no formally recommended use regimen. This position is likely to continue because the original patentees have stated that they do not intend to make any application for aquacultural use.While significant usage has resulted from this approach with all development expenditure being borne by other bodies, it may not continue for much longer; an analogue compound, ememectin, is now being developed specifically for aquaculture.

(a) Dose rate

As noted above, 0.2 mg/kg was originally used because that is the standard dose rate for mammals. It was used in two trials in water at 13-16°C, and in the first it appeared to be effective against all stages of sea-lice. In the second only adults were eliminated but the juveniles appeared not to survive to become adults. The drug acts in a profoundly different way from organo-phosphorus compounds and hydrogen peroxide in that it reaches the parasite from within the fish. So while residues at an adequate concentration remain in the fish it will continue to be active against the parasites. The dose of 0.2 mg/kg appeared to remain active for about 20 days.

This dose rate has also been used successfully in Canada for the treatment of Atlantic salmon infested by *Ergasilus labracis*, another copepod ectoparasite.

Kilmartin *et al.* (1997) observed that many Irish salmon farmers were believed to use 0.05 mg/kg once or twice a week - presumably dosing more frequently in the summer than in the winter. Two doses of 0.05 mg/kg at an interval of 3 days has become the normal procedure where ivermectin is used. It is efficacious against chalimus larvae as well as pre-adults and adults, but takes about 2 weeks to reach maximum cumulative effect. In consequence it is used as a routine preventive measure rather than as a treatment of established infestations.

(b) Administration

Johnson and Margolis (1993) tested the efficacy of 0.05 mg/kg given every third day for three or six occasions and found both regimens significantly to reduce sea-lice burdens on Atlantic salmon. They used a commercially available 1% oral drench, diluted it in deionized water and sprayed it onto feed pellets. Assays of samples of the pellets a month after the treatments showed only about 60% of the targeted concentrations of ivermectin. As it is not known whether the losses occurred during medication of the pellets or during subsequent storage the actual dose rate used is not known, but it would appear that the use of a gelatin binder is worthwhile.

14.4.4 TARGET SPECIES TOLERANCE

(a) Single treatments in Atlantic salmon

The original efficacy trial showed that 0.2 mg/kg was satisfactory. There was a suggestion that at 0.4 mg/kg Atlantic salmon were prone to lethargy. This was not surprising in view of later work by the same team at Galway in which Kilmartin *et al.* (1997) using a wide range of doses and concentrations determined the following parameters:

Oral LD_{50} in Atlantic salmon at 11°C:	0.5	mg/kg
Immersion 96 h LC_{50} in Atlantic salmon at 11°C:	17	µg/l
Intraperitoneal injection LD_{50} in brown trout at 14°C:	0.3	mg/kg

(b) Multiple treatments in Atlantic salmon

Johnson *et al.* (1993) reported an extensive series of studies in four salmonid species. They used various daily doses which were given every second day for 50 days (25 doses) or, as frequently happened in Atlantic salmon, until all the fish stopped feeding. Their observations were recorded against the nominal dose rates although, as in Johnson and Margolis (1993) above, the actual recoveries of ivermectin from medicated feed were often little over 50% of nominal.

For Atlantic salmon the water temperature was in the region of 9°C. At doses of 0.2 mg/kg or higher feeding ceased after six or seven doses (12-14 days), with cumulative mortalities of 80% or more. At 0.1 mg/kg feeding also stopped after seven doses but the cumulative mortality was only 14%. Mortality ceased when the fish stopped feeding, apart from an additional 4% at 0.5 mg/kg and 6% at 0.1 mg/kg.

At 0.05 mg/kg feeding did not cease and cumulative mortality at the end of the trial (25 doses) was 10%. However, after six doses the feed intake dropped to about 66% of the initial level, so the medication rate would have self-limited at about 0.033 mg/kg every 2 days.

Toxicity symptoms consisted of a darkening of the skin and loss of equilibrium. This occurred after the first dose at the highest rates and after six doses at the lowest. Except at the lowest dose rate (0.05 mg/kg) the eyes of the fish rolled down so that the lenses were no longer visible; they returned to normal after death.

(c) Multiple treatments in chinook salmon

The water temperature was in the region of 10°C and only doses of 0.05 or 0.1 mg/kg on every other day were used. There was no mortality at the lower dose rate. At the higher rate the skin darkened, feed intake was reduced and there was 10% cumulative mortality. The eye symptoms seen in Atlantic salmon did not occur.

(d) Multiple treatments in coho salmon

The water temperature was in the region of 12°C; dose rates of 0.05, 0.1 or 0.2 mg/kg were used. At the two lower rates cumulative mortality was less than in the untreated controls and all the offered food was taken. The fish darkened after 16 doses. At 0.2 mg/kg cumulative mortality was 20%; feed intake was only about 68% after the 5th dose (day 11) and the fish darkened after two doses. The eye symptoms seen in Atlantic salmon did not occur.

(e) Multiple treatments in steelhead trout

The water temperature was in the region of 11°C; dose rates of 0.05 and 0.1 mg/kg were used. Anomalously, the cumulative mortality at the lower dose rate (68%) was significantly higher than at the higher dose rate (48%); however the results were confounded by an intercurrent bacterial infection. After eight doses feed intakes fell - to 72% at 0.05 mg/kg and to 57% at 0.1 mg/kg.

(f) Conclusions

The bacterial infection made the results in steelhead trout impossible to compare with those in the other species. Of the other three, coho salmon appeared to be the most tolerant of ivermectin, followed by chinook salmon with Atlantic salmon the most susceptible to toxicity. 0.05 mg/kg every 2 days is safe in coho and chinook salmon; but continuous treatment at this rate and frequency in Atlantic salmon would have an adverse effect on feed intake and consequently on growth. Feed intakes by Atlantic salmon on twice weekly dosing need to be carefully monitored.

14.4.5 PHARMACOKINETICS IN ATLANTIC SALMON

Høy *et al.* (1990) studied the pharmacokinetics of tritium-labelled ivermectin administered as a single dose by gavage to Atlantic salmon.

At 2 hours after dosing the radio-activity was still in the stomach and pyloric caeca. At 6 hours there were low levels in most organs and at 12 hours these had increased considerably, especially in the liver and kidney, although there was still a high level in the intestines. At 24 and 48 hours there were further increases in most organs, with the highest concentrations in the liver, kidney and fat, and, although the total weight was small, in the retina. There was a high level in the bile which was obviously an important route of excretion.

4 days was T_{max} for most organs; by this time there was a particularly high level in the fat and a significant level in the central nervous system. 29% of the administered radio-activity was distributed between the blood, muscle, liver and kidney. This was still the picture after 7 days. Subsequently there were gradual falls in all organs but levels remained high in the bile. At 28 days 19% of the administered radio-activity was in the blood, muscle, liver and kidney.

Although the musculature represents a high proportion of the total body weight the concentration in it was lower than in other organs; this includes the brain from day 2. The concentration in the bile ranged from 8 to 37 times that in the muscle.

77% of the administered dose was unchanged after 24 hours and 44% after 3 days. Thereafter metabolism was much slower, presumably due to the movement of the drug into such metabolically inert sites as the bile, fat and central nervous system.

It was concluded that:

1. Absorption is slow; T_{max} for most organs is 4 days.
2. Elimination is slow; the drug accumulates in fat and is released from it only slowly (as is the case in mammals).
3. There is an entero-hepatic circulation.
4. The concentration is higher in the brain than in the muscle within 2 days (unlike the position in mammals). This may account for the low tolerance by Atlantic salmon.
5. Excretion is mainly through the bile.

14.4.6 LEGAL STATUS

Ivermectin has no market authorization for use in the world. In the EU it should theoretically not be used unless it can be shown that the authorized drugs, azamethiphos and hydrogen peroxide, are ineffective or unusable. Conceivably this might be the case for fish in a cage too big for a bath. In this situation a withdrawal period of 500 degree-days would apply (see Section 3.3.1). However, various authors have pointed out the persistence of ivermectin residues in Atlantic salmon, and it has been suggested that were a market authorization to be granted at some time in the future the withdrawal period would probably lie in the range 700-1200 degree-days. This would effectively preclude its use in this species in the year in which the fish were to be harvested.

14.5 Cypermethrin

14.5.1 DEVELOPMENT OF PYRETHROIDS AS SEA LICE TREATMENTS

(a) Chemical nomenclature

Some confusion has arisen from the use, particularly in N. American fish farming journals, of the word "pyrethrin" for what is in fact the pyrethroid, cypermethrin. In scientific literature the word, pyrethrin, is confined to natural compounds extracted from plants of the *Pyrethrum* (Compositae) genus. There are four such insecticides, two stereo-isomers of each of two cis:trans isomers, and commercial extracts are mixtures of all four. The mixture is usually synergised with piperonyl butoxide. Pyrethrins are very safe for mammals and birds but they have the fundamental defect of being extremely photo-labile.

A wide variety of chemical analogues of pyrethrins have been synthesized; many have insecticidal properties and as a group they are known as pyrethroids. Photo-stable pyrethroids may be less safe for mammals than pyrethrins but many are now in commercial use as insecticides.

(b) Safety of pyrethroids

While being generally safe for homoiothermic animals, pyrethrins and pyrethroids vary in their toxicities, and in general the more potent insecticides are the more toxic to mammals. All these compounds are in varying degrees toxic to fish as well as to invertebrates. The selection of a pyrethroid for any particular purpose from the many available has to be based on considerations of its efficacy against the arthropod and its safety to the target species and the environment. Safety to the consumer is not usually a determining factor.

(c) Pyrethrins

At one time efforts were made to develop an oil-based formulation of pyrethrins for the control of sea lice. The proposed method of administration was to suspend a wide-bore cylinder half in and half out of the water, and float the oily solution on the surface of the contained water. Fish dropped through the the oily solution would acquire a surface layer of medication. The advantage propounded for the system was that very little insecticide was used and much of the unused material could be recovered from the environment. The idea has been abandoned due to high labour costs and poor efficacy.

(d) Cypermethrin

This pyrethroid is efficacious against sea lice at very low concentrations in seawater. It was originally tested for in-feed administration, but while it appeared to be efficacious it was found that the sea lice were actually being killed by drug which had leached out of the food into the water. Owing to the toxicity of cypermethrin to both fish and invertebrates further develoment has been focused on a formulation for bath administration which is safe in the environment.

14.5.2 EXCIS®

(a) Efficacy

Hart *et al.* (1997) reported on the efficacy of a formulation of cypermethrin devised specifically for use in sea-lice control. It is a 1% alcoholic solution with a bio-degradable surfactant; and the cypermethrin used has a low cis content, this being more toxic than the trans isomer. They demonstrated that at 5 µg/l (=0.005 ppm) in seawater at 10°C for 1 hour it produced profound reductions in in the counts of chalimus stages III and IV as well as pre-adults and adults. Counts of chalimus stages I and II were less affected; it was claimed that this was not due to inefficacy of cypermethrin against these stages but to the rapid replacement of killed sea-lice by new infestations.

(b) Administration

Since Excis is a 1% solution the bath concentration of 5 µg/l is achieved with 0.5 ml per m^3. With the bath tarpaulin set up in the normal way the volume of enclosed water and hence the requirement of Excis should be calculated. The Excis should be mixed into 40 l. seawater; oxygen must be diffused into the bath and the diluted Excis added to the bath at several points including the diffuser positions to ensure rapid dispersion. The recommended exposure time is 1 hour.

Strict attention should be given to operator precautions as with other sea-lice treatments administered by baths. In the case of Excis the alcoholic nature of the solution necessitates particular attention to precautions against fire in storing the drug as well as in using it.

(c) Environmental and consumer safety

The bio-degradable surfactant in Excis enhances the diffusion and distribution of the cypermethrin when the diluted solution is first added to a bath. However when the tarpaulins are dropped at the end of the exposure time the further dilution (and bio-degradation) of the surfactant means that it can no longer hold the hydrophobic cypermethrin in suspension. The cypermethrin becomes adsorbed onto particulate, and to a lesser extent dissolved, organic matter. Muir *et al.* (1994) have shown that this profoundly reduces the proportion of the cypermethrin in the water phase and in turn reduces the uptake by rainbow trout. It may reasonably be assumed to have the same effect in invertebrates.

The half-life of cypermethrin in rainbow trout (at 10°C) was estimated to be 78 hours. However the Excis formulation means that only a very low level of the drug is absorbed and a very short withdrawal period is being proposed for it.

14.6 Benzyl-ureas

14.6.1 MODE OF ACTION

While the organo-phosphorus compounds, ivermectin and pyrethroids all act on the parasite's nervous system, the benzyl-ureas, the latest chemical group to be used for sea-lice control, have an entirely different mode of action. They interfere with the synthesis of chitin, the protein from which the exoskeleton of arthropods is formed. The exoskeleton is rigid, and the only way an arthropod can grow is by shedding it (ecdysis) and forming a new larger one. In sea-lice ecdysis marks the division between one stage and the next, *e.g.* between the four chalimus stages, two pre-adult and adult stages.

Benzyl-ureas have no effect on arthropods in a stage between ecdyses, but in the case of copepods at ecdysis benzyl-ureas render them incapable of shedding old exoskeletons completely. To be effective benzyl-ureas have to be administered over a period long enough to ensure the inclusion of an ecdysis.

14.6.2 CHEMICALS IN USE

Diflubenzuron has been in commercial use for a number of years, particularly as a pesticide for use on plants. It has also been used in veterinary medicine as a sheep dip, and is used in the USA for infestations of *Argulus, Lernaea* and *Ergasilus* in koi carp. Teflubenzuron has been developed particularly for sea-lice. Both compounds are in use in aquaculture in Norway.

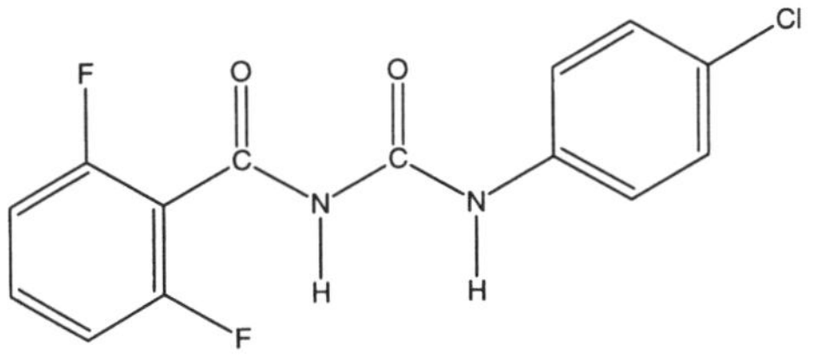

Diflubenzuron

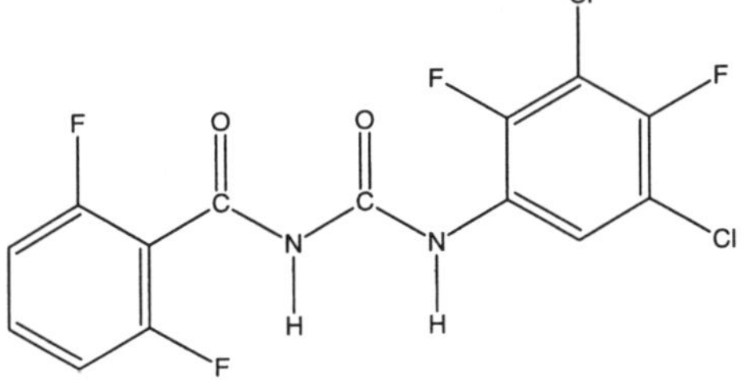

Teflubenzuron

Figure 14.3.

14.6.3 ADMINISTRATION AND DOSAGE OF TEFLUBENZURON

Teflubenzuron is supplied as a 100% powder for surface-coating onto feed pellets. It is mixed into pellets at 2 kg per tonne and oil is added as binder. The pellets are fed at 0.5% for 7 days, which gives a medication rate of 10 mg/kg/day. When, as will normally be the case during the season of severest sea lice challenge, the fish are feeding at more than 0.5% per day, the medicated feed should be given as the first feed of the day, and care must be taken to ensure that all the fish in a netpen receive some medicated pellets. After this unmedicated pellets may be given to satiety.

14.6.4 PHARMACO-KINETICS OF DIFLUBENZURON

Horsberg and Høy (1991) found that absorption of diflubenzuron by Atlantic salmon was slow and incomplete. At 12 hours after a single oral dose of 75 mg/kg radio-labelled drug they found only 3.7% in the major tissues. This was the highest level they observed although higher levels could have been reached as their next observations were not until 36 hours after dosing. The moiety which had been absorbed was mainly in the liver, kidney, brain, bile and cartilage; activity was also detectable in the cutaneous mucus.

At 2 days the level in bile and the continuing high levels in the intestines indicated an entero-hepatic circulation. At this time cutaneous mucus contained radio-activity equivalent to 4 μg/g diflubenzuron. Other tissues had lower levels than at 12 hours. Diflubenzuron is rapidly metabolized, and as early as 24 hours after dosing more than half the absorbed radio-activity is in the form of polar compounds.

Further reading

Burridge, L.E. and Haya, K. (1993) The lethality of ivermectin, a potential agent for treatment of salmonids against sea lice, to the shrimp *Crangon septemspinosa*. *Aquaculture* **117**, 9-14.

Hart, J.L., Thacker, J.R.M., Braidwood, J.C., Fraser, N.R. and Matthews, J.E. (1997) Novel cypermethrin formulation for the control of sea lice on salmon (*Salmo salar*). *Veterinary Record* **140**, 179-181.

Hodneland, K, Nylund, A., Nilsen, F. and Midttun, B. (1993) The effect of Nuvan, azamethiphos and hydrogen peroxide on salmon lice (*Lepeothpheirus salmonis*). *Bulletin of the European Association of Fish Pathologists.* **13** 203-206.

Horsberg, T.E. and Høy, T. (1991) Tissue distribution of ^{14}C-diflubenzuron in Atlantic salmon (*Salmo salar*). *Acta Veterinaria Scandinavica* **32**, 527-533.

Høy, T., Horsberg, T.E. and Nafstad, I. (1990) The disposition of ivermectin in Atlantic salmon. *Pharmacology and Toxicology* **67**, 307-312.

Johnson, S.C. and Margolis, L. (1993) Efficacy of ivermectin for control of the salmon louse *Lepeophtheirus salmonis* on Atlantic salmon. *Diseases of Aquatic Organisms* **17**, 101-105.

Johnson, S.C., Kent, M.L., Whitaker, D.J. and Margolis, L. (1993) Toxicity and pathological effects of orally administered ivermectin in Atlantic, chinook and coho salmon and steelhead trout. *Diseases of Aquatic Organisms* **17**, 107-112.

Kilmartin, J., Cazabon, D. and Smith, P. (1997) Investigations of the toxicity of ivermectin for salmonids. *Bulletin of the European Association of Fish Pathologists* **17**, 58-61.

Mattson, N.S., Egidius, E., Kryvi, H. and Solbakken, J.E. (1987) Fate of radioactivity in Atlantic salmon (*Salmo salar*) following intragastric administration of (methyl-^{14}C)-trichlorfon. *Journal of Applied Ichthyology* **3**, 61-67.

McAndrew, K.J., Sommerville, C., Wootten, R. and Bron, J.E. (1998) The effects of hydrogen peroxide treatment on different life-cycle stages of the salmon louse, *Lepeophtheirus salmonis* (Krøyer, 1837). *Journal of Fish Diseases* **21**, 221-228.

McHenery, J.G., Turrell, W.R. and Munro, A.L.S. (1991) Control of the use of the insecticide dichlorvos in Atlantic salmon farming. In *Chemotherapy in Aquaculture: from theory to reality* (eds. Alderman, D.J. and Michel, C.) O.I.E. Paris. pp. 179-186.

Messager, J.L. and Esnault, F. (1991) Traitement par le dichlorvos des copépodoses de la truite arc-en-ciel élevée en mer: modalités de traitement adaptées aux conditions environnmentales françaises. In *Chemotherapy in aquaculture: from theory to reality* (eds. Alderman, D.J. and Michel, C.) O.I.E, Paris. pp. 195-205.

Muir, D.C.G., Hobden, B.R. and Servos, M.R. (1994) Bioconcentration of pyrethroid insecticides and DDT by rainbow trout: uptake, depuration and effect,of dissolved organic carbon. *Aquatic toxicology* **29**, 223-240.

O'Halloran, J. and Hagans, W.E. (1996) First use in North America of azamethiphos to treat Atlantic salmon for sea lice infestation: procedures and efficacy. *Canadian Veterinary Journal* **37**, 610-611.

Roth, M., Richards, R.H., Dobson, D.P. and Rae, G.H. (1996) Field trials on the efficacy of the organophosphorus compound azamethiphos for the control of sea lice (Copepoda: Caligidae) infestations of farmed Atlantic salmon (*Salmo salar*). *Aquaculture* **140**, 217-239.

Samuelsen, O.B. (1987) Aeration rate, pH and temperature effects on the degradation of trichlorfon to DDVP and the half-lives of trichlorfon and DDVP in seawater. *Aquaculture* **66**, 373-380.

Southgate, P. (1993) Disease in Aquaculture. In *Aquaculture for veterinarians* (ed. Brown, L.A.) Pergamon Press, Oxford. pp. 91-129.

Thomassen, J.M. (1993) Hydrogen peroxide as a delousing agent for Atlantic salmon. In *Pathogens of Wild and Farmed Fish: Sea Lice* (eds. Boxshall, G.A. and Defaye, D.) Ellis Horwood, Chichester. pp. 290-295.

15. ANTHELMINTICS

15.1 Worm parasites of fish

15.1.1 TREMATODES

(a) Monogenean flukes

The most important worm parasites of fish are the monogenean flukes (*i.e.* flukes having the entire life cycle in a single host species). Of these the two most important genera are *Dactylogyrus* which are gillflukes and *Gyrodactylus* which are found mainly on skin and fins. Both genera are capable of infesting a wide range of fish host species. Other gill flukes which parasitize more specific groups of fish include *Pseudodactylogyrus* spp. in eels and *Diplozoon paradoxum* on chubs and bream. Gill and skin flukes are external parasites and are therefore most easily treated by immersion.

(b) Digenean flukes

While there are digenean fluke species whose life cycles include an immature (usually metacercaria) stage in fish, there are few of any economic significance in which the adults parasitize fish. One such is *Sanguinicola inermis* which lives in the blood vessels of carp. The adults are not in fact pathogenic, but the eggs may form emboli in the gill blood vessels causing necrosis, and death due to respiratory failure.

There are several species known to fish farmers as eye flukes, *Diplostomum spathaceum* being one of the commonest. Fish are invaded by free-swimming cercariae; the metacercariae encyst in fish eyes, and the adult flukes are in fish-eating birds. Eye fluke in festations cause cataract and eventual blindness. *Diplostomum* spp. infest a wide range of freshwater fish species.

The "yellow grub" is the metacercaria of *Clinostomum* spp. which have a similar life-cycle to *D. spathaceum*. The cysts vary from 1 to 5 min diameter and are found in the skin, fins, musculature and viscera of many warm-water fish, notably the channel catfish. A heavy infestation of large cysts will render the fish unsaleable.

15.1.2 CESTODES

(a) Strobila stages

There are few tapeworms found in the intestines of fish. Where they do occur they are not pathogenic unless present in sufficient numbers to cause physical blockage. Species which may have this effect include *Bothriocephalus* spp. in cyprinids and turbot, *Proteocephalus* spp. in bass and *Caryophyllaeus* spp. in carp

(b) Plerocercoids
This is the cestode infestation usually seen in fish. The fish is the second intermediate host, the first having been a planktonic invertebrate. The final host is a predator of the fish and may be either another fish species, a bird or a mammal.

Triaenophorus spp. plerocercoids are pathogenic in salmonids, producing cysts in the viscera, particularly the liver. *Diphyllobothrium* spp. can have the strobila stages in man and hence render infested fish unsaleable due to their zoonotic potential.

15.1.3 NEMATODES

There are few nematode species of any significance in fish. One genus which has been of sufficient pathogenicity for research to be conducted into its control is *Anguillicola* which infests the swim-bladder of the European eel.

15.1.4 THE LEGAL POSITION

No anthelmintics carry market authorizations in either the EU or North America for use in fish, despite the range and severity of helminth parasitisms which occur. Disease has been the stimulus for a number of studies of efficacy in relation to target species safety, but there has been virtually no consideration of the consumer or environmental safety issues.

15.2 Flukicides

15.2.1 BENZIMIDAZOLES

(a) Active members of the range
Nine commercially available benzimidazoles were screened for activity as immersion treatments against *Gyrodactylus* spp. skin flukes; the majority had no activity at concentrations below the no observable effect concentration (NOEC) for rainbow trout. Mebendazole, fenbendazole, parbendazole and triclabendazole were active, and in each case exposure time rather than the initial concentration was the determining factor for efficacy. None had significant efficacy in 3 hours exposure. Mebendazole and parbendazole at 25 mg/l for 12 hours produced over 90%, but not total, eliminaton of flukes. Fenbendazole was 100% effective at 25 mg/l and showed a high degree of activity at as little as 1.5 mg/l; its mode of action against flukes is obviously different from that against nematodes in mammals where it is believed to be converted to the active oxfendazole. Triclabendazole, which alone among benzimidazoles is used in mammals as a flukicide rather than a nematocide, was nearly 100% effective at 25 mg/l for 12 hours.

(b) Mebendazole and trichlorfon
Before the development for aquacultural use of drugs with specific anthelmintic action, gill and skin fluke infestations were controlled with general biocides such as formalin. Trichlorfon, introduced to aquaculture for sea-lice control, was also used for gill flukes but

parasite resistance developed to the extent that doses toxic to the fish became necessary for efficacy. In consequence, when dichlorvos displaced trichlorfon for sea-lice it was not seriously considered as a flukicide.

In a study on skin and gill flukes (*Gyrodactylus elegans* and *Dactylogyrus vastator* respectively) of goldfish it was found that *G.elegans* were susceptible to mebendazole but resistant to trichlorfon (see section 14.2.2). In contrast, mebendazole had no effect on gill flukes but trichlorfon was effective. A combination of 0.4 ppm mebendazole and 1.8 ppm trichlorfon for 24 hours was effective against both types of parasite. The combination was less potent than mebendazole for skin flukes but more potent than trichlorfon alone for gill flukes. The greater activity of mebendazole against *Gyrodactylus* spp. than against *Dactylogyrus* spp. has been observed in several tropical ornamental fish species as well as goldfish.

The mebendazole-trichlorfon mixture has not been followed up by a mebendazole-dichlorvos mixture because more specific flukicides have now become available.

(c) Mebendazole for eels

The activity of mebendazole against gill flukes has been applied by Buchmann and Bjerregaard (1990) to pseudodactylogyrosis in European eels. There are two species of flukes, *Pseudodactylogyrus bini* and *P. anguillae*, and they differ in that *P. bini* provokes more tissue reaction.

As with *Dactylogyrus* spp., duration of exposure appears to be important for efficacy; and a 72 hour exposure was used with the drug being flushed out over the next few days. The drug used was a 5% suspension commercially available as a drench for mammals. This was initially diluted 1:1000 (50 ppm mebendazole) and then added to the water over a period of 45 minutes at a rate of 1:50 giving a final 1 ppm mebendazole. All gill flukes were eliminated within the 72 hour exposure and the fish remained uninfested for 2 weeks. The persistence of the action was attributed to the mebendazole killing the fluke eggs, as benzimidazole carbamates are known to do to nematode eggs.

P. anguillae was more resistant than *P. bini in vitro*, but the reverse was the case *in vivo*. It was suggested that the latter effect may have been due to the protective effect of the tissue reaction round *P. bini*.

Up to 100 ppm mebendazole was tolerated by pigmented eels for 72 hours, but 500 ppm was 100% lethal by 52 hours. Glass eels were much more susceptible, 1 ppm causing 20% mortality and 100 ppm causing 40% mortality in 72 hours.

It was noted that at the concentrations used mebendazole had less adverse effect on environmental microfauna than formalin.

(d) Fenbendazole for carp

Bothriocephalus acheilognathi is a significant parasite of cyprinids for which mebendazole has been reported to be inactive. However, fenbendazole has been found to be active in common carp; it is given in feed at a rate of 40 mg/kg twice at an interval of four days.

15.2.2 PRAZIQUANTEL

(a) Mode of action

Praziquantel was originally developed as a cestodicide and it has become a standard treatment for tape worms, both strobila and cyst stages, in birds and mammals including Man. The drug is believed to impair the neuromuscular system of the parasite such as it is, and thus to inhibit the hooks and suckers on the scolex. It may also affect the permeability of the integument of the parasite leading to osmotic and nutritional imbalances. Mammal and bird hosts actively excrete the drug from the blood into the lumen of the intestine.

In fish praziquantel has proved useful against trematodes. Optimum dose regimens and use conditions have not been determined in most cases but it is known that for immersion administration it is more efficacious when dissolved in dimethyl-sulphoxide (DMSO) than in ethanol.

(b) In-feed administration for eye flukes

Prior to the development of praziquantel, several anthelmintics had been tested for activity against eye flukes, and while a few showed activity against metacercariae *in vitro* none had any activity *in vivo.* Control of the parasites depended on control of the molluscan hosts of the rediae and water treatment to kill free-living cercariae.

Bylund and Sumari (1981) found that metacercariae in the lenses of rainbow trout were affected by oral doses as low as 50 mg/kg/day but that this was reversible. 330 mg/kg/day for 7 days appeared to be the optimum therapeutic dose regimen. Feed pellets were medicated at 1.2% so this dose rate would have been achieved by feeding medicated pellets at 2.75% per day. This high feeding rate was achievable because the drug had no effect on the fishes' appetite. The destruction of the metacercariae initally exacerbated the lens opacities but in several fish a gradual restoration of transparency occurred over a period of time.

Ad lib. feeding of pellets medicated at the same rate for 15 minutes to uninfested fish gave virtually complete protection against artificial challenge for 48 hours and very substantial protection for a further 24 hours. It was suggested that while long term medication of fish would be uneconomic, the therapeutic regimen might be useful for brood fish and the prophylactic regimen where acute outbreaks of disease could be predicted from weather conditions.

(c) Immersion for eye flukes

Plumb and Rogers (1990) used immersion administration of praziquantel in a population of channel catfish many of which were infested with both eye flukes and yellow grubs. At a concentration of 2 ppm for 2 or 4 hours they found a rapid kill of a substantial number of eye flukes but the process continued at a decreasing rate for as much as 3 weeks.

(d) Skin flukes

Using *Gyrodactylus aculeatus* on sticklebacks as an experimental model, praziquantel has been shown to be active with exposure time rather than concentration being important for efficacy. The mode of action may well be similar to that on cestodes in mammals since the affected gill flukes dropped off the hosts.

In vitro tests against *Pseudodactylogyrus bini* showed 100% efficacy immediately with 600 ppm, and after 8 hours exposure at 120 ppm; and 60% efficacy after 8 hours at 1 ppm. *In vivo* both the 600 and 120 ppm regimens paralysed the eel hosts. Eels tolerated concentrations of 10-30 ppm for 15 hours and by 8 hours 80-90% of *P. bini* were immobilized. As 10 and 30 ppm appeared to be equally efficacious, 10 ppm was recommended as the optimum concentration. In view of the importance of avoiding the development of parasite resistance to anthelmintics it has been suggested that for pseudodactylogyrosis in eels mebendazole should be alternated with praziquantel.

Praziquantel was found to be ineffective for *Amphibdelloides*, a gill fluke of electric rays. In this instance the drug was given orally and by injection and it seems conceivable that its pharmacokinetics meant that there was no contact with the parasites. Positive reports of the efficacy of praziquantel against gill flukes have invariably involved immersion administration.

(e) Yellow grubs

Lorio (1989) tested the efficacy of praziquantel and ivermectin injections and praziquantel and trichlorfon by immersion on yellow grubs in channel catfish. Trichlorfon had no action. Praziquantel as a single intramuscular injection of 25 mg/kg or by immersion at 0.65 ppm, and ivermectin injected at 220 μg/kg showed some but not 100% efficacy. Praziquantel was considered superior to ivermectin. Plumb and Rogers (1990) used immersion at the higher concentration of 2 ppm and obtained greater efficacy but it was still only about 80% and they considered praziquantel to be more useful for eye flukes.

15.2.3 OTHER FLUKICIDES

Such is the economic importance of flukes, particularly monogeneans, in fish culture that a wide variety of other anthelmintics has been tested for activity against them. While several are active at dose rates toxic to the fish few show any useful activity.

Niclosamide (dissolved in DMSO) is active against *Gyrodactylus* spp. at 0.1 ppm, and appear to have a greater effect than praziquantel on immature parasites. In view of its lower use rate than mebendazole it might well be considered for alternating with mebendazole or praziquantel in culture systems where these flukes occur. Nevertheless it has negligible activity against *Pseudodactylogyrus* spp. at this concentration; 1 ppm is necessary and this causes 36% mortality in eels.

Nitroscanate has been shown to be 100% effective against *Gyrodactylus* spp. in rainbow trout at the very low concentration of 0.07 ppm for 3 hours. The drug is very insoluble in water and the observed efficacy was attributed to particles adhering to the fish. Large scale use would depend on maintaining the drug in suspension during the exposure period.

15.3 Tapeworm and roundworm treatments

15.3.1 TAPEWORM TREATMENTS

Treatment of fish for tapeworms is normally done with one of two drugs, niclosamide and praziquantel, both originally developed for this use in mammals. In both cases their efficacy against flukes depends on immersion administration, whereas for the strobila stages of tapeworms they are given orally, normally in feed. For niclosamide the normal dose is 40 mg/kg/day and for praziquantel 5 mg/kg/day; both are given daily for 3 days. Mebendazole is also active against many tapeworms but only at very high doses, such as 150 mg/kg/day.

As noted in Section 15.2.1(d) fenbendazole is active against *Bothriocephalus acheilognathi* in common carp. This tapeworm is also an important parasite of grass carp (*Ctenopharyngodon idella*) and niclosamide as well as mebendazole is reported to be inactive. Praziquantel is effective but doses of at least 35 mg/kg/day are needed. *Proteocephalus amblyoptilis* in bass is another tapeworm which is insensitive to normal dose regimens; praziquantel at 100 mg/kg/day for 4 days achieves about 90% elimination.

15.3.2 NEMATODE TREATMENTS

Taraschewski *et al.* (1988) investigated the use of five anthelmintics for the treatment of air bladder worms in eels; all were administered orally, three by immersion, and ivermectin was also given by intramuscular injection. The only treatment considered suitable for further investigation was levamisole by immersion where 1 ppm for 24 hours was curative and the lethal dose was 250 ppm for 24 hours.

Further reading

Buchmann, K. (1987) The effect of praziquantel on the monogenean gill parasite *Pseudodactylogyrus bini. Acta Veterinaria Scandinavica* **28**, 447-450.

Buchmann, K. and Bjerregaard, J. (1990) Mebendazole treatment of pseudodactylogyrosis in an intensive eel-culture system. *Aquaculture* **86**, 139-153.

Bylund, G. and Sumari, O. (1981) Laboratory tests with Droncit against diplostomiasis in rainbow trout, *Salmo gairdneri* Richardson. *Journal of Fish Diseases* **4**, 259-264.

Lorio, W. (1989) Experimental control of metacercariae of the yellow grub *Clinostomum marginatum* in channel catfish. *Journal of Aquatic Animal Health* **1**, 269-271.

Plumb, J.A. and Rogers, W.A. (1990) Effect of Droncit (praziquantel) on yellow grubs *Clinostomum marginatum* and eye flukes *Diplostomum spathaceum* in channel catfish. *Journal of Aquatic Animal Health* **2**, 204-206.

Taraschewski, H., Renner, C. and Mehlhorn, H. (1988) Treatment of parasites: 3. Effects of levamisole HCl, metrifonate, fenbendazole mebemdazole and ivermectin on *Anguillicola crassus* (nematodes) pathogenic in the air bladder of eels. *Parasitology research* **74**, 281-289.

PART FOUR

PHARMACODYNAMIC AGENTS

16. ANAESTHETICS

16.1 General considerations

16.1.1 STAGES OF ANAESTHESIA

The following stages of anaesthesia in fish are generally recognized:

1. Light sedation	- Slight loss of reactivity to external stimuli, equilibrium normal:
2. Deep sedation	- Total loss of reactivity to external stimuli except strong pressure; slight increase in opercular ventilation rate; equilibrium normal;
3. Partial loss of equilibrium	- Partial loss of muscle tone, erratic swimming; reaction only to strong tactile and vibrational stimuli;
4. Total loss of equilibrium	- Total loss of muscle tone and equilibrium; rapid opercular ventilation (slow with some agents); reaction only to deep pressure stimuli;
5. Loss of reflex reactivity	- Total loss of reactivity; opercular movements very shallow, heart rate very slow;
6. Medullary collapse	- Opercular movements cease immediately after gasping, followed by cardiac arrest.

Stage 4 is adequate for most purposes in fish culture. As in mammalian anaesthesia the aim is to induce this as rapidly as possible to limit stress, but in practice it is not possible to induce anaesthesia in fish as rapidly as is achieved with injectable barbiturates in mammals. When a fish is placed in an anaesthetic solution there is an initial phase of excitement followed by erratic swimming; then the fish becomes inactive and sinks to the bottom of the tank, resting on its side or back.

Sedation (stage 1-2 anaesthesia) is being increasingly used to reduce stress during transport and this may last for several hours. Drugs are used for this, but where suitable vehicles are available chilling is a safer method.

16.1.2 PHARMACOLOGY

(a) Effects of anaesthetics on the central nervous system

Anaesthetic agents depress different parts of the central nervous system sequentially. As the concentration of the drug builds up in the body the more sensitive parts, which are the phylogenetically newer parts, are depressed before the others. Depression affects successively:

1. The cerebral cortex - leading to analgesia and some sedation;
2. The basal ganglia and cerebellum - leading to delirium and excitement;
3. The spinal cord - producing surgical anaesthesia;
4. The medulla - causing vaso-motor paralysis and death.

(b) Effects of anaesthesia on blood chemistry

With the exception of carbon dioxide, the anaesthetic agents now used produce very similar changes in the blood chemistry of fish. The surgical stage (5) of anaesthesia produces a rapid fall in oxygen tension and a corresponding rise in carbon dioxide. During recovery the oxygen tension rises to above the base level. The blood pH falls during anaesthesia and this too returns during recovery. The adrenalin level rises in prolonged anaesthesia and returns on recovery. All changes are referrable to hypoxia resulting from reduced opercular movements and hence a reduced water flow across the gills.

Where a fish is to be given a surgical operation arrangements must be made to maintain its respiration if it is to be out of water for more than 4 minutes. The gills can be sprayed or flooded with water or anaesthetic solution; or the head and gills can be immersed in a plastic bag of the solution.

(c) Effects of anaesthesia on stress responses

Blood cortisol falls slowly through both anaesthesia and recovery. In some reports rises have been noted but they are probably attributable to stresses of chasing, netting and handling and not to the anaesthetic drugs.

Thomas and Robertson (1991) investigated the stress responses of red drum fry and the effects of anaesthesia on them. They used transference with restraint and exposure to air as a stressor, and rises in plasma cortisol and glucose as measures of stress response. They concluded that short-term anaesthesia blocks stress responses but that long-term anaesthesia is itself a stressor.

16.1.2 DRUGS USED IN FISH

(a) Criteria of suitability

In the early 1970s it was suggested that for an anaesthetic agent to be suitable for use in fish it should:

1. Induce stage 4 anaesthesia within 3 minutes;
2. Be safe to the fish when used for 30 minutes;
3. Allow full recovery within 20 minutes.

These are still formally regarded as the criteria of suitability but they are requirements for surgery rather than commercial aquaculture. Gilderhus and Marking (1987) pointed out that the major uses to which anaesthetics are put in aquacultural practice include in particular injecting, sorting broodstock and measuring and tagging fish. For these purposes the fish must be 'handleable' - it must be possible to hold it on a table for a few moments without undue flopping or jumping. While handleability is not a classic stage of anaesthesia (and was found to be inadequate for fin clipping), it is subjectively recognizable, and the authors suggested that this should be achievable in 3 minutes.

For any given concentration of an agent the induction time will vary between individual fish. This is because of variations in the rate of water flow over the gills - which in turn depends on the oxygen requirement. Thus the concentration, or even the agent used, may have to vary according to the fish species and size.

(b) Obsolete agents

Gilderhus and Marking (1987) surveyed a range of anaesthetic agents using their own criteria of suitability. Apart from handleability in 3 minutes they lowered the target species safety criterion to 15 minutes on the grounds that surgery is rarely conducted on fish other than for experimental purposes, and fish will not normally need to be anaesthetized for 30 minutes. At the same time, with the development of anaesthetic agents more specifically suited to fish the time for full recovery can now be expected to be about 10 minutes.

Of the 16 drugs they surveyed, 12 were found to be unsatisfactory. Nicotine and salt were ineffective; halothane, metofane and sodium thiamylal were lethal at or below the effective concentrations, and methyl-pentynol and chlorbutanol were toxic to small fish. Etomidate, metomidate, propanidid and piscaine had extended recovery periods which did not meet the set criterion; but the first two have nevertheless been considered for use as sedatives in North America, and piscaine is used as a sedative in Japan. Carbon dioxide was also classified as having too long a recovery period but it has the advantage of being legally classified as "generally regarded as safe" (GRAS) in the USA.

Other drugs not included in this survey but which have been used in fish include chloral hydrate, diethyl-ether, propoxate and urethane.

(c) Suitable drugs

The four drugs which met the criteria of Gilderhus and Marking (1987) were:

- MS-222 (Tricaine methane-sulfonate)
- Benzocaine (ethyl *para*-amino-benzoate)
- Phenoxyethanol (phenoxetol, 2-phenoxy-ethanol)
- Quinaldine sulfate

The first two are chemically related to the local analgesic agents used in mammals, and this group is in fact generally more suitable for fish than either the general anaesthetics such as halothane and ether or the hypnotics such as barbiturates. General anaesthetics have a narrow safety margin in fish, whereas the local analgesic analogues are safer and are rapidly absorbed from immersion solutions and distributed within fish.

A new anaesthetic agent for fish, AQUI-S®, has been developed in New Zealand in the mid-1990s. This is a mixture of clove oil and polysorbate-80, and is already in use, like carbon dioxide in other parts of the world, to anaesthetize fish prior to bleeding out when harvesting for the table.

16.2 MS-222

16.2.1 LEGAL STATUS

Tricaine methane-sulfonate was originally produced in a search for local analgesic alternatives to cocaine, and has been used as such in Man. However its value as an anaesthetic for aquatic animals was soon recognized and its further development has been exclusively for this purpose. In many countries it is the only anaesthetic with a market authorization for fish; and in the EU member states where it is authorized the use of any other anaesthetic for fish is, strictly speaking, illegal.

The name MS-222 was originally a trade mark of the discoverers, Sandoz of Basle, Switzerland. The registration has now been abandoned and the name is in general use.

16.2.2 DOSAGE AND ADMINISTRATION

(a) Chemistry

MS-222 is supplied as a white powder which is 100% pure drug. It is intended for dissolution in water and one of its advantages is its very high solubility. It also has a very high lipid solubility. The compound is subject to degradation in light forming methyl sulfate; so containers should be re-sealed after use to exclude moisture, and the compound should be stored in the dark. Solutions should be kept out of direct sunlight before use and they should be used only on the day they are prepared.

Tricaine, the pharmacologically active moiety of MS-222 is a weak base, but solutions of MS-222 are acidic due to the formation of methane-sulfonic acid, which has a pK_a of 3.5. Where species sensitive to low pH are to be anaesthetized the preliminary solution can be buffered with sodium bicarbonate before addition to the tank holding the fish. An example is rainbow trout which are visibly irritated in an MS-222 solution at pH 4.

(b) Use concentration

MS-222 may be used in both freshwater and seawater.

The stage of anaesthesia reached is determined mainly by the concentration used, but also by the duration of exposure because absorption continues throughout the period of immersion. Table 16.1 shows concentrations recommended for various types of fish together with maximum exposure times. In each case a range of concentrations is given because the stage of anaesthesia actually reached at any given moment depends on other factors as well as concentration. Potency and hence toxicity rise with temperature, not least because of the reduced oxygen capacity of the water. Salinity and hardness of the water reduce toxicity and increase the concentration required. Small fish are more sensitive than large ones. The density of the biomass of fish also affects anaesthesia because as fish absorb the drug they reduce the true concentration in the water. A limit of 80 g/l is recommended for biomass of fish in a solution of MS-222.

Table 16.1. After Bové, F.J. (1965) with permission

	MS222 CONCENTRATION mg/litre OF WATER	IMMERSION TIME minutes
Trout Species (7-17degrees C)		
Sedation	10-30	up to 480
Anaesthesia: light	30-80	up to 30
Anaesthesia: deeper	80-180	up to 10
Salmon Species		
Sedation	7-30	up to 240
Anaesthesia: light	30-80	up to 10
Anaesthesia: deeper	80-100	up to 5
Bass Species		
Sedation	8-30	up to 480
Anaesthesia: light	30-70	up to 20
Anaesthesia: deeper	70-100	up to 4
Carp Species		
Sedation	20-30	up to 1440
Anaesthesia	30-200	up to 8
Fresh Water Tropical fish		
Sedation	30-50	up to 1440

16.2.3 PHARMACOKINETICS

(a) Absorption and excretion

Due to its high lipid solubility tricaine passes across gills very rapidly in either direction. In rainbow trout the blood concentration can reach approximately 75% of the bath concentration in 2.5 minutes. The gills are the main route of excretion and Hunn and Allen (1974) point out that since at the pH of fish blood only 0.01% of the compound is ionized it is the lipid-soluble free base which will be excreted. Houston and Woods (1972) estimated the half-life in the blood of brook trout to be 20 minutes, which is consistent with the observation in rainbow trout that blood levels returned to background levels in about 8 hours.

The withdrawal period in the UK is 10 days, irrespective of species or temperature. In the USA it is 21 days.

(b) Effects on the cardio-vascular system

In many fish species MS-222 causes tachycardia within 30 seconds of immersion; this is transient and is followed by a prolonged bradycardia. The tachycardia is attributed to the rapid absorption of the drug, its early penetration into the central nervous system and its having an immediate effect on the autonomic nervous system producing a form of 'vagus escape'. The later bradycardia is probably due to a direct effect on the myocardium. Whole hearts perfused with MS-222 solution accumulate up to three times the concentration in the perfusate, and perfused ventricular strips show a progressive decline in their contractile force.

One of the effects of MS-222 is to produce significant vaso-constriction in the gills. So although the initial absorption of the drug is rapid, the rate soon falls off; and the rapid achievement of stage 5 anaesthesia requires the use of a higher concentration of drug in the water than is in the blood at that stage. So after stage 5 has been reached absorption continues, albeit slowly. This is a recognized disadvantage of MS-222, imposing strict limits on the duration of anaesthesia in order to avoid lethal overdosage.

Studies on blood parameters of rainbow trout anaesthetized with MS-222 show increases in haematocrit and haemoglobin concentration and a decrease in mean corpuscular haemoglobin content (MCHC). Taken together these findings indicate swelling of the red cells, a condition which in itself will reduce the blood flow through the gill lamellae.

Although blood glucose levels are maintained during MS-222 anaesthesia, lactate tends to rise, indicating hypoxia. MS-222 solutions should be aerated during induction and maintenance of anaesthesia.

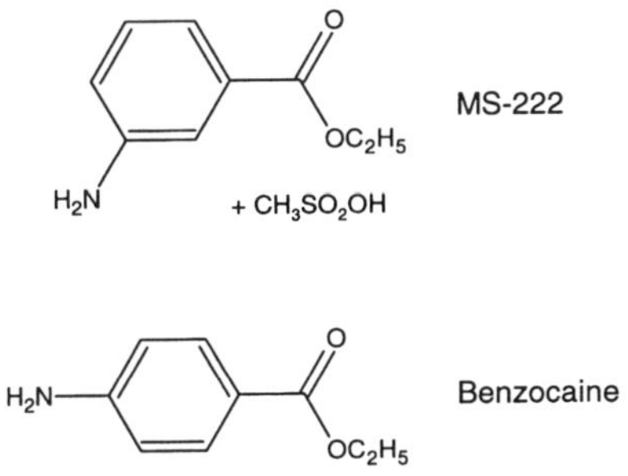

MS-222: H_2N, OC_2H_5, + CH_3SO_2OH

Benzocaine: H_2N, OC_2H_5

Figure 16.1.

16.3 Benzocaine

16.3.1 CHEMISTRY

Benzocaine is an isomer of tricaine which has probably been used as an anaesthetic for fish since before the introduction of tricaine. However because there has never been any patent protection on it, it has never been studied in the same depth and is not the subject of a market authorization anywhere in the world. For the same reason it is much less expensive than MS-222.

Benzocaine has been studied and used as the free base, as the hydrochloride and as neutralized hydrochloride, *i.e.* hydrochloride solution adjusted to pH 7 with sodium bicarbonate. Benzocaine itself forms colourless crystals or a white crystalline powder and is very sparingly soluble in water. It can be used as an anaesthetic only after initial solution in ethanol or acetone, both of which compounds are irritants to fish. Thus in practice the hydrochloride salt is usually used as it is more soluble in water. However this is also irritant due to its acidity; hence the use of neutralized hydrochloride.

Benzocaine, like tricaine, is highly lipid-soluble. The cation in the hydrochloride is considerably less so, and the neutralized hydrochloride has an intermediate lipid-solubility. These properties affect the rate of absorption across biological membranes: the more lipid-soluble molecules are absorbed faster.

Benzocaine, the hydrochloride and solutions of them are photo-labile and should be stored in the dark.

16.3.2 USE CONCENTRATION

As there is no market authorization for benzocaine hydrochloride there are no formal recommendations as to use concentration.

Benzocaine hydrochloride is a rather more active anaesthetic than MS-222; thus lower concentrations are required or, at the same concentrations, benzocaine acts more quickly. For example, in tests on tilapia and carp at 19°C it was found that at 50 mg/l benzocaine hydrochloride induced anaesthesia (stage 5) in 6.5 minutes in tilapia and 3.9 minutes in carp. These are both undesirably long induction periods, but in comparison MS-222 did not achieve anaesthesia even after 15 minutes; it was only a tranquillizer. At 80 mg/l the induction time with MS-222 was twice as long as with benzocaine hydrochloride in tilapia and three times as long in carp. At 100 mg/l the induction time was about 1.5 times as long with MS-222 as with benzocaine in both species.

As with MS-222, the duration of exposure to benzocaine hydrochloride has an effect on the depth of anaesthesia reached.

Concentration requirements vary widely between species. In the above study carp were induced more quickly than tilapia at all concentrations, and the 3 minute criterion would probably be met by 65 mg/l for carp and 80 mg/l for tilapia. Other recommendations which have been made include 25-45 mg/l for salmonids and 100-200 mg/l for northern pike. Gilderhus (1990) working with spawning-phase chinook and Atlantic salmon found that small changes in concentration had little effect on induction times but considerable effect on recovery times. He also noted a profound difference in handleability times between the two species.

The activity of benzocaine is unaffected by water hardness or pH. However higher doses are required in warmer water; this has been attributed to metabolism (acetylation) of the drug by the fish. Target species tolerance is also lower at higher temperatures so the margin of safety is narrower.

16.3.3 PHARMACOLOGY

Benzocaine has a similar effect on the cardio-vascular system to MS-222. Solutions should be aerated similarly while in use.

There is evidence that the species variation in response to benzocaine solutions depends to a significant extent on differential rates of absorption. Absorption through perfused gills is faster in rainbow trout than in carp and faster in carp than in tilapia. This is the same order as the lipid content of the gills, and the same order as the rate of induction.

Absorption also occurs across skin. The extent is probably not sufficient to affect induction times, but it may become relevant when the absorption rate through the gills has become reduced due to vaso-constriction and erythrocyte swelling. In a comparative study carp showed the fastest absorption through the skin, followed by tilapia and then rainbow trout. This was the reverse order of the mass per unit area of skin.

Radio-labelled benzocaine injected into the dorsal aorta of rainbow trout is detectable in the water 1 minute later; by 3 hours 59% has been excreted. There is also slow excretion in the urine. The compound excreted through the gills is mainly benzocaine at 3 and 15 minutes after injection, but mainly the acetyl conjugate by 1 hour. Benzocaine is also hydrolysed in fish to para-amino-benzoic acid (PABA) and acetyl-PABA; these being polar compounds are not excreted through the gills to any extent, but they are excreted in the urine.

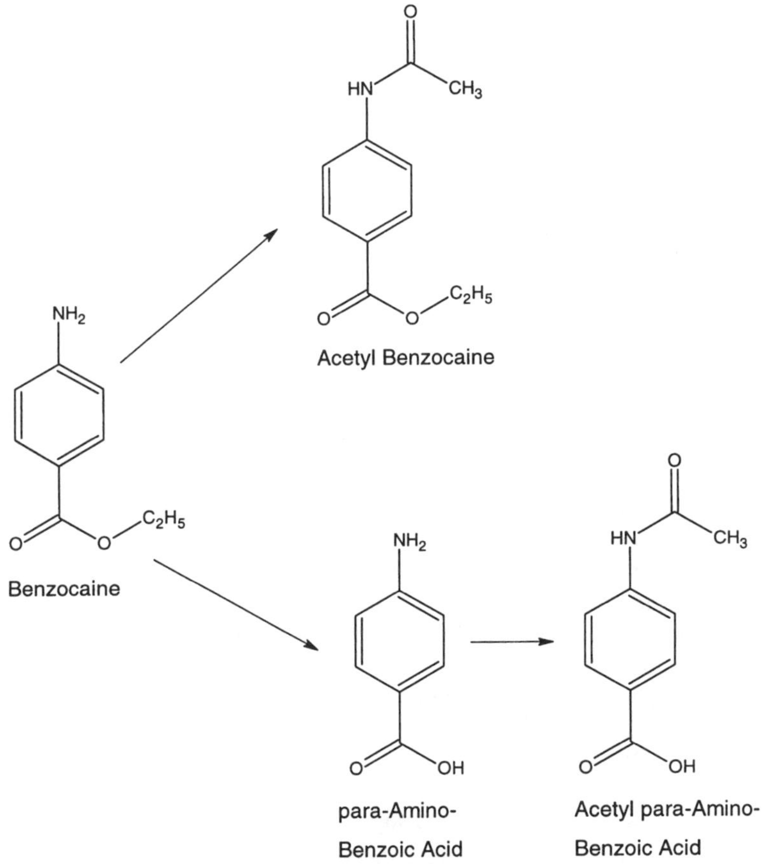

Figure 16.2. Benzocaine metabolism.

16.4 Other anaesthetic agents

16.4.1 QUINALDINE SULFATE

(a) Uses

Quinaldine sulfate is a rather expensive anaesthetic, but it meets the criteria of rapid induction and recovery. It has the distinctive advantage over MS-222 and benzocaine that it can be used for long periods. The disadvantage of it is that fish often retain some reflex responsiveness; thus while it is useful for transporting fish and for short interferences such as injections, it would be unsuitable for prolonged surgical intervention.

Use concentrations vary between species; recommendations include:

- salmonids 25 mg/l
- striped bass 40 mg/l
- channel catfish 25 mg/l

Largemouth bass (*Micropterus salmoides*) are particularly sensitive to quinaldine (96 h LC_{50} = 6.8 mg/l) and the drug should not be used for this species.

(b) Chemistry

Quinaldine is an oily liquid which is practically insoluble in water and therefore unsuitable for use in fish anaesthesia. It is a tertiary amine and the sulfate salt, a yellow crystalline powder, is very soluble in water. The solution is acid ($pK_a = 5.4$) and should be buffered with sodium bicarbonate before administration to freshwater fish. (Seawater is more than adequately buffered already). This is not only to reduce the irritancy of the acid but also because the unbuffered salt is ionized and therefore poorly absorbed by fish. At pH 5, 72.5% of the drug is in the form of the quinaldinium ion; at pH 7, 97.4% is free base.

(c) Pharmacokinetics

Absorption is rapid at pH 7 or above; this is mainly through the gills but in some species a small but significant proportion may pass through the skin. The drug concentrates in tissues with a high lipid content such as the central nervous system. It is also found in the bile very soon after first administration. Concentration in the plasma remains low, and on prolonged administration the bile:plasma ratio rises but the muscle:plasma ratio does not.

Quinaldine is excreted unchanged. This is also mainly through the gills, but in channel catfish 5% was found to be excreted in the urine. There is also probably a small amount in the faeces as would be expected with a drug concentrated in the bile. Elimination is rapid: in striped bass a half-life in muscle of 1 hour was determined and in rainbow trout the drug is undetectable in muscle within 24 hours.

16.4.2 PHENOXYETHANOL

Although phenoxyethanol (phenoxetol, 2-phenoxy-ethanol) has been recognized as an anaesthetic agent for fish for a long time there has been little definitive work done on its pharmacology. It is a colourless oily liquid with a low solubility in water. Like quinaldine

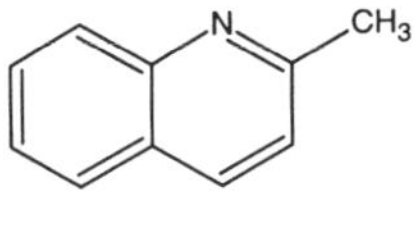

Figure 16.3.

sulfate it has an advantage over MS-222 or benzocaine in that it does not accumulate in the fish after anaesthesia has been induced; so it can be used for long periods. At the normal use concentrations of 0.1-0.5 ml/l it is less expensive than MS-222, benzocaine or quinaldine sulfate. It is safe and efficacious in most species but in a few, *e.g.* cod, induction is too slow.

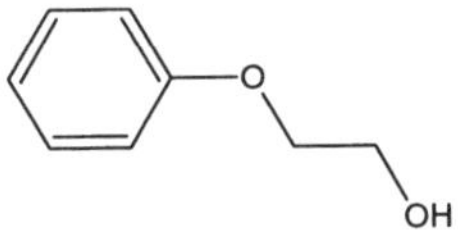

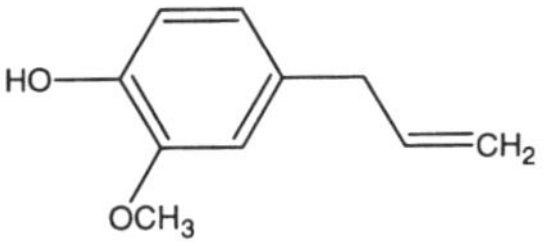

Rainbow trout exposed to 0.25 ml/l for 30 minutes had heart rates reduced to a mean of 54% of normal. Rates began to return to normal as soon as the fish were returned to unmedicated water.

Recommended concentrations include:

- Pacific salmon 0.1-0.5 ml/l
- spotted grunter 0.2-0.4 ml/l (Deacon *et al.*, 1997)
- carp 0.6 ml/l for surgery
  0.2 ml/l for handling, *e.g.* in spawning operations

16.4.3 CARBON DIOXIDE

(a) Uses

The value of carbon dioxide as a fish anaesthetic lies in its safety - to the operator, the environment and above all to the consumer. As a compound occurring naturally in all animals and which can have no effect on consumers other than possibly a very transient increase in respiration rate, it is acceptable with no withdrawal period. It is used in harvesting operations at many fish farms to render fish insensible (the equivalent of 'stunning' mammals) prior to bleeding them out. In North America it has other uses resulting from the paucity of fish anaesthetics with market authorizations: in Canada there are no authorized chemical anaesthetics and in the USA only MS-222 is authorized, with a withdrawal period of 21 days.

The disadvantages of carbon dioxide lie in the equipment needed for its administration. For large scale use large and therefore heavy cylinders are needed, together with pressure regulators and porous tubing or other means of dispersing the gas in the water.

(b) Dosage and administration

Prince *et al.* (1995) estimated that in sockeye salmon 150-200 mg/l were needed to produce a loss of equilibrium. Induction of a surgical plane of anaesthesia was not possible with carbon dioxide alone; nor was it achievable with sodium bicarbonate in the water as the pH was too high. 40 g sodium bicarbonate and 15 ml glacial acetic acid in 30 l water induced a surgical plane of anaesthesia in 7 minutes. 30 l is an appropriate volume of solution in which to anaesthetize one adult sockeye salmon; and the same solution can be used for three fish in succession. The carbon dioxide concentration was in the range 195-325 mg/l (pH 6.9-7.6) depending on the pH of the water before medication. After induction, dilution of the solution with two volumes of water (*e.g.* from 30 to 90 l) provides a satisfactory maintenance solution.

Fish hyperactivity, at least among sockeye salmon, was common and could be minimized by avoiding handling stress. Movement in water is less stressful than movement in air for many species, and if a such a fish is lifted in water the chemicals can be dissolved and then added to that water without handling the fish at all.

16.4.4 AQUI-S®

AQUI-S is an equal-part mixture of eugenol, a colourless or pale yellow liquid which is virtually insoluble in water, and the solubilizing agent, polysorbate 80. It is already in use in New Zealand, and because it has no withdrawal period it is being used particularly for anaesthetizing farmed fish prior to bleeding out at harvest. Use of AQUI-S is claimed virtually to eliminate bruising during harvest leading to both improved carcase quality and better bleeding out. The drug is also useful for general husbandry operations, with the fish not only recovering but returning to feeding very rapidly.

Both constituents are legally classified as 'generally regarded as safe' (GRAS) in the USA; and the International Association of Fish and Wildlife Agencies (IAFWA) is considering submissions to the Food and Drug Administration for an NADA approval (see Section 3.7.3(c)) for it as a fish anaesthetic.

At 6 ppm it will produce sedation and at higher concentrations up to 17 ppm it will slowly induce anaesthesia. Between 17 and 25 ppm induction is increasingly rapid. 17 ppm is normal for surgical anaesthesia, and concentrations above 30 ppm are not recommended.

16.5 Hypnotics and sedatives

16.5.1 METOMIDATE

(a) Pharmacodynamics

Metomidate and etomidate are chemical analogues of propoxate, an obsolete fish anaesthetic. They are strictly hypnotics rather than anaesthetics, having virtually no analgesic action (c/f short-acting barbiturates in mammalian anaesthesia).

Thomas and Robertson (1991) showed that metomidate blocked the cortisol response to ACTH; it is known to inhibit cortisol synthesis in several mammal species. In dose/

response studies by other workers using Atlantic salmon, 2 ppm was found to allow a slight cortisol rise but 4 or 8 ppm prevented any rise. Prolonged use of metomidate may therefore be stressful to fish even though the endocrine response normally used as the measure of stress is negative. Metomidate is useful for sedation but not for surgery.

(b) Use concentrations

In Atlantic salmon in seawater 40 mg/l metomidate was found to give almost immediate anaesthesia and the fish could be maintained in 10 mg/l for 20 minutes. Parr kept at 5°C reached the handleable stage:

- in 12 minutes in 1 ppm
- in 5 3/4 minutes in 2 ppm
- in 3 3/4 minutes in 3 ppm
- in 2 1/2 minutes in 5 ppm

Adults responded much more quickly than parr.

- 7 mg/l is an immobilizing dose in red drum.
- 5 mg/l produces rapid hypnosis in cod.

In most species recovery is slower than from MS-222 or benzocaine.

16.5.2 ETOMIDATE

(a) Pharmacodynamics

Etomidate is very similar to metomidate, but whereas the latter is alleged to cause haemolysis in some fish species etomidate does not.

In rainbow trout exposed to 0.1 ppm etomidate for 30 minutes the mean heart rate fell during exposure and continued falling for 10 minutes after fresh water flow had been restored to flush away the drug. The minimum heart rate was only 45% of the pre-treatment rate. Some irregular and missed beats were observed during exposure but ceased immediately on restoration of the water flow. The mean heart rate rose from 10 minutes after restoration of the water flow, in due course exceeding the pre-treatment rate and staying high for over 90 minutes.

Merits which have been claimed for etomidate are:

1. A low effective concentration;
2. A long safe exposure period;
3. Absence of colour or odour;
4. Operator safety.

(b) Use concentrations

Etomidate is available as 1% free base in 55% propylene glycol. Various species of aquarium fish have been immobilized with 1-4 mg/l. In channel catfish 3 mg/l can be used safely for at least 40 minutes; however 0.8 mg/l is satisfactory for long term sedation. Striped bass are very sensitive to etomidate: while up to 1.6 mg/l have been used for short procedures the 96 h LC_{50} is 0.68 mg/l. 0.1 mg/l is satisfactory for long term use, for example for transportation.

16.5.3 PISCAINE

Piscaine is a sedative which can safely be used on fish for long periods. A concentration of 6 g/l is recommended for rainbow trout. The drug is conjugated to the glucuronide in rainbow trout and carp, and both parent drug and metabolite are excreted through the gills. On transfer of the fish to fresh water the elimination half-life is 40 minutes for the first 4 hours, rising to 22 hours from 9 hours after transfer. Detectable residues are still present at 48 hours, so the drug is not suitable for transport to slaughter.

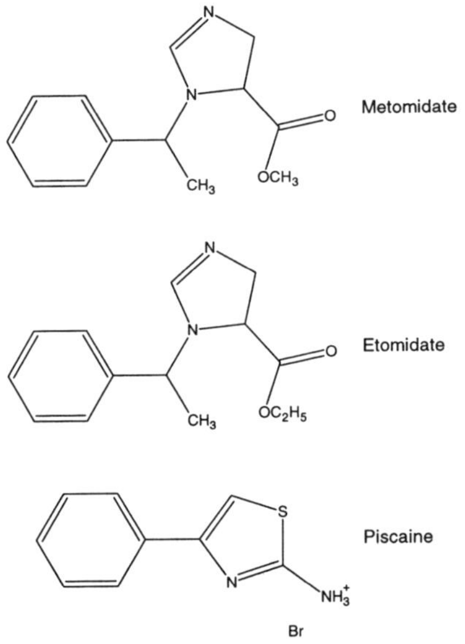

N
N
O
CH_3
OCH_3
Metomidate

N
N
O
CH_3
OC_2H_5
Etomidate

S
N
NH_3^+
Br
Piscaine

Figure 16.4.

Further reading

Bové, F.J. (1965) *MS-222 Sandoz: the Anaesthetic and Tranquillizer for Fish, Frogs and Other Cold-blooded Organisms.* Thomson & Joseph Ltd., Norwich, UK 24 pages.

Deacon, N., White, H. and Hecht, T. (1997) Isolation of the effective concentration of 2-phenoxyethanol for anaesthesia in the spotted grunter, *Pomadasys commersonnii*, and its effect on growth. *Aquarium Sciences and Conservation* **1**, 19-27.

Ferreira, J.T., Schoonbee, H.J. and Smit, G.L. (1984) The uptake of the anaesthetic benzocaine hydrochloride by the gills and skin of three freshwater fish species. *Journal of Fish Biology* **25**, 35-41.

Gilderhus, P.A. (1990) Benzocaine as a fish anaesthetic: efficacy and safety for spawning phase salmon. *Progressive Fish-Culturist* **52**, 189-191.

Gilderhus, P.A. and Marking, L.L. (1987) Comparative efficacy of 16 anaesthetic chemicals in rainbow trout. *North American Journal of Fisheries Management* **7**, 288-293.

Houston, A.H. and Woods, R.J. (1972) Blood concentrations of tricaine methane sulphonate in brook trout, *Salvelinus fontinalis*, during anaesthetization, branchial irrigation and recovery. *Journal of the Fisheries Research Board of Canada* **29**, 1344-1346.

Hunn, J.B. and Allen, J.L. (1974) Movement of drugs across the gills of fishes. *Annual Reviews of Pharmacology* **14**, 47-55.

Prince, A.M.J., Low, S.E. and Lissimore, T.J. (1995) Sodium bicarbonate and acetic acid: an effective anaesthetic for field use. *North American Journal of Fisheries Management* **15**, 170-172.

Thomas, P. and Robertson, L. (1991) Plasma cortisol and glucose stress reponses of red drum (*Sciaenops ocellatus*) to handling and shallow water stressors and anaesthesia with MS-222, quinaldine sulfate and metomidate. *Aquaculture* **96**, 69-86.

17. BREEDING INDUCTION AGENTS

17.1 Reproductive physiology

17.1.1 THE PROBLEM

Few if any species of finfish reproduce naturally in conditions of intensive culture. This may be presumed to be a result of intensivism because goldfish will reproduce in garden ponds, and many species of tropical ornamentals will reproduce in tanks, in both cases the livemass of fish per unit volume of water being less than economic for food production. Carp would have spawned naturally in the ponds of mediaeval monasteries but their keepers would not have faced the economic constraints of modern fish farmers and it may be assumed that the monks used low stocking densities.

In female fish the reproductive dysfunction may simply be in the voiding of fully developed eggs ("spawning"), or it may be in the later stages of oocyte maturation and ovulation. In males the usual problem is reduced sperm production. In Japanese eels, which are catadromous, the gonads are immature at the beginning of the sexual migration; and if migration is blocked maturation does not occur.

17.1.2 NON-MEDICINAL SOLUTIONS

Without resort to drugs there are three methods which fish culturists use to obtain young fish to grow on for harvesting for food. The method used depends on the species of fish but all methods have distinct demerits.

(a) Catching juveniles from the wild

Eels, which spawn in a marine environment and migrate to freshwater to grow are caught as elvers and placed in freshwater tanks. This is the normal method of stocking eel farms; but it is unsatisfactory in that the supply of juveniles is seasonal and unreliable, and the stress of transporting them to the freshwater farms often causes high mortality.

(b) Stripping

Some species, notably salmonids, will develop fertile gametes but will not void them ("spawn") in intestive culture. This may be associated not only with the stocking density but also with the fishes' difficulty in fulfilling their reproductive instincts of making a redd in which to deposit the eggs. The normal method of breeding such fish is to catch them individually and manually express eggs from females and then milt from males and mix them manually in a plastic bucket. The fish are more easily handled under anaesthesia, or at least deep sedation, but all traces of the drug must be rinsed off the fish before it is stripped because virtually all anaesthetics and sedatives reduce sperm motility.

The procedure is labour-intensive, and results in a supply of fertilized eggs to hatcheries which is seasonal.

(c) Catching wild broodstock

For very many farmed species the only source of new stock is wild fish. For successful breeding they must be caught when they are ready to spawn ("ripe"). This again means that the supply of fertilized eggs is seasonal and unreliable; but the technique is also wasteful because fish caught at not quite the right time may well be so stressed that they will never spawn. Moreover there are a number of species which take several years to mature but will then spawn more than once per year; even if ripe adults of such species are caught and stripped at the right time they will not spawn again.

17.1.3 STAGES OF GAMETOGENESIS

(a) In the female

Four stages are recognized:

- Oogenesis - oogonia are formed by meiosis from the germinal epithelium of the ovary; they acquire a covering layer of follicular cells;
- Vitellogenesis - the follicular cells produce the yolk from vitellogenin synthesized in the liver;
- Ovulation - the eggs are shed from the ovary into the peritoneal cavity;
- Spawning - the eggs are voided.

Ovulation in any species cannot be stopped once it has started; and eggs once ovulated must be spawned or stripped immediately as otherwise they become over-ripe and unfertilizable.

(b) In the male

Three stages are recognized:

- Spermatogenesis - proliferation of spermatogonia and then meiosis to spermatocytes;
- Spermiogenesis - maturation of spermatids to spermatozoa;
- Spermiation - voiding of spermatozoa in milt

17.1.4 ENDOCRINE CONTROL OF REPRODUCTION

(a) Gonadotrophin-releasing hormones

In most temperate fish species reproduction is seasonal, and the seasonal changes which stimulate reproduction may be:

- Change in day length;
- Water temperature;
- Rainfall - possibly affecting water pH;
- Food supply;
- or any combination of these.

The seasonal influences are sensed in the hypothalamus and stimulate the secretion of gonadotrophin-releasing hormone (GnRH). As in higher vertebrates, fish GnRH is a decapeptide; the chemical structure has not been worked out for all fish species, but those for Pacific salmon (*Oncorhynchus* spp.) and tilapia have. They are called sGnRH and tGnRH, and are known to differ from mammalian luteinizing hormone releasing hormone (LHRH) only in two amino-acids, at positions 7 and 8, see Table 17.1

The main, but probably not the only, target organ for GnRH is the anterior pituitary gland which secretes gonadotrophin (GtH). In addition to the seasonality of hypothalamic GnRH secretion there is, at least in rainbow trout, a seasonality in the responsiveness of the pituitary to GnRH. Synthetic analogues of GnRH which are more active than the natural hormone also act directly on the ovary inducing ovulation; it may be assumed that natural GnRH has the same action, and probably also stimulates the earlier stages of gametogenesis.

In addition to GnRH the brain secretes a gonadotrophin-releasing inhibiting factor (GRIF) which has been identified as the catecholamine, dopamine. Dopamine antagonists promote the pituitary secretion of GtH, and it seems possible that dopamine secretion may normally be continuous (*i.e.* non-seasonal), preventing GtH secretion except under the seasonal influence of GnRH.

(b) Gonadotrophins

In several unrelated species of teleostean fish there are two gonadotrophins, known as GtH I and GtH II. In rainbow trout they are secreted at different stages in the sexual cycle. GtH I is secreted first and it stimulates oestrogen secretion by the ovaries; this in turn causes vitellogenin production by the liver. A few months later there is a much greater secretion of GtH II causing final maturation of oocytes and ovulation in the female and spermiation in the male. In cyprinids GtH II is important for spawning. In European eels the differentiation of GtH into two types has not specifically been shown but GtH does have each of the above two activities.

It was noted in Section 17.1.1 that in Japanese eels gonadal maturation does not occur if migration is blocked. Pituitary extract (of either carp or salmon) induces gonadal development, ovulation and spermiation; similarly human chorionic gonadotrophin (HCG) will induce spermiation in European eels. However, such treatments are not used commercially because the resulting larvae are not viable.

(c) Oestrogens

There are two naturally occurring oestrogens in the rainbow trout, oestradiol and oestrone. Their main action appears to be stimulation of vitellogenin secretion by the liver, and the same action of oestrogens is recognized in European eels.

In eels oestradiol stimulates the gonadotrophic cells of the pituitary (in contrast to the mammalian situation where steroids have a negative feed-back to the anterior pituitary). There is a dose-dependent increase in pituitary GtH content; but the GtH is not released except in the presence either of both GnRH and a dopamine antagonist or of testosterone.

(d) Androgens

The teleostean testis secretes three hormones:

- Testosterone,
- 11-keto-testosterone (11-KT),
- 17α,20β-dihydroxy-4-pregnene-3-dione (20-P).

Only the first two of these are, strictly speaking, androgens, and testosterone is the most important. It may subsequently be 'aromatized' to oestrogens in the central nervous system and possibly the head kidney. 11-KT is a less important androgen; it is probably produced from testosterone and is not aromatizable to oestrogen.

Testosterone is, in fact, the normal gonadal hormone secretion in fish of both sexes, and much higher concentrations of it are produced in the female rainbow trout than in the male. Its function in the female is not clear but peak concentrations occur shortly before spawning and it is possible that some is excreted into the environment to act as a pheromone.

Testosterone stimulates gonadal development in juvenile rainbow trout of either sex; 11-KT does not show this action. In the male testosterone levels peak before spermiation but 11-KT peaks at spermiation. All three hormones stimulate spermiation, the action of testosterone possibly being due either to its conversion to 11-KT or to its stimulating the secretion of 20-P.

In female Japanese eels testosterone in slow-release pellets (50 mg/kg at 15 day intervals) stimulates ovarian development. This is in contrast to European eels where the effect is achieved better with oestradiol and a dopamine antagonist.

(e) Progestagens

Among several progestagens produced in the female close to spawning the two most important are 17α-hydroxy-progesterone and 20-P. The latter in particular is important in stimulating maturation of eggs; GtH is also needed for ovulation in most species but in the African catfish 20-P will induce ovulation.

As noted in the previous section, 20-P is also present in the male at spermiation. Secretion of it is stimulated by GtH II, testosterone or both. Like the androgens it stimulates spermiation, and in salmonids it is the most important hormone for this action. 20-P also promotes sperm motility.

(f) Thyroid hormones

In female Pacific salmon (*Oncorhynchus* spp.) thyroid hormone levels are high in early vitellogenesis, falling as the fish approach final maturation; in Atlantic salmon (*Salmo* spp.) the levels rise just prior to ovulation. Tri-iodo-thyronine (T_3) enhances the actions of GnRH in inducing ovulation and of GtH in inducing steroid secretion by the gonads. Immersion of fish larvae in moderate concentrations of thyroid hormones accelerates absorption of yolk, promotes growth, increases fin differentiation and improves survival. However high concentrations adversely affect growth and survival rates. It is postulated that physiological concentrations of thyroid hormones enter the yolk at vitellogenesis and thence pass to the larvae.

17.2 Drugs used in artificial breeding

17.2.1 GONADOTROPHIN-RELEASING HORMONES

(a) The range of compounds

Table 17.1. Amino-acid structures of natural and synthetic gonadotrophin-releasing hormones

Position	1	2	3	4	5	6	7	8	9	10
Natural Hormones										
Mammalian LHRH:	p-Glu –	His –	Trp –	Ser –	Tyr –	Gly –	Leu –	Arg –	Pro –	Gly amide
sGnRH	p-Glu –	His –	Trp –	Ser –	Tyr –	Gly –	Trp –	Leu –	Pro –	Gly amide
tGnRH	p-Glu –	His –	Trp –	Ser –	Tyr –	Gly –	Leu –	Gln –	Pro –	Gly amide
Synthetic Analogues										
LHRH-A	p-Glu –	His –	Trp –	Ser –	Tyr –	D-Ala –	Leu –	Arg –	Pro –	des-Gly ethyl amide
sGnRH-A	p-Glu –	His –	Trp –	Ser –	Tyr –	D-Arg –	Leu –	Arg –	Pro –	des-Gly ethyl amide

p-Glu = pyroglutamic acid; His = histidine; Trp = tryptophane; Ser = serine
des-Gly = desoxy-glycine; Tyr = tyrosine; Gly = glycine; Leu = leucine
Glu = glutamic acid; Gln = glutamine; Arg = arginine; Pro = proline
Ala = alanine

sGnRH is used, but in addition there are two synthetic decapeptides which are analogues of mammalian LHRH and sGnRH - shown by '-A' after the acronym for the corresponding hormone. Both analogues end with desoxy-glycine ethyl-amide ($-NH-CH_2-CH_2-NH-C_2H_5$) and have a dextro-amino-acid at position 6.

The relative potencies of sGnRH, sGnRH-A and LHRH-A vary with fish species. While the mammalian LHRH shows activity in some species, LHRH-A shows sufficiently higher potency to justify the cost of synthesizing it. The analogue is more resistant to degradation by tissue enzymes in fish and has a prolonged binding to the pituitary. In goldfish C_{max} in plasma for LHRH-A is about 100 times C_{max} for LHRH suggesting that the apparently higher potency of the analogue may in fact be a result of its much greater persistence in plasma. In coho salmon it is about 50-fold more potent than LHRH and in grass carp about 100-fold. As might be expected sGnRH-A is highly active in salmonids; it is also highly active in African catfish, and is about 10 times as potent as LHRH-A in the common carp and goldfish.

(b) Formulation and administration

As polypeptides these compounds are susceptible to proteolytic enzymes in the gastro-intestinal tract of fish; they therefore have to be injected. They are sufficiently soluble in water for adequate doses to be injected in aqueous solution; but single aqueous injections rarely give satisfactory results because they are cleared from the circulation too quickly. Multiple injections are stressful, and the stress may counteract the hormonal action.

Long-term treatment is needed, and several slow-release formulations have been found efficacious.

Sherwood *et al.* in Crim *et al.* (1988) described the production of pellets for implantation each containing 30 mg LHRH-A, and having matrices consisting of varying proportions of cholesterol and cellulose. They were named after the percentage content of cholesterol; and the cellulose merely acted as a binder for the cholesterol, reducing the brittleness of the pellets. Pellets containing up to 75% cholesterol were found to release 90% of the LHRH-A within 24 hours whereas 95-100% pellets released only 18-20% in this time. The authors pointed out that the formulation technique provided for adjustment of total dose and release rate according to the species of fish.

Sato *et al.* (1995) described an injectable water-in-oil-in-water emulsion for intramuscular injection which delayed T_{max} of the GtH response to LHRH-A from 3 to 24 hours. Mylonas *et al.* (1995) used microspheres with a matrix of a copolymer of fatty acid dimer and sebasic acid.

17.2.2 GONADOTROPHINS

(a) Carp pituitary extract

Carp pituitary extract (CPE) or carp pituitary homogenate (CPH) was probably the first gonadotrophin to be developed for induction of breeding in fish. Its use is called hypophysation. Woynarovich and Horvath (1980) illustrate a technique for obtaining fish pituitary glands (Fig. 17.1) but point out that the GtH content of a gland varies with season and the age of the fish. Dosage is expressed either in terms of mg of acetone-dried gland or, where the recipient is a carp, in units given by the ratio:

$$\frac{\text{Body weight of donor}}{\text{Body weight of recipient}}$$

It should be noted that this convention is used in Table 17.1.

Acetone-dried, pulverized gland is available on the market in some countries but it may be contaminated with brain tissue so the potency is unreliable. The user can obtain a more reliable product from live fish in the spawning season; the weights of donor fish should be noted. Physiological saline is a suitable solvent for dried gland. Where a solvent is prepared at the farm care should be taken to use iodine-free salt. Injections are normally given by the intramuscular route.

In female fish hypophysation is used when vitellogenesis is complete and the eggs are "dormant", *i.e.* ready for spawning. It is normal to give two or more doses especially in high temperatures, the interval being 24 hours. The preparatory doses should be only 10% of the total, the remainder being given in the final ("decisive") dose; higher preparatory doses may cause premature partial spawning. Males are given a single decisive dose at the time of the decisive dose for the females. The dose rate for females depends on body weight:

For fish over 5 kg - 2.5-3.0 mg/kg in total
2-5 kg - 1.5 mg/kg
under 2 kg - 0.75 mg/kg

For males the dose rate is 1.0-1.5 mg/kg.

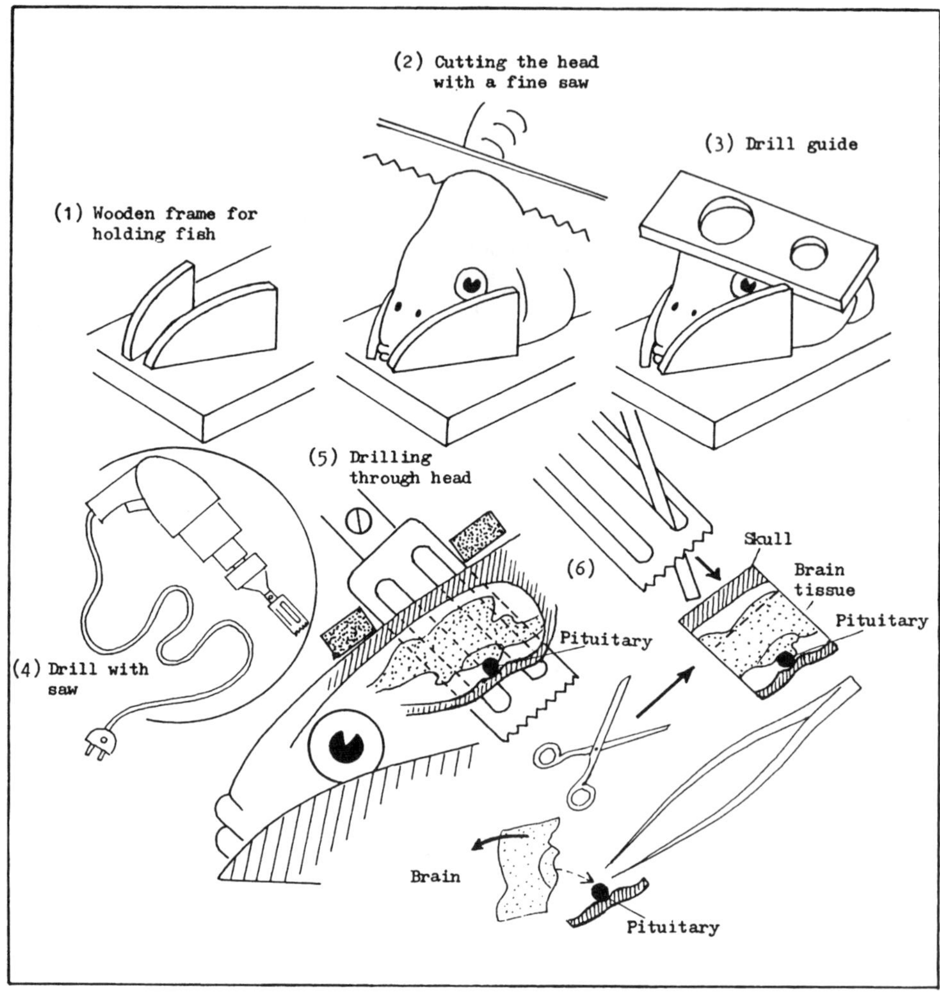

Figure 17.1. The extraction of pituitary gland from fish.

(b) Human chorionic gonadotrophin

Human chorionic gonadotrophin (HCG) or Chorionic Gonadotrophin Ph Eur is extracted from the urine of pregnant women. In mammals it has a similar action to pituitary luteinizing hormone (LH) and has been used for many years in both human and veterinary medicine as the routine source of luteinizing hormonal activity. Its activity is measured in international units (IU) and samples are bio-assayed in comparison with an international standard. The bio-assay is a measure of the effect in increasing the weight of the seminal vesicles or prostate glands of immature rats. The pharmacopoeial requirement is that potency should be within 80% to 125% of that stated, but such is the insensitivity of the test that the fiducial limits of error are 64 to 156%.

In many freshwater species HCG can be use to accelerate the pre-ovulatory phase of egg maturation in which the nucleus moves towards the micropyle and the egg absorbs water. Actual ovulation cannot always be induced but in general carnivorous species respond better than herbivorous ones. Spawning can be induced with HCG in channel catfish, striped mullet and the Chinese major carps. Dose rates recommended by Woynarovich and Horvath (1980) are:

- Channel catfish - 700-2000 IU/kg by intraperitoneal injection depending on the maturity of the fish;
- Striped mullet - 6000 IU/kg by intramuscular injection, usually as two doses at an interval of 24-48 hours;
- Chinese major carps - 800-1000 IU/kg in divided doses at an interval of 8 hours.

Chinese major carps are a group which Woynarovich and Horvath (1980) do not define; presumably it includes:

- Bighead carp (*Aristichthys nobilis*)
- Mud carp (*Cirrhinus molitorella*)
- Grass carp (*Ctenopharyngodon idella*)
- Silver carp (*Hypothalamichthys molitrix*)

The authors recommend that the females of this group are given 60% of the dose in the first injection and 40% in the second. Males are dosed at the time of the second dose for the females.

Although widely used, HCG has no market authorization in north America and concerns have been expressed about the consumer safety of using a human hormone in food fish. Apart from the fact that broodstock are rarely if ever food fish, it has been shown that HCG disappears extremely rapidly from cooked fish flesh. HCG is the hormone detected by pregnancy test kits and such kits are a sensitive, although not quantitative, test for the drug in fish flesh.

(c) Fish pituitary gonadotrophins

On the assumption that pituitary gonadotrophins may be species specific in their chemical composition some workers have experimented with the use of homologous pituitary extracts in breeding induction. It appears that gonadotrophins of fish origin are much more potent in fish than those of mammalian origin, such as HCG, but that fish gonadotrophins are not species specific in potency. A partially purified salmon gonadotrophin has been produced commercially under the name SG-G100; it is rather more potent than CPE and 1 mg is equivalent to 2000-2250 IU HCG.

17.2.3 DOPAMINE ANTAGONISTS

(a) The range of active compounds

Six drugs, each representative of a different chemical group of dopamine antagonists, have been tested with and without LHRH-A for their ability to stimulate a GtH response in goldfish. In the absence of LHRH-A only pimozide and domperidone had any effect and domperidone was considerably more potent. When used with LHRH-A, metoclopramide and fluphenazine showed some action but it was poor compared to pimozide or domperidone and the latter two drugs had comparable potency.

While much of the early research on the use of dopamine antagonists in breeding induction in fish used pimozide, domperidone is now regarded as having the wider safety margin, not least because in at least some fish species, as in mammals, it is known not to cross the blood-brain barrier. Hence domperidone is the drug used commercially in breeding induction in fish.

(b) Pharmacology of domperidone

After intraperitoneal injection of radio-labelled domperidone into goldfish T_{max} for radioactivity in serum is 6 hours. Although it does not cross the blood-brain barrier it has a high affinity for the pituitary gland and from 12 hours after injection it has a higher concentration in the pituitary than in the serum.

In the goldfish both domperidone alone and sGnRH-A alone produce dose-related increases in serum GtH, and within limits each drug synergizes the responses stimulated by the other. When used in combination with 10 mg/kg sGnRH-A, doses of domperidone in the range 1-5 mg/kg produce ovulation. Lower doses produce GtH resonses without ovulation. Doses of sGnRH-A in the range 3.3-330 mg/kg are potentiated by domperidone but no potentiation of 1000 mg/kg sGnRH occurs. These responses by goldfish are unusual: in most fish species, including a majority of cyprinids, there is a GtH response to domperidone alone but the fish do not ovulate. In consequence domperidone is rarely used alone in practice.

(c) Formulation of domperidone

Domperidone has little activity by itself in fish but it potentiates gonadotrophin-releasing hormones. The injectable formulation is usually a 0.1% suspension in fish physiological (0.7%) saline containing 0.1% sodium metabisulphite. It is also available commercially for use in fish as a combination with GnRH-A.

Small quantities of domperidone have been formulated for experimental purposes in fish by pulverizing the tablets available for human medicine (Motilium® - Janssen). The drug is extracted with dimethyl sulfoxide (DMSO) at 0.2 ml per tablet and the solution suspended in 0.5 ml propylene glycol. Undissolved particles of the tablet excipient are removed by centrifugation.

(d) The Linpe method

Combined use of a gonadotrophin-releasing hormone analogue and a dopamine antagonist has been called the 'Linpe' method of breeding induction after its developers, Lin and Peter (see Peter *et al.* in Crim *et al.* (1988)). It is particularly useful in cyprinids. Previously CPE or HCG were used; the advantages of the Linpe drugs are known potency leading to accuracy in dosage, and greater stability on storage.

17.2.4 SYNTHETIC STEROIDS

(a) Methyltestosterone

17α-methyl-testerosterone or Methyltestosterone Ph Eur (MT) differs from other androgens in being active by mouth in mammals. It is therefore normally formulated as tablets. It is, however, obtainable as a powder which is practically insoluble in water but soluble in alcohols, acetone and arachis oil. For use in induction of breeding in fish it is normally formulated in slow-release pellets for implantation.

(b) Deoxycortone acetate

Deoxycortone acetate (DOCA) acts as a mineralocorticoid in mammals, promoting sodium retention and potassium excretion. In females of some fish species it has shown a gonadotrophic action but it is not widely used.

17.2.5 SPERM-ACTIVATING SOLUTIONS

(a) Uses

In some cultured fish species, notably salmonids, sperm motility and fertility depends on ovarian fluid voided by the female with her eggs. This ovarian fluid varies in quantity and quality between individual females. Under natural conditions this may not be a significant constraint on the fish population but in farming the aim is to harvest as many fish as possible for the table. It may therefore be advantageous to supplement or replace ovarian fluid and the solutions used are known as sperm-activating solutions. They prevent sperm deterioration; prolong sperm motility time, and delay the egg cortical reaction which prohibits subsequent sperm penetration.

(b) Composition

Sperm-activating solutions are usually saline solutions adjusted to a slightly alkaline condition. Potassium ions are inhibitory to fish sperm. The pH is critical in some species, such as the rainbow trout whose sperm are inactive at pH below 7.5, and irrelevant in others such as the Atlantic salmon. Osmolarity should be in the range 150-250 mOsm. An adequate volume of fluid to enable sperm to reach all the eggs is always important, and 20 ml per 300 eggs is recommended.

Billard (1992) recommends 250 mM sodium chloride buffered to pH 9 with glycine 30 mM and Tris 20 mM.

Steyn *et al.* (1989) recommended a borax-boric acid buffer which can be prepared from a concentrated stock solution. They used:

- Solution A: 61 g boric acid in 5 litres distilled water
- Solution B: 76 g disodium tetraborate in 1 litre distilled water
- Stock buffer solution: 1000 ml Solution A + 1180 ml Solution B
- The activator solution contains: 2 litres stock buffer solution
7.5 litres distilled water
36 g sodium chloride

17.3 Procedures for individual species

17.3.1 SALMONIDS

(a) Rainbow trout

Salmonids are not routinely medicated to induce breeding because they can be spawned manually and normally die after spawning rather than breeding again. Nevertheless, because they are widely farmed, salmonids in general and rainbow trout in particular are extensively used as experimental animals.

It is recognized that there is a gonadotrophin-releasing hormone inhibitory factor (GRIF) in most salmonids, but while dopamine antagonists will stimulate GtH secretion they do not, by themselves, induce ovulation. In rainbow trout a combination of LHRH-A at low dosage (*e.g.* 1 µg/kg) and domperidone will induce ovulation. This effect can be exploited to achieve some degree of breeding synchronization, since in populations of rainbow trout as of most salmonid species individual females ovulate at different times over several months. Groups of females ready to ovulate within 2-3 weeks can be medicated to synchronize them, and this will facilitate an "all in, all out" system of hatchery hygiene. There is an economic limit to the size of a group of females because those which will not ovulate naturally for several weeks can be induced but ovulation of immature oocytes results in low fertility of the eggs.

(b) Brown trout

LHRH-A at 10 µg/kg will advance ovulation but only at the onset of spawning. Lower doses and slow-release formulations are ineffective. Domperidone synergizes this action.

(c) Arctic charr

Charr tolerate a high stocking density and have higher market value and faster growth than Atlantic salmon or rainbow trout. Constraints on farming them include the limited supply of brood females and poor fertility. The poor spawning is due to blockage at the ovulation stage rather than delays in gonadal development and maturation. At low temperatures the spawning period is long and hatchability is low; GnRH-A induces spawning and therefore shortens the spawning period. Ovulation does not occur naturally at all at temperatures above 10°C but sustained release LHRH-A will induce it and domperidone synergizes the LHRH-A. Significant improvements in ovulation have been demonstrated with the use of Ovaprim®, a combination of 2% sGnRH-A and 1% domperidone, at 0.1 ml/kg followed 2 days later by 0.4 ml/kg.

17.3.2 MILKFISH

Quite apart from the fact that milkfish often do not mature sexually in captivity, under normal conditions males do not mature until they are at least 4 years old and females not until they are at least 5 years old, but even at these ages they do not spawn without hormone induction. Farmers rely for their replacement stock on catches of fry from seawater. Nevertheless once maturity is reached milkfish will spawn more than once in a year. Medicinal induction of breeding is problematic because the species is particularly sensitive to handling stress. Breeding induction requires prolonged elevation of hormone levels but multiple injections are stressful and hence counter-productive.

Slow release formulations of LHRH-A or methyltestosterone (MT) or a combination of both will produce mature males. The combination is the most efficacious and the LHRH-A the least, but implantation of slow-release formulations of LHRH-A can stimulate and maintain spermiation. LHRH-A and the combination also induce maturation in females. For LHRH-A alone the stage of maturity most receptive to treatment is the presence of oocytes averaging 750 μm or more with a unimodal size distribution; the dose rate should be in the range 1-5 μg/kg. MT alone appears to advance and support vitellogenesis.

Spawning in mature females has been induced with SG-G100 (see Section 17.2.2(c)) or a combination of fish pituitary homogenate and HCG, but the eggs so produced have very low fertility. The most successful spawning regimen uses a coho salmon pituitary homogenate (SPH) together with HCG. The dose rate is 70 mg SPH and 10,000 IU HCG repeated after 12 hours. This induces ovulation but since the oocytes do not hydrate synchronously stripping time is critical - the ideal is 10.5 hours after the second hormone dose.

17.3.3 EELS

(a) European eels

Anguilla spp. are catadromous and the artificial creation of a breeding environment similar to the Sargasso Sea is virtually impossible. The normal source of fish for farming is catching elvers as they migrate up rivers. This is expensive, unreliable, confined to specific seasons and offers no possibility of genetic improvement, so there is a clear need for artificially induced breeding. A number of medications has been tested but no rcliable regimen has so far been developed.

In European eels, spermiation can be induced with a single injection of HCG at 250 IU, this being appropriate for fish with body weights in the range 40-100 g. However, spermiation is not sychronized between individual fish - it may occur at any time over the following 3 months, a point which illustrates the long persistence of HCG in eels.

Oestradiol stimulates the proliferation of pituitary gonadotrophic cells in the female European eel. There is a dose-dependent increase of GtH content but no release of it. A combination of GnRH-A and pimozide has been shown to stimulate GtH release but neither drug worked alone. Pre-treatment with oestradiol increased the response to the combination. It may be presumed that domperidone could be used instead of pimozide but it does not appear to have been specifically tested.

In European eels GtH has important actions in vitellogenesis. It stimulates the secretion of steroids, including oestrogens, by the ovary, and these in turn promote the synthesis of vitellogenin in the liver. GtH also promotes the incorporation of vitellogenin into the yolk.

(b) Japanese eels

Spermatogenesis and spermiation can be induced in some fish with GtH or with either a single dose of HCG at 5 IU/g or repeated doses of HCG at 1 IU/g/week. The problem is that the maturity induced varies widely between individual fish. Maturity is usually measured in terms of the gonado-somatic index (GSI):

$$GSI\ (\%) = \frac{\text{Weight of the gonad}}{\text{Total weight of the fish}} \times 100$$

Immature Japanese eels given 8 IU/g HCG (a relatively high dose) showed a range of GSI after 35 days from 0.27-8.2%. Of 18 fish only the three with the highest GSIs spermiated on manual pressure. The drug was given intraperitoneally and C_{max} in serum was at 1 day. Thereafter serum levels declined with a half-life of about 2 days but the differences in levels between individual fish increased with time.

(c) American eels

Spermiation can be induced with weekly doses of HCG, usually at 500 IU per fish. Once induced a male will produce viable sperm for 2 weeks.

A technique for induction of spawning in females is less well established. A course of twice weekly injections of a combination of CPE and HCG for 5 weeks has been shown to induce maturation. In ripe females 1-2 mg of 4-pregnene-17α,20β-diol-3-one, a synthetic analogue of 20-P, was shown to induce a release of eggs but it was a small proportion of the total ovulated. Under natural conditions males nudge the females to induce swimming and the female's body movement causes the voiding of eggs; females spawn limited numbers of eggs repeatedly over a short period.

(d) Ricefield eels

In ricefield eels (*Monopterus albus*) it is the availability of males which is the greatest breeding problem because it is a protogynous hermaphrodite. That is, adults alternate their sex annually, starting as females; so fish caught as juveniles will all be female on initial maturation. Administration of androgens does not cause sex inversion - the gonadal steroids secreted are the result, not the cause, of gender.

Mammalian luteinizing hormone produces limited spermatogeny so it is possible that the pituitary is involved in the sex inversion process. An attempt has been made to induce sex inversion with sGnRH-A, with or without domperidone, to induce secretion of endogenous GtH, or with CPH as a source of exogenous GtH. sGnRH-A at 0.1 μg/g twice weekly for 9 weeks induced sex inversion in eight out of nine female fish. Domperidone had no effect. CPH induced ovarian development.

17.3.4 CYPRINIDS

There is a particular problem with cyprinids in that ovulation cannot be stopped once it has started; and once it has taken place the eggs must be spawned or stripped forthwith because within an hour or so they become over-ripe and impenetrable to spermatozoa. Even in species which do spawn spontaneously after induced ovulation, controlled spawning and fertilization may be desirable: under natural conditions the eggs of some species clump and all except the outer layer die from lack of oxygen.

(a) Goldfish

The goldfish is rarely used for food and breeds readily in the conditions under which it is normally kept. For the latter reason it is frequently used as an experimental animal, although experiments on breeding induction have shown it to be to some extent anomalous (see Section 17.2.3(b)).

In the male goldfish spermiation can be induced with CPE, carp GtH, HCG, LHRH-A or progesterone. In the female sGnRH-A as a single dose has some 10 times the potency of LHRH-A, but GtH responses to two doses of LHRH-A are very much higher than to a single dose. As little as 0.01 μg/g sGnRH-A will stimulate significant rises in serum GtH in pre-spawning females but without ovulation. Domperidone synergizes LHRH-A and sGnRH-A and either combination can induce ovulation.

Suzuki *et al.* (1988a and b) showed that spermiation and ovulation could be induced with salmon pituitary extract. The interesting finding was that although the assumed active ingredient of the extract (GtH) was a glycoprotein, it was active when given orally. The authors noted that this was not true for all fish species, and they presumed their results were associated with the absence of a stomach in the goldfish.

(b) Common carp

The common carp female will not spawn in the presence of fish of other species but she does require the presence of a male. However, even where natural spawning can occur it should not be allowed because the species produces sticky eggs which clump together. Even where the inner eggs of a clump are fertilized they tend to die from lack of oxygen. The use of CPE to induce controlled breeding is a long established practice. In male fish the normal dose is 125 μg with spermiation occurring 2 days later, and it is possible to collect milt in this way weekly for four collections. However, some workers have claimed that up to 4 mg CPE will give dose-related spermiation, and that daily dosage at 125 μg will stimulate spermiation for up to 5 days. In contrast to the position in goldfish, HCG has no action.

A commonly used sperm-activating solution consists of 4 g sodium chloride and 3 g urea per litre. Tap water may be used as the solvent but it must be chlorine-free; table salt may be used for the sodium chloride but it must be iodine-free. A more concentrated solution also known as the fertilizing solution and consisting of 40 g sodium chloride and 160 g urea per litre is used later.

Eggs and milt are manually stripped from the broodfish and mixed gently but thoroughly. After that the sperm-activating solution is added, a small quantity (say 5 ml) at a time, with constant gentle stirring to a total of 50 ml per 300 eggs. After 2-3 minutes 40 ml of the fertilizing solution is added and the mixture swirled into the eggs. As the eggs absorb the solution further aliquots of fertilizing solution are added so that the surface of the egg mass is always covered. Within 90 minutes absorption of liquid is complete and the eggs separate. The sticky material on the surface may be washed off with three or four changes of the fertilizing solution and they may be hardened with a solution of tannin at 8 g in 10 l.

(c) Chinese cyprinids

Tables 17.2 and 17.3 list traditional Chinese and Linpe methods of inducing spawning in the females of a number of cyprinids farmed for food. The males are normally given a single dose of CPE or HCG when the females are given the second injection.

Since the publication of these tables there have been one or two further developments in the technology.

For silver carp Table 17.2 shows traditional use of either CPE or HCG; however broodfish which have been spawned repeatedly with HCG lose their sensitivity to it. LHRH-A at 10 μg/kg with domperidone at 5 mg/kg gives more consistent results.

The bighead carp is a riverine species adapted to life in large volumes of fast-flowing water. It reaches sexual maturity in 2 years but does not spawn spontaneously in captivity. A 2-dose regimen using a 1:200 combination of LHRH-A and domperidone has been found more efficacious than a single injection: the doses recommended are initially 7.5 μg/kg LHRH-A and 1.5 mg/kg domperidone, followed by 9 times this dose. The fish can be induced to spawn 3 times per year but the survival rate is poor in the third brood.

The Chinese loach (*Paramisgurnus dabryanus*) requires a sudden surge in GtH for ovulation. Minimum doses for 100% ovulation are 10 μg/kg sGnRH-A and 5 mg/kg domperidone. LHRH-A has little activity in this species.

(d) Indian major carps

Indian major carps have traditionally been induced to spawn with a two injection regimen of initially 250 IU/kg HCG and then 6 mg/kg CPE. However, the mrigal (*Cirrhinus mrigala*) can be induced to spawn by a single dose of 10 μg/kg LHRH-A but not by 5 μg/kg. Dopamine inhibitors alone will induce ovulation but not spawning.

(e) Thai carp (Puntius goniotus)

Induced ovulation in this species has been achieved with a combination of sGnRH-A and domperidone, and notably this has been done by oral administration (gavage) as well as by injection. An effective dose rate by injection was 20 mg/kg of each drug; of several dose rates tested by gavage the best results were found with 100 mg/kg sGnRH-A and 50 mg/kg domperidone.

The activity of the drugs by gavage was attributed to the anti-emetic action of domperidone preventing any egestion; and the absence of a stomach in Thai carp preventing any proteolytic digestion until at least some of the sGnRH-A had been absorbed.

17.3.5 CATFISH

(a) African catfish (Clarias gariepinus)

DOCA (see Section 17.2.4(b)) at 50 mg/kg causes maturation but not ovulation, which must be evoked mechanically. A surge in GtH secretion is necessary for spontaneous ovulation and the action may be mediated by the GtH stimulating ovarian secretion of 20-P. Ovulation as well as maturation can be stimulated by CPE at 4 mg/kg or HCG at 200 IU/kg. LHRH-A is of limited efficacy but better in the spring breeding season. LHRH-A in combination with domperidone is efficacious but not normally used.

Table 17.2. Traditional methods of induced spawning of cultured freshwater fish in China[a]. After Peter R.E. *et al*, 1988

Species	Water temp. (°C)	First injection	Interval (h)	Second injection	Time to ovulation (or spawning) following last injection (h)
Common carp	20-30	CPE[b] (2-4/kg)	-	-	14-12
		hCG[c] (800-1000 IU/kg)	-	-	14-12
Silver carp	20-30	CPE (2-4/kg)	5-6	CPE (10-20/kg)	8-6
		hCG (100-200 IU/kg)	5-6	hCG (700-1000 IU/kg)	8-6
		hCG (100-200 IU/kg)	5-6	hCG (400 IU) + LHRH-A (10μg/kg)	8-6
Mud carp	22-28	CPE (2-4/kg)	4-5	CPE (16-30/kg)	6-4
		CPE (2/fish) + LHRH-A (100 μg/kg)	4-5	CPE (10-16/kg) + LHRH-A (100 μg/kg)	6-4
Bream	22-30	CPE (2-4/kg)	5-6	CPE (12-24/kg)	10-8
		CPE (2/fish) + LHRH-A (10 μg/kg)	5-6	CPE (6-10/kg) + LHRH-A (50μg/kg)	10-8
Grass carp	18-30	CPE (1-2/kg)	5-6	CPE (6-12/kg)	8-6
		CPE (1/kg)	5-6	CPE (4-10/kg) + LHRH-A (10 μg/kg)	8-6
		LHRH-A (10 μg/fish)	5-6	CPE (2-4/kg) + LHRH-A (10 μg/kg)	8-6
Bighead carp	20-30	CPE (2-4/kg)	5-6	CPE (10-20/kg)	8-6
		hCG (100-200 IU/kg)	5-6	hCG (700-1000 IU/kg)	8-6
		hCG (100-200 IU/kg)	5-6	hCG (400 IU/kg) + LHRH-A (10 μg/kg)	8-6
Black carp	20-30	CPE (2-4/kg)	6-8	CPE (16-30/kg)	10-8
		CPE (2/kg) + LHRH-A (100 μg/kg)	4-5	CPE (10-16/kg) + LHRH-A (100 μg/kg)	10-8
Loach	18-30	CPE (25-30/kg)	-	-	12-10
		hCG (1000 IU/kg)	-	-	12-10

[a]Information gathered from a number of fish hatcheries in Guangdong Province, P.R. China.
[b]CPE = Carp pituitary extract.
[c]hCG = Human chorionic gonadotropin.

Table 17.3. Linpe method of induced spawning of cultured freshwater fish in China. After Peter R.E. *et al*, 1988

Species	Water temp. (°C)	First injection	Interval (h)	Second injection	Time to ovulation (or spawning) following last injection (h)
Common carp	20–25	Domperidone (5 mg/kg) + LHRH-A (10 μg/kg)	-	-	16–14
		Domperidone (1 mg/kg) + sGnRH-A (10 μg/kg)	-	-	16–14
Silver carp	20–30	Domperidone (5 mg/kg) + LHRH-A (50 μg/kg)	-	-	12–8
		Domperidone (5 mg/kg) + sGnRH-A (10 μg/kg)	-	-	12–8
Mud carp	22–28	Reserpine (5 mg/kg) + LHRH-A (50 μg/kg)	4	LHRH-A (50 μg/kg)	6
Bream	22–30	Pimozide (1 mg/kg) + LHRH-A (100 μg/kg)	-	-	10–8
Grass carp	18–30	Domperidone (5 mg/kg) + LHRH-A (10 μg/kg)	-	-	12–8
Bighead carp	20–30	Domperidone (5 mg/kg) + LHRH-A (50 μg/kg)	-	-	12–8
Black carp	20–30	Pimozide (10 mg/kg) + LHRH-A (100 μg/kg)	6	Pimozide (5 mg/kg) + LHRH-A (50 μg/kg)	6–8
Loach	18–30	Pimozide (1 mg/kg) + LHRH-A (50 μg/kg)	-	-	14–11
		Domperidone (5 mg/kg) + sGnRH-A (1 μg/kg)	-	-	14–11

No medication has been found which will induce spawning and after the hormone injection the females have to be stripped manually. The interval between injection and stripping is critical; 14 hours appears to be ideal.

(b) Asian catfish (Clarias batrachus)
CPE is routinely used at 6 mg/kg. This is followed by stripping 17 hours afterwards, the time being critical also in this species.

(c) Hito (Clarias macrocephalus)
Spawning is normally induced with homogenates of hito pituitaries. The proportion of treated fish which actually spawns varies widely, probably due to inconsistencies in the potency of the material injected.

In a search for other more reliable drugs HCG, ACTH, hydrocortisone, DOCA and dehydrotestosterone actetate were tested. Only HCG had any action, suitable doses being 450-500 IU per female and 150-250 IU per male.

17.3.6 BASS

(a) Lates calcarifer
This species of sea bass first matures as males at 3 years of age and become females as 4-5 year-olds. GnRH-A has been used to induce spawning, and both injections and cellulose-cholesterol pellets (Sherwood *et al.*, in Crim *et al.*, 1988) are effective in females. They spawn only once per injection, but four injections of 100 µg per fish at 24-48 hour intervals have been shown to induce four spawnings. A single pellet is unsatisfactory but one each containing 80% and 95% cholesterol causes the release of very large numbers of eggs with very satisfactory fertility. At least two males per female are needed. Spermiation can be induced with an injection of 10 µg GnRH-A per male fish.

(b) Dicentrarchus labrax
Female sea bass of this species can be induced to spawn with either HCG or LHRH-A. The latter in doses of 5-10 mg/kg or 100 µg per fish produces larger numbers of eggs with a very high viability. Eggs induced by HCG tend to have poor viability. The latency period between dosage and spawning is short at high temperatures but prolonged with considerable variation between individual fish at lower temeratures. The regimen can be used successfully up to 6 weeks before the normal spawning season, thus extending the period for which hatcheries can be supplied.

(c) Striped bass
The usual method of obtaining eggs for hatching is to catch mature wild broodfish and induce ovulation with HCG. The dose rate is 330 IU/kg for females and 150 IU/kg for males, injections normally being made into the dorso- median sinus (see Section 1.4.2(c). When ovulation has occurred the fish are stripped manually, four males being used with each female.

Although this technique is effective the breeding season is short and so unsatisfactory use is made of hatchery capacity. Furthermore the sign that a female is suitable for HCG treatment is the presence of oocytes with coalescing lipid droplets; and this involves ovarian biopsy which is stressful if not, on occasions, lethal. Experiments with GnRH in cellulose-cholesterol pellets have shown that maturation before the normal season can be induced. As in *D. labrax*, an 80% pellet is ineffective; an 80% and a 95% pellet in each female induced maturation but with variable latency periods and unreliable ovulation, fertility and hatching rates. Care is necessary to avoid hormonal overdose when the breeding season starts as this causes the production of over-ripe and infertile eggs.

17.3.7 OTHER SPECIES

(a) Walleye

Stocks of walleye (*Stizostedion vitreum*), either for the table or for restocking, have traditionally been produced at hatcheries. Obtaining fertilized eggs has been a problem, the normal source being wild fish caught after ovulation and just before natural spawning. Since ovulation is poorly synchronized in walleye populations, a large number of fish have to be caught to obtain a satisfactory supply of eggs.

HCG at 500 IU/kg and CPE will induce ovulation in walleye. Pankhurst *et al.* (1986) showed that LHRH-A will advance (and presumably synchronize) ovulation in fish which would ovulate anyway, but it does not increase the proportion of fish in a population which ovulate. The action is not synergized by dopamine inhibitors. The authors classified oocytes as:

Stage 1 - Central germinal vesicle (CVG)
2 - Germinal vesicle off-centre (GVOC)
3 - Germinal vesicle migration (GVM)
4 - Germinal vesicle breakdown (GVBD)
5 - Ovulated

They suggested that fish at stage 1 could not be induced and at stage 4 would ovulate anyway; the advancing action affected only fish at stage 2 or 3.

(b) Pacu

The pacu (*Piaractus mesopotamicus*) has traditionally been induced to spawn with acetone-dried pituitary from wild, pre-spawning *Prochilodus lineatus*. The drug is used only in fully ripe females identifiable by having soft swollen abdomens and protruding genital papillae. However, it has been shown to be possible to induce spawning with LHRH-A. As a single dose 100 μg/kg is needed, but two doses of 25 μg/kg at an interval of 6-12 hours are equally satisfactory.

(c) Rabbitfish

Induction of spawning in this species normally involves pairing males with females and inducing the females with HCG. This will achieve a single spawning but is unsatisfactory as a means of obtaining a regular supply of fertile eggs.

In females implanted pellets of LHRH-A prolong the availability of vitellogenic oocytes. In males injections of LHRH-A at 200 μg/kg intramuscularly will induce a peak of milt production 24 hours later. This can be repeated weekly, and for at least five weeks results in a progressive increase in the milt volume so long as the fish is stripped 24 hours after dosing. However there is a true increase in sperm production only for the first 3 weeks; thereafter there is increased volume but reducing spermatocrit.

(d) Grey mullet

The grey mullet is cultured in fresh, brackish and seawater. However, in freshwater it will mature but not spawn. Fish grown to maturity in freshwater can be acclimated to seawater for only 24 hours and then induced to spawn with CPE or HCG, so salinity is an important condition for spawning. Daylength is also important: natural spawning is in the winter and an artificial optimum condition for spawning in females can be created with 6 h light, 18 h dark.

Spermatogenesis and spermiation can be induced with HCG but a more reliable drug is methyltestosterone (MT). This can be given in feed, a satisfactory dose being 12.5 mg/kg bodyweight/day. With a suitable diet medicated in this way males can be maintained in spawning condition for as long as a year and will spawn spontaneously with spawning females. Although MT promotes egg maturation it does not promote growth of testes in immature fish.

Subject to the physical conditions mentioned above, females have been induced to spawn with:

- mullet pituitary homogenate;
- carp pituitary homogenate;
- salmon pituitary homogenate;
- HCG;
- Synahorin® (a combination of HCG and fish pituitary homogenate);
- SG-G100 (see Section 17.2.2(c)).

Of these SG-G100 is the most reliable. A total dose in the range 12-21 μg/kg (50-100 μg/fish) is given to fish with eggs of mean diameter in the range 600-700 μm. Within these ranges the dose is inversely proportional to the mean size of egg. For females with smaller eggs the treatment will increase the size of the eggs, and it should be followed by a second treatment in the range 5-13 μg/kg given in divided doses of 1/3 and, after an interval of 24 hours, 2/3. The priming injection induces hydration of the eggs resulting in a noticeable increase in the body weight of the fish. One variant recommendation is that the second injection should be delayed until this process is complete and the centres of the oocytes have become clear - at about 48 hours. The latency period before spawning is 10-15 hours.

Further reading

Billard, R. (1992) Reproduction in rainbow trout: sex differentiation, dynamics of gametogenesis, biology and preservation of gametes. *Aquaculture* **100**, 263-298.

Crim, L.W., Lee, C.-S., Peter, R.E. and Sherwood, N.M. (editors) (1988) Induced Spawning of Asian Fishes: *Proceedings of a Fish Breeding Workshop, Singapore 7-10 April 1987. Aquaculture* **74(1/2)**, special issue.

Krise, W.F., Hendrix, M.A., Bonney, W.A. and Baker-Gordon, S.E. (1995) Evaluation of sperm-activating solutions in Atlantic salmon *Salmo salar* fertilization tests. *Journal of the World Aquaculture Society* **26**, 384-389.

Lee, C.-S., Tamaru, C.S. and Kelley, C.D. (1986) Technique for making chronic-release LHRHa and 17α-methyltestosterone pellets for intramuscular implantation in fishes. *Aquaculture* **59**, 161-166.

Mylonas, C.C., Sullivan, C.V. and Hinshaw, J.M. (1994) Thyroid hormones in brown trout (*Salmo trutta*) reproduction and early development. *Fish Physiology and Biochemistry* **13**, 485-493.

Mylonas, C.C., Tabata, Y., Longer, R. and Zohar, Y. (1995) Preparation and evaluation of polyanhydride micr-spheres containing gonadotrophin-releasing hormone (GnRH), for inducing ovulation and spermiation in fish. *Journal of Controlled Release* **35**, 23-34.

Mylonas, C.C., Magnus, Y., Gissis, A., Klebanov, Y. and Zohar, Y. (1996) Application of controlled-release GnRHa-delivery systems in commercial production of white bass x striped bass hybrids (sunshine bass) using captive broodstocks. *Aquaculture* **140**, 265-280.

Pankhurst, N.W., van der Kraak, G. and Peter, R.E. (1986) Effects of human chorionic gonadotrophin, DES-GLY^{10} (D-ALA^{6}) LHRH-ethylamide and pimozide on oocyte final maturation, ovulation and levels of plasma sex steroids in the walleye (*Stizostedion vitreum*). *Fish Physiology and Biochemistry* **1**, 45-54.

Peter, R.E., Chang, J.P., Nahorniak, C.S., Omeljaniuk, R.J., Sokolowska, M., Shih, S.H. and Billard, R. (1986) Interactions of catecholamines and GnRH in regulation of gonadotrophin secretion in teleost fish. *Recent Advances in Hormone Research* **42**, 513-545.

Sato, N., Kawazoe, I., Shiina, Y., Furakawa, K., Suzuki, Y. and Aida, K. (1995) A novel method of hormone administration for inducing gonadal maturation in fish. *Aquaculture* **135**, 51-59.

Steyn, G., van Vuren, J. and Grobler, E. (1989) A new sperm diluent for the artificial insemination of rainbow trout (*Salmo gairdneri*). *Aquaculture* **83**, 367-374.

Suzuki, Y., Kobayashi, M., Aida, K. and Hanyu, I. (1988) Transport of physiologically active salmon gonadotrophin into the circulation in goldfish, following oral administration of salmon pituitary extract. *Journal of Comparative Physiology B* **157**, 753-758.

Suzuki, Y., Kobayashi, M., Nakamura, O., Aida, K. and Hanyu, I. (1988) Induced ovulation of the goldfish by oral administration of salmon pituitary extract. *Aquaculture* **74**, 379-384.

Woynarovich, E. and Horvath, L. (1980) The Artificial Propagation of Warm-Water Finfishes - A Manual for Extension *Fisheries Technical Pape*r **201** FAO, Rome.

18. SEX CONTROL

18.1 The Objectives of sex control

18.1.1 ADVANTAGES OF SINGLE-SEX POPULATIONS

(a) Male tilapia

The obvious aim in the husbandry of fish for food is to use the fastest growing types and to harvest them at the end of their period of fastest growth and most economic food conversion ratio (FCR). In some species of fish, notably tilapia and salmonids, there are problems relating to sexual maturation and there are economic advantages in growing single-sex populations. Tilapia have high fecundity but somatic growth can be stunted when they are kept at high stocking densities. It is virtually essential to have single-sex populations to produce fish in excess of 300 g liveweight. Although tilapia males mature at low body weights they grow significantly faster than females, and so a single-sex population should preferably be male.

(b) Female rainbow trout

In salmonids sexual maturation is associated with poor FCR, poor flesh quality and increased susceptibility to infections; furthermore in rainbow trout and channel catfish the broodfish may contain tapeworms and should not be eaten even if the flesh appears to be satisfactory. Since rainbow trout males usually mature at 2 years of age and females at 3; and since harvesting for the table is normally at 2 years, these problems apply only to males, and an all-female population is desirable.

(c) Female anadromous salmonids

For anadromous salmonids the ideal harvesting time is some 2.0 to 2.5 years after they have been put into seawater cages as smolts, and there is a problem in that a few fish mature early. In the case of Atlantic salmon this can occur after a little over a year, and such fish are called grilse; but the problem is well recognized not only in other salmonids such as coho salmon but also in unrelated species such as tilapia. In salmonids the proportion of the population varies with the water temperature in spring and with the genetic strain of fish, but is typically about 20%. Fish showing signs of impending maturation have to be harvested immediately, yielding a low return on the investment in producing them.

There are more male than femalc grilse, and while oestrogens could be used to increase the proportion of females in a population, this would mean the medication of fish ultimately destined for the table. This is generally undesirable, and experience with mammals suggests that it would be banned in the EU

(d) Production of eggs for the table

In some countries there is a different commercial advantage in the production of all-female populations of fish. This is that fish eggs of any species are regarded as a delicacy comparable to caviare. Some farmers find it more profitable to allow the females to mature and spawn, and to discard the carcasses, than to harvest the fish at the optimum time for their flesh.

(e) Ornamental males

In many ornamental fish species the males are more colourful and therefore command a higher price than the females.

18.1.2 ADVANTAGES OF STERILE FISH

In mature carp of either sex the gonads can reach 25% of bodyweight and are not used for food; thus there is an economic advantage in producing sterile fish for harvesting.

Another reason for developing sterile fish is the possibility of escapes of any age. The possibility of escaping salmonids breeding with wild populations is of concern in Europe and north America where angling interests allege that it would produce progeny of low sporting potential. The salmon farming industry, anxious to avoid political opposition to its development, is concerned to find ways of preventing such inter-breeding. In British Columbia attempts are being made to farm Atlantic salmon. This species is of a different genus from the indigenous Pacific salmon and will not interbreed; but there is concern that escapees of the foreign species might compete with the indigenous ones and exterminate one or more of them.

18.2 Induction of Triploidy

18.2.1 PRESSURE SHOCK

Techniques have been developed in parallel in British Columbia for Pacific salmon and in Scotland for Atlantic salmon which result in the production of all-female populations or of triploids. The all-female diploid populations have a reduced production of grilse; and while male triploids (XXY or XYY) may develop gonads, and hence hormones, female (XXX) triploids are sterile. Both single-sex populations and triploids can be achieved by treating eggs with either pressure-shock or drugs, and triploids have been produced by temperature shock. The production of all-female (diploid) populations by pressure-shock means the fusion of the egg nucleus with the second polar body, and this would lead to in-breeding. So modern practice is to confine the pressure-shock technique to the production of triploids.

18.2.2 MEDICINAL INDUCTION

Although it is rarely, if ever done, it has been shown to be possible to induce triploidy by drugs; and those known to be efficacious for this purpose are the anaesthetics.

They are used on eggs immediately after fertilization to cause fusion of the nuclei of the eggs, spermatozoa and second polar bodies. A comparative test of six such drugs in Atlantic salmon, both as gases or vapours under normal use conditions and under pressure, showed nitrous oxide to be the best. None of the drugs had any useful level of activity under normal use conditions and it was found necessary to use nitrous oxide at 11 atmospheres to achieve over 90% triploidy. The superiority of nitrous oxide was attributed to the fact that the eggs were in water and it was the least hydrophobic drug tested. However it was probably the only one of the six drugs tested which it was physically possible and safe to use at that pressure.

The use of nitrous oxide for 30 and 60 minutes made no significant difference in the conversion to triploids; but with a higher mortality after 60 minutes exposure, 30 minutes at 11 atmospheres was clearly the optimum regimen.

18.3 Medicinal production of all-female populations

18.3.1 AVAILABLE METHODS

There are two distinct methods: in one eggs or alevins are immersed in baths of oestrogen to induce genetic males to develop as phenotypic females; and in the other an androgen is administered, by immersion or in-feed or both, to induce genetic females to develop as phenotypic males. In the former method the fish are intended to be harvested for the table, and since they have been treated with sex hormones, even though this was at a very young age, the procedure is regarded by consumers as undesirable. In the latter method the treated fish are not intended for the table but as broodstock for table fish. Its disadvantage is in the delay of a generation in the production of the initial single-sex population.

The latter technique is applicable where, as in mammals and birds, the male is heterozygous (XY) and the female homozygous (XX); this does not apply to all taxonomic groups of fish but it is true for salmonids. The medicated fish are known as gynogenetic males; they are not sold for the table but are used to breed with normal females. Being genotypically female, gynogenetic males produce only X spermatozoa and only female progeny, the latter being unmedicated and saleable for the table. Once such an all-female population has been established it is possible to use it not only to harvest for food but also as the source of fry for further production of gynogenetic males. It cannot also be used as the source of brood females as this would result in severe in-breeding.

18.3.2 DRUGS AND ADMINISTRATION

(a) Oestradiol

The oestrogen normally used is oestradiol. This is a natural hormone and so is regarded as environmentally safer than synthetic steroids. For immersion treatment of eggs or alevins it is formulated as a solution of 10mg in 10 ml ethanol (0.1%) and this is added to water at 2 ml in 5 l. This gives a final concentration of 400 μg/l in the bath. Typically the exposure time is 60 minutes.

For administration in fry food it is dissolved in the rather less volatile iso-propanol, at the rate of 20 mg in 200 ml. This is sufficient for 1 kg fry food. It is thoroughly mixed into the food and the mixture is then spread out on trays and left overnight for the iso-propanol to evaporate. The hormone stays in the lipid fraction of the food and medicated food may be stored in a refrigerator for up to 48 hours. A rather more dilute initial solution in ethanol can be used to medicate diets for juvenile salmonids: the recommended final concentration is 20 mg/kg diet.

(b) Androgens

The androgen usually used is methyltestosterone (MT) (see Section 17.2.4(a)). It is practically insoluble in water but is soluble in alcohols. It is obtainable as a powder and for immersion use it may be formulated in the same way and at the same concentration as described above for oestradiol. It may also be used to medicate feed in the same way as oestradiol, but the usual concentration in the diet is 3 mg/kg and hence the initial concentration in iso-propanol should be 3 mg in 200 ml.

A large number of synthetic and natural oestrogens and, more particularly, androgens and have been tested for this use. One such androgen, 11-β-hydroxy-androstenedione (OHA), has the advantage claimed for it that it is a natural hormone, but it has been found generally inferior to MT and has not so far been adopted commercially. As a general rule the synthetic androgens have proved both cheaper and more slowly eliminated. Among the latter, 17-α-ethinyl-testosterone and mibolerone are notable.

Some non-steroidal compounds have sex controlling effects in fish. Acriflavine, which is a mixture of acriflavin and proflavin, will masculinize *adult* mosquito fish and has an androgenic action on tilapia at doses equimolar with MT. Tamoxifen, which is an anti-oestrogen rather than an androgen, is also active in tilapia.

18.3.3 RAINBOW TROUT

(a) All-female populations

It was noted in Section 18.1 that it is economically desirable to farm all-female populations of rainbow trout. In this species immersion of alevins in oestrogen has produced a low and variable conversion of genetic males to phenotypic females. A more consistent result is obtainable with NN-dimethyl-formamide ($HCON(CH_3)_2$), but this drug has never been adopted commercially. A more expensive but more efficacious approach is to give juvenile fish a diet containing 20 mg/kg oestradiol from first feeding. This is continued for a month; shorter periods produce some hermaphrodites. During treatment the fry are more susceptible to adverse environments and grow more slowly than untreated fish. However they catch up on cessation of the treatment and are indistinguishable from untreated fish at 150 days. This procedure means the medication of fish ultimately intended for the table; but only juvenile fish are treated, a natural hormone is used and it is very rapidly excreted - the half-life of radio-oestradiol is less than 12 hours.

(b) Gynogenetic male production

In contrast to the problems of chemical induction of all-female populations, the procedure for producing gynogenetic males is well established. Starting from a normal population of fry with an approximately 50:50 distribution of XX and XY genotypes the medication will produce, respectively, gynogenetic males and normal males at maturity and they will be indistinguishable. In the early years of the technique the procedure was to medicate feed with 3 mg/kg MT and feed it for 718-1180 degree-days. This produced a stock half of which was strippable and half of which had occluded efferent testicular ducts. Progeny testing showed the latter to be the gynogenetic males. Ductless testes were removed surgically, macerated and used, with or without a saline diluent, as milt.

Recently extensive studies on different immersion regimens have shown that acceptable conversions of females to gynogenetic males can be achieved by immersion treatment of alevins. The optimum appears to be a single immersion in MT seven days after 50% hatch; this gives about 75% conversion with about 90% of the converted males having patent ducts. Although these gynogenetic males are strippable they are indistinguishable from genetic males, so until an all-female population has been established the progeny of each male must be kept separate; and the fry for further production of gynogenetic males must be taken only from all-female families. Hence the technique of producing gynogenetic males with occluded ducts continues to be used.

18.3.4 OTHER SALMONIDS

(a) Chinook salmon

For this species the use of MT both as a bath and in feed is recommended. A 2 hour bath is given to alevins in the period 4 to 11 days after hatching, and this is followed by giving medicated feed to the fry for the first 6 weeks. In contrast to the results in rainbow trout, this rather high dose regimen does not produce gynogenetic males with occluded efferent testicular ducts. The problem of distinguishing between gynogenetic and genotypic males therefore exists, but recently a polymerase chain reaction test has been developed to detect Y chromosomes in chinook salmon. Androgen treated stock, which are phenotypically all-male, may be tested and the positives rejected for breeding.

(b) Coho salmon

In this species the use of MT as a bath for eyed eggs or alevins produces sterility. In older fish still in freshwater it has no significant sex inversion effect; but at the abnormally high feed medication rate of 10 mg/kg it has been shown to cause initial testicular hypertrophy and subsequent degeneration. Ovaries were unaffected. Feed medicated over a long period at a rate in the range 1-10 mg/kg had a significant growth promoting effect although this was partially compensated by reduced growth after transfer to seawater.

Oestradiol used as a bath for eyed eggs or alevins has been shown to produce a significant increase in the proportion of females. A single immersion for 2 hours has been used on eyed eggs and on alevins; optimum results were obtained with alevins at 6 days post-hatch. This procedure is not widely used as the fish to be marketed for the table are hormone-treated.

(c) Amago salmon

Only immersion studies have been reported for this species. For a single 2 hour treatment of alevins with MT at 400 μg/L it has been shown that limited conversion of females to gynogenetic males can be achieved at any time up to 22 days after hatching. The optimum time was found to be 17 days when 55% of females converted. Those which did convert had normal testes with efferent ducts.

(d) Atlantic salmon
Feeding MT to fry causes some sex inversion but the regimen is inefficient - over half the fish are sterilized rather than inverted. Of those that do become gynogenetic males none can be stripped normally; they have to be treated as described above for rainbow trout with occluded efferent testicular ducts.

Immersion treatment has also been studied in this species because it can be administered at a more appropriate time earlier in development. Hatching normally occurs at about 500 degree-days after fertilization, and sets of three immersions at 50 degree-day intervals have been tested beginning with the second immersion at the time of hatching. A regimen of immersions at 750, 800 and 850 degree-days after fertilization was the most efficacious.

18.4 Sex control in tilapia

18.4.1 PHYSIOLOGY

More extensive studies have been made into the manipulaton of sex in this genus than in most other fish because of the commercial incentive (see Section 18.1.1).

It appears that in cichlids generally only oestrogens have any physiological role in determining an individual's sex; if oestrogens are present a female is produced; if not, a male develops. High proportions of females can be produced by similar techniques to that described previously for coho salmon, and this is used in some tilapia species where precocious maturation of males is the main problem. A further effect of this physiology is that an all-male population can be produced with the anti-oestrogenic drug, tamoxifen.

An important behavioural point to note about tilapia is that they are mouth-brooders. It is therefore impossible to medicate the water in which the eggs are incubated; and the juveniles can only be treated after their release from maternal care, by which time water medication would be uneconomic. So in practice medication is usually in-feed.

18.4.2 MEDICINAL PRODUCTION OF ALL-MALE TILAPIA

The physiological role of oestrogens in tilapia does not mean that androgens are inactive (although some species such as the golden tilapia may be relatively insensitive to them). MT and 17-α-ethinyl-testosterone (17αET) have been used in several tilapia species to sex-reverse females to males. For example in Mozambique tilapia MT has been used at 40 mg/kg in the feed for 40 days starting as soon as the juveniles are released from maternal care; and the daily feeding rate is gradually reduced from 36% to 6%. Another synthetic androgen, mibolerone, has been used in golden tilapia and found to have about 100 times the potency of 17αET.

Some interesting work has been done on the induction of all-male populations of hybrids of Nile and golden tilapia. In-feed use of acriflavine, tamoxifen and MT were studied. In all cases the medicated feed was given at 15% of bodyweight daily for 6 weeks. Acriflavine produced a male-skewed population but the effect was not dose-dependent and there were a number of progeny of indeterminate sex. MT at 15 ppm in feed and tamoxifen at 100 ppm in feed each produced 100% males. Lower doses of either of the latter two drugs produced dose-dependent sex-reversal including a proportion of fish of indeterminate sex. Survival in the groups of fish given any of the drugs at any of the dose rates was as good as in untreated controls; but the dose of MT needed to induce 100% sex-reversal (15 ppm in feed) produced gross deformity in 10% of the fish treated. Lower doses of MT and all doses of acriflavine and tamoxifen produced only 3% deformity. Of the three drugs, MT was the only steroid: its induction of deformity was postulated to be an example of an effect of androgenic steroids on non-gonadal physiological processes. Economics aside, tamoxifen would appear to be the best of these drugs for sex-inversion, at least in this hybrid of tilapia.

Mibolerone is a synthetic androgenic steroid which has shown value in golden tilapia, a species which is of particular value for farming because it is more cold-tolerant than other tilapia. In this species mibolerone is some 100 times as potent as 17αET; it has been shown to be effective in food, and, although there are demerits in the procedure, it has also shown efficacy in water if the fish are under 11mm in length. The recommended in-feed regimen is to medicate the feed at 1 ppm and feed it at 10% of body weight daily for 4 weeks. Survival is very good; growth is promoted, and the resulting population develops as about 85% male, 11% indeterminate sex and only 4% females.

Water medication with mibolerone is of interest less for its commercial potential than because it has been shown to work at all. Groups of experimental golden tilapia were kept in aquaria at 27±2°C and the water was medicated with 0, 0.3, 0.6 or 1.0 ppm mibolerone. In each case the water also contained 0.0017 ppm dimethyl sulphoxide (DMSO), a solubilizing agent which promotes the passage of drugs through mammalian skin and may be expected to do the same with fish skin and more particularly gills. Water medication was continued for 5 weeks with fresh medicated water being provided weekly. The lowest medication rate produced a population mainly of males and intersexes but still containing 0.7% females; such a result may be acceptable for commercial purposes. The two higher dose rates produced all male or intersex populations but at the expense of significant dose-related reductions in survival rates. Furthermore although in-feed mibolerone had a positive effect on growth the immersion procedure had a negative effect. An interesting aspect of the experiment was that at the end, analysis of the fish showed that all groups contained higher concentrations of mibolerone than was in the water, showing that the fish had actively absorbed the drug.

18.4.3 BREEDING ALL-MALE TILAPIA

It has been noted that the medication of fish ultimately destined for the table is believed to be unacceptable to the market. This is in spite of the fact that any sex hormone administered to eggs, alevins or fry would deplete beyond the limit of detection within a few weeks; and no fish would be marketable in less than 18 months. In the procedures described above with the exception of those for coho salmon and tilapia the broodstock were hormone-treated but the all-female fish to be marketed were bred. Desirable as it may be in tilapia, it is not in fact possible to breed an all-male population of a species which has homozygous (XX) females and heterozygous (XY) males. However the tilapia are a remarkable family in that some species have the XY system of sex determination and some have the opposite system - heterozygous (WZ) females and homozygous (ZZ) males. The golden tilapia is one of the latter species and, as noted above, is of particular value for farming. It appears to be rather insensitive to androgens but a method has been developed to produce broodstock which will breed all-male progeny.

Initially a cohort of golden tilapia fry were medicated in-feed with ethinyloestradiol at 60 mg/kg feed for 30 days. The feed was given at 20% of bodyweight per day, making a dose of 12 mg/kg bodyweight per day. Histology on a sample of this cohort showed 82% were phenotypically female. The rest of the cohort were allowed to grow, and at 40 g liveweight when the sexes could be differentiated the males were culled. This meant that in the remaining all-female population there would have been about 39% sex-reversed genotypic males. The group was grown on to maturity, bred with normal males and progeny tested. Those producing 50% male progeny were obviously heterozygous (genotypic females) and were culled. Those producing all-male progeny were bred again and their fry, known to be genotypically all-male, were oestrogen-treated. Any developing as females, in this or subsequent oestrogen-treated generations, must be sex-reversed genotypic males and useable to breed all-male progeny.

18.5 Production of sterile carp

The drug used is MT, as for other fish groups, but the medication rate is two orders of magnitude higher and the treatment regimen is longer. Sterilization has been shown by giving feed medicated at 400 ppm for 30 days starting with day-old hatchlings. 300 ppm has been found to result in the maturation of a few fish although with a gross effect on the sex ratio. As noted at the beginning of this chapter, the weight loss on evisceration of mature carp may be over 25%; in carp medicated as hatchlings at 400 ppm in the feed it can be as little as 5-6%.

As with the treatment of salmonids, an alcoholic solution of the MT is sprayed on to the food. However a much more concentrated solution is required - 800 mg/l is typical; 500 ml/kg food will then provide the 400 ppm medication rate. Ethanol can be used as the solvent and the drying process can be expedited by keeping the sprayed food at room temperature for 3 hours and then putting it in a hot air oven for an hour at 65-70°C.

Some Chinese carp species have specialized feeding habits which preclude the use of medicated feeds and late gonadal differentiation which precludes administration by immersion. For them slow release oral formulations have been developed, for example with the drug in Silastic®. Using this, results may be achieved with a single dose force fed to the fish.

18.6 Induction of sex change in hermaphrodite species

In fish the word hermaphrodite is not normally used to mean individuals having gonads of both sexes at the same time, but those which change sex as a normal physiological process. The groupers (*Epinephelus* spp.) are a genus of fast growing marine fish in many ways highly suitable for farming; however breeding them presents severe economic problems. When they first mature, typically at 3 years old, they are all female; 4 years later they turn into males ('protogynous hermaphroditism'). Thus natural breeding cannot take place until some at least of the broodstock are 7 years old. In practice they have been farmed in the past by catching juveniles from the wild.

It has been shown possible in both common groupers (*E. tauvina*) and blue-spotted groupers (*E. fario*) to produce males at the time of initial maturation by a regimen of oral 3-methyl-testosterone. The hormone is disolved in 95% ethanol and given in feed pellets. In common groupers 1 mg/kg/day for 145 days has been found efficacious; in blue-spotted groupers this regimen and 0.5 mg/kg/day for the same period have been found equally effective. The males so developed will produce milt without further medication; untreated fish which develop as females need to be induced to spawn.

In general the dose of a steroid given by injection can be as little as 10% of the dose required by the oral route. Injections are obviously not possible in the eggs and alevins normally treated but changing the sex of hermaphrodites would appear to be an activity where the technique should be tried.

18.7 Safety aspects

18.7.1 ENVIRONMENTAL SAFETY

While waste medicated water will rapidly be diluted below the level of pharmacological activity, the environmental safety aspect of the medication of feed must be borne in mind: lipophilic compounds will leach out of food only slowly and uneaten feed may become available to wildlife.

18.7.2 CONSUMER SAFETY

Studies have been conducted in Mozambique tilapia and rainbow trout of the depletion of radio-labelled MT given in the diet. In both species initial high radio-activity was found in the viscera with the probability that the drug was concentrated in the liver and gall bladder. Between 20 and 50 hours after dosing there was a rapid decline in total radio-activity suggesting elimination in the faeces. At 50 hours only about 2.5% of the original activity remained and at 100 hours it was down to 1%. Straight line semi-logarithmic plots could not be constructed, presumably due to metabolism of the radio-carbon. It was clear however that negligible levels of drug would remain in full grown fish.

18.7.3 TARGET SPECIES SAFETY

Where androgenic or oestrogenic steroids are being used to achieve sex changes any other pharmacodynamic effects whether desirable or otherwise must be regarded as side effects. Three are generally recognized:

1. An increase in mortality;
2. A change in growth rate - usually an increase in juveniles, and a decrease where the technique is used in adult fish such as hermaphrodites. Taxonomic groups of fish vary in the degree of growth response they make to steroid treatment; but in all cases there is an optimum dose rate and overdosage reduces responses.
3. An increase in the number of deformed fish, especially where androgenic steroids are used.

Further reading

Pandian, T.J. and Sheela, S.G. (1995) Hormonal induction of sex reversal in fish. *Aquaculture* **138**, 1-22.

Yamazaki, F. (1983) Sex control and manipulation in fish. *Aquaculture* **33**, 329-354.

19. IMMUNO-STIMULANTS

19.1 Immuno-modulation

19.1.1 ADVANTAGES OF IMMUNO-STIMULATION

Since 1990 considerable research effort has been concentrated on immuno-stimulants for fish. Immuno-stimulants are compounds which enhance the resistance of the treated animal to infections in general. They are effective against both overt and sub-clinical disease. As medicinal agents for the control of overt disease they have important safety advantages in that they have none of the negative effects which antibacterial drugs and live vaccines have on the environment; and being natural compounds they produce no undesirable residues.

Immuno-stimulants can be given in feed over long periods, and in this way they have distinct advantages over vaccines in farmed fish. The immune system of fish is compromised by stress, which may result from high stocking rates and handling. Furthermore vaccinal immunity in fish is often short-lived, and some vaccines are difficult to administer to large numbers of fish.

By controlling sub-clinical infections immuno-stimulants improve the general health of the medicated animal, and this in turn produces the economic advantages of enhanced growth and improved food conversion ratio.

19.1.2 DEFINITIONS

Immunity systems in the animal body are normally described as specific or non-specific. Specific systems are so-called because they increase the body's resistance to a particular chemical, which may be exuded (as an exotoxin) by an invading parasite or be on its surface. Such parasites may vary in size from a virus to a helminth. Non-specific systems, once stimulated, increase the body's resistance not only to the stimulant but also to every other foreign chemical.

Antigens stimulate specific immunity systems. The term, immuno-stimulants, should theoretically include antigens but in practice it is confined to substances stimulating non-specific systems. Adjuvants are used with artificial antigens to enhance specific immune responses; although any one adjuvant will potentiate a range of different antigens they are not normally regarded as immuno-stimulants because they have no action in the absence of an antigen and they do not stimulate non-specific systems. Nevertheless immuno-stimulant compounds are on occasions used as adjuvants. In *in vivo* tests using *Aeromonas salmonicida* bacterin in rainbow trout it was found that:

- the bacterin alone produced a low humoral response;
- the bacterin with immuno-stimulant produced a good humoral response (and delayed

death after furunculosis challenge);

- the immuno-stimulant alone enhanced non-specific immune responses.

19.1.3 THE RANGE AND IMPORTANCE OF IMMUNO-STIMULANTS

A wide variety of different types of compound has been found to show immuno-stimulant properties in fish. A majority are active only by injection but some are active orally or by immersion. Those most extensively studied include glucans, certain polypeptides, levamisole and vitamin C.

Non-specific immune systems, and therefore immuno-stimulants, are more important in fish that live in cold environments and have slow specific immune responses. In its initial encounter with a pathogen a fish is almost entirely dependent on its non-specific immunity. For this reason a glucan, MacroGard®, has been included in fish feeds in Europe; but only very limited claims for it can be made in the EU because it is not authorized under Directive 70/524 (see Section 3.4.2).

19.1.4 ADMINISTRATION OF IMMUNO-STIMULANTS

Anderson and Siwicki (1994) studied the use of a glucan and chitosan in brook trout before challenge by immersion in a culture of *A. salmonicida*. The immuno-stimulants were given by injection at 100 μg/kg or immersion for 30 minutes in 100 μg/ml. While there was no difference between the two immuno-stimulants, the injection regimen was more efficacious than immersion, and by either route of administration the protection was less at 14 or 21 days after medication than at 7 days.

Where glucans, and presumably other types of immuno-stimulants, are given in feed there is a limit inclusion rate above which there is no further enhancement of non-specific immunity. Using feed at or near this limit medication rate for 2 weeks is long enough for the immuno-stimulation to last about 6 weeks, so a regimen of 2 weeks medicated feed alternating with 6 weeks unmedicated feed is the economic optimum. The exception is fry from first feeding to about 15 g body weight. For them a maximum non-specific response is desirable and is best achieved by continuous use of medicated feed. This is because the specific immune system does not develop until the fish reaches about 4 g body weight.

19.1.5 IMMUNO-SUPPRESSION

Although deliberate medicinal immuno-suppression is used in transplant surgery in Man, there is no indication for it in fish. Nevertheless, chemical immuno-suppression does occur in fish due to pollutants in the water which may be particulate or soluble. The latter include pesticides and heavy metals (although cadmium in sub-lethal concentrations may be immuno-stimulant). In addition, a few substances used medicinally have immuno-suppressive side effects; the most widely used of these is oxytetracycline.

It is probable that the immuno-suppressive effects of physical stressors, pollutants and oxytetracycline are all mediated by the secretion of cortico-steroid hormones, particularly hydrocortisone. These hormones have a suppressive effect on leucopoiesis.

19.1.6 DETECTION OF IMMUNO-STIMULATION

The following subsections describe experimentally measurable characteristics of non-specific immunity. An increase in any characteristic in medicated fish over unmedicated controls is evidence of immuno-stimulation. Experimental methods are given by Kajita *et al.* (1990), Jørgensen *et al.* (1993) and Siwicki *et al.* (1994).

(a) Haematocrit and leucocyte count

Non-specific immunity is mediated by leucocytes, so a raised leucocyte count with an essentially unchanged haematocrit is evidence of immuno-stimulation.

(b) Oxidative radical production

A major way in which neutrophil granulocytes contribute to non-specific immunity is by the production of oxidative radicals. Nitro-blue tetrazolium (NBT) reacts with oxidative radicals producing a dark blue colour and is used to identify neutrophils actively producing them.

(c) Phagocytic activity

Phagocytosis can be assayed by incubating blood with a killed bacterial (*e.g. Staphylococcus aureus*) culture and examining stained smears for phagocytes containing bacteria.

(d) Bactericidal activity

Phagocytosed bacteria are not necessarily killed. Bactericidal activity can be assayed by incubating macrophages with a live bacterial culture and then washing off the supernatant liquid, lysing the macrophages and examining the residue for live bacteria.

(e) Myelo-peroxidase production

Activated neutrophils also produce myelo-peroxidase. The level of activation can be assessed by incubating blood smears in an indicator reagent and examining cells under the microscope for the degree of staining.

(f) Immuno-globulin concentration

Although some serum immuno-globulins are humoral antibodies and therefore mediate specific immunity, many others are non-specific in action.

(g) In vitro measurement

Jeney and Anderson (1993) have described an *in vitro* method for screening substances for immuno-stimulation. In essence finely divided pieces of rainbow trout spleen are maintained in a tissue culture medium with the test substance, and after 4 days cell suspensions are prepared. For neutrophils the cell suspensions are treated with NBT and examined by spectrophotometry; for phagocytes aliquots of the cell suspension are shaken for 15 minutes with a suspension of fixed sheep erythrocytes, and then smears are made for microscopy.

(g) In vivo measurement

As specific immunity develops slowly in fish it is possible to assess immuno-stimulation by a challenge infection which rapidly produces high mortality. Any delay or reduction in mortality in treated fish compared to untreated controls may reasonably be attributed to non-specific immunity systems.

19.2 Glucans

19.2.1 MACROGARD®

(a) Chemistry

Glucans are polysaccharides composed exclusively of glucose units. MacroGard is a glucan with β-1.3 and β-1.6 linkages between glucose units obtained from the cell walls of yeast (*Saccharomyces cerevisiae*). Glucans are known to be important stuctural compounds in the cell walls of all fungi and MacroGard is not the only glucan obtainable from yeast.

MacroGard is insoluble in water but can be suspended in it by ultra-sonication.

(b) Mode of action

MacroGard stimulates fish macrophages to secrete the hormone interleukin. Interleukin causes the proliferation of T-lymphocytes and stimulates them to secrete interpherone; interpherone acts back on the macrophages increasing their phagocytic action and causing them to secrete antibacterial enzymes such as lysozyme. Interpherone thus enhances non-specific immunity; and insofar as MacroGard leads to secretion of interpherone it is acting indirectly as an immuno-stimulant. It also has a direct immuno-stimulant action on macrophages causing increased production of oxidative radicals. However interleukin also enhances T-lymphocyte production of antibodies, so MacroGard can have an indirect adjuvant action, a point which has been confirmed experimentally (see Table 19.1 from Rørstad *et al.*, 1993).

The direct immuno-stimulant action of MacroGard is subject to its concentration. *In vitro* the maximum production of oxidizing radicals by macrophages is by 0.1-1 μg/ml MacroGard; 10 μg/ml has no effect, and 100 μg/ml is inhibitory.

(c) Use by injection

Robertsen *et al.* (1990) showed that MacroGard enhances the resistance of Atlantic salmon presmolts to three different bacterial pathogens. They used intraperitoneal injections of a suspension in physiological saline at 0.2 ml per 20 g fish and obtained different dose rates by varying the concentration. It may be noted that the intraperitoneal route would be the ideal method of injecting a drug of this type because it was a suspension, not a solution, and because the peritoneal cavity drains directly into the lymphatic system. Using a dose of 2mg per fish it was found that maximum protection developed at 3 weeks. Using a challenge at 1 week the optimum dose rate was found to be in the range 50-200 μg per fish (2.5-10 mg/kg).

Jørgensen *et al.* (1993) tested MacroGard in rainbow trout with *in vivo* medication and *post mortem in vitro* assay of its effect. Fish weighing 300-400 g were given 1 ml of 1% MacroGard in 0.9% saline (= 25-33 mg/kg) by intraperitoneal injection. Groups were slaughtered at 1, 2 and 3 weeks after injection and head kidney macrophages assayed for oxidative radical production, bactericidal activity and lysozyme production. A significant increase in macrophage bactericidal activity over unmedicated controls was found only in fish slaughtered 2 weeks after injection; and there were similar findings for oxidative radical production. Serum lysozyme activity was significantly raised in all medicated fish but it was highest in those slaughtered 1 week after injection and then fell progressively. The authors point out that these observations on head kidney macrophages show that MacroGard does not only affect cells in the peritoneum. It is now thought that particles of the insoluble glucan are phagocytosed by macrophages in the peritoneum and so transported to the head kidney.

(d) In-feed use

MacroGard has also been shown to be efficacious when given by mouth. The manufacturers have tested it in Atlantic salmon at 1 g/kg in the diet for 5 weeks. They showed significant protection against *Vibrio anguillarum* and *V. salmonicida* infections. Siwicki *et al.* (1994) fed a diet containing 2 g/kg to rainbow trout for 1 week and showed some protection against an intraperitoneal challenge with *Aeromonas salmonicida*. They also studied several of the characteristics of non-specific immunity listed in Section 19.1.6 and found that most were elevated in medicated fish but the haematocrits and leucocyte counts were virtually unchanged.

19.2.2 OTHER GLUCANS

(a) Other yeast glucans

There have been a number of reports of tests of "glucans" or "β-glucans" in various fish species including turbot, common carp and sea bream, but in many cases the substance used has not been adequately characterized. Glucans do not all have the same activity. Another yeast cell wall glucan, designated DL-glucan, has been compared with MacroGard and shown to be significantly inferior as an immuno-stimulant in Atlantic salmon although not entirely inactive (Robertsen *et al.* (1990)).

(b) Laminaran

Laminaran is another glucan described as β(1,6)-branched units of β(1.3)-dextrose; it is obtained from the brown alga, *Laminaria hyperborea*. It has shown immuno-stimulatory properties *in vitro* but was found to be inactive in common carp challenged with *Edwardsiella tarda*. Unlike most other glucans laminaran is soluble in water, and it is absorbed by fish from the intestine.

Tritium-labelled laminaran has been given intravenously to Atlantic salmon for liquid scintillation counting and whole-body auto-radiography. It was found that the highest concentration remained in the blood for some hours but that by day 2 the highest concentration was in the kidney and therafter also in the spleen and liver - organs rich in immuno-competent cells. Throughout the study there was a high concentration in the urine but only a trace in the bile, so the urine is probably the main route of excretion.

(c) Fungal glucans

A number of glucans extracted from different fungi have been tested for immuno-stimulatory activity in fish. Most have proved inactive, but significant activity has been found in:

- Schizophyllan extracted from *Schizophyllum commune*
- Scleroglucan extracted from *Sclerotium glucanicum*

Both are (1-6)branched (1-3)β glucans. They are known anti-tumour agents and immuno-stimulants in mammals.

Suspensions of each of these glucans in physiological saline have been shown to be active when injected intraperitoneally into common carp. Two doses at an interval of 2 days conferred significant protection against an artificial *Edwardsiella tarda* infection which was lethal to all unmedicated fish within 3 days. 2 mg/kg at each dose was adequate with schizophyllan but 5 mg/kg was necessary with scleroglucan.

The same two glucans have been tested in yellowtail of 38-50 g body weight. Two intraperitoneal injections in the range 2-10 mg/kg at an interval of 3 days, with challenge 3 days after the second dose, showed dose-related protection against strepto-coccosis. Enhancement of the activity of head kidney cells and, to a lesser extent, enhancement of lysozyme production were shown. However, the same regimen gave no protection against pasteurellosis. This was attributed to the ability of *Pasteurella piscicida* to replicate in phagocytes.

19.3 Other immuno-stimulants

19.3.1 PEPTIDES

(a) FK-565

FK-565 is a synthetic tetrapeptide produced from a lactoyl-tetrapeptide extract of *Streptomyces olivaceogriseus*. Unlike MacroGard, which acts on macrophages and so indirectly on T-lymphocytes, FK-565 stimulates T-lymphocytes directly. When given to rainbow trout at 1 mg/kg by intraperitoneal injection before challenge with *Aeromonas salmonicida* it enhanced the survival of treated fish; best results were from treatment only 1 day before challenge. Enhanced bactericidal activity of the phagocytes was shown.

FK-565 also shows adjuvant activity with bacterins. At 20-25 mg/kg it enhances the antibody response especially to low doses of bacterin. It appears to have no activity by immersion administration.

(b) KLP-602

KLP-602 is a dimer of lysozyme, a mediator of non-specific immunity, and is consider-ably less toxic than the natural monomer. Rainbow trout given intraperitoneal injections of 100 μg/kg once, twice or 3 times daily for 2 days were bled 7, 14, 21 or 28 days after treatment and the blood samples tested for various characteristics of immuno-stimulation (see Section 19.1.4). All characteristics were positive, the phagocytic and myelo-peroxidase results being higher in those fish given KLP-602 twice or three times daily than in those treated only once. An *in vivo* challenge with *Aeromonas salmonicida* showed that

KLP-602 gave short-term protection, and here also the twice daily injection regimen was more efficacious.

(c) ISK

ISK is a short chain polypeptide derived from fish by-products. It is produced in Russia where it is used for health and growth promotion in poultry, and it has been similarly used in common carp in Hungary.

Rainbow trout weighing 20-30 g given 1 μg ISK (33-50 mg/kg) by injection showed elevated phagocytic activity *post mortem*. In conjunction with O antigen from an *Aeromonas salmonicida* bacterin (which is itself an immuno-stimulant as well as an antigen) ISK enhances all non-specific immunity characteristics and in particular leucocyte counts. *In vivo* ISK delays mortality from *Aeromonas salmonicida* challenge, and combination with specific bacterin extends the delay.

19.3.2 LEVAMISOLE

(a) In rainbow trout

As well as being an anthelmintic widely used in ruminants, levamisole has immuno-stimulatory properties and was at one time used in sheep vaccines for its anthelmintic and adjuvant action. *In vitro* tests on rainbow trout spleen sections have shown low concentrations to be immuno-stimulant to both non-specific and specific immune responses; higher doses suppress specific responses, and very high doses suppress both types of response.

Kajita *et al.* (1990) showed that it has immuno-stimulant action *in vivo* in rainbow trout, the optimum dose by intraperitoneal injection being 0.5 mg/kg. On challenge with *Vibrio anguillarum* 5 days after dosing, the LD_{50} in treated fish was 50 times that in untreated fish. Other immuno-stimulation characteristics were also enhanced.

(b) In common carp

In common carp levamisole is immunostimulant both by injection and when given in feed. The optimum oral dose is in the range 3-8 mg/kg body weight daily. When given at 5 mg/kg/day every 3rd day for 15 days (6 doses), levamisole raised the leucocyte count, enhanced phagocytic activity and increased the level of serum lysozyme. A notable feature of these effects was that they lasted for about 3 months.

A dose titration using the intraperitoneal route showed that 5 and 10 mg/kg three times at 3 day intervals was immuno-stimulant but 15 and 20 mg/kg at this frequency were immuno-supressive.

(c) In sea bass

In sea bass levamisole is immuno-stimulant when administered by immersion. A concentration of 1 mg/l for 30 minutes developed the characteristics of immuno-stimulation and reduced mortality due to artificial infection with *Pasteurella piscicida*.

19.3.4 ASCOGEN

(a) Composition
Ascogen is an extract from a yeast which has been described as, "... a biologically active complex of metabolites of nucleotide metabolism ... Pyrimidines are the most important item among them." (Adamek, 1994)

(b) Mode of action
Commercial literature for Aquagen®, a brand of ascogen formulated and marketed for fish, claims that it, "improves the performance of vaccines but has a different mode of action from immuno-stimulants." Instead of stimulating leucocyte activity it enhances leucopoiesis by the provision of a ready source of pyrimidine and purine groups, which cannot be synthesized by these cells. This means that it can be used to advantage in combination with other immuno-stimulants or with vaccines.

Ascogen is on the market as a growth promoter for pigs and poultry in several countries including the UK and USA. In the USA it is claimed to lead to better growth and food conversion ratios (FCR) specifically as a result of "strengthening the immune system". As an extract from yeast it is not regarded as requiring registration as a feed additive.

(c) Feed medication rates
For fish recommended inclusion rates in feed vary between 1 and 5 kg/tonne. In common carp ascogen produces dose-related performance enhancement at up to 5 kg/tonne, although 2.5 kg/tonne depresses growth in goldfish. Adamek (1994) cited a finding of a positive effect of 5 kg/tonne in the European catfish (*Silurus glanis*) but found 2.5 kg/tonne to be optimum for rainbow trout, with 5 kg/tonne giving worse results than in unmedicated controls.

In mullet and silver carp ascogen at 2 kg/tonne improves growth and reduces mortality, and in tilapia it has been shown to improve growth and FCR with 2 kg/tonne being nearly as good as 5 kg/tonne. However the higher feed medication ıate gave a significantly higher antibody response to *Aeromonas hydrophila* vaccine given either by intramuscular injection or by immersion.

19.3.5 SODIUM ALGINATE

Fujiki *et al.* (1994) extracted two water-soluble fractions from the edible brown alga, *Undaria pinnatifida*. They showed that one fraction, which was soluble in strong acid, had a protective action against artificial *Edwardsiella tarda* challenge in the common carp. Chromatography showed the fraction to be an alginate, that is, a mixture of β-D-mannuronate and α-L-glucuronate, a chemically distinct group from the glucans.

In later work the same group studied the mode of action of the alginate and showed that it was different from that of the glucans. Sodium alginate did not activate carp complement (whereas scleroglucan does), and it has little effect on head kidney phagocytes. Peritoneal exudate phagocyte (PEP) counts were found to rise in medicated carp and the phagocytic activity of the PEP cells was also raised. It was concluded that sodium alginate acts primarily by enhancing the chemotactic migration of head kidney phagocytes to the peritoneal cavity. It does so by stimulating resident peritoneal cells to secrete chemotactic factors.

Further reading

Adamek, Z. (1994) Effect of ascogen probiotics supplementation on the growth rate of rainbow trout (*Oncorhynchus mykiss*) under conditions of intensive culture. *Zivocisna Vyroba* 247-253.

Anderson, D.P. (1992) Immunostimulants, adjuvants and vaccine carriers in fish: Applications in aquaculture. *Annual Review of Fish Diseases* **2**, 281

Anderson, D.P., van Muiswinkel, W.B. and Roberson, B.S. (1984) Effects of chemically induced immune modulation on infectious diseases of fish. In *Chemical Regulation of Immunity in Veterinary Medicine* (*eds. Kende, Gainer and Chirigos*). Alan R. Liss Inc., New York, pp. 187-211.

Anderson, D.P. and Siwicki, A.K. (1994) Duration of protection against *Aeromonas salmonicida* in brook trout immunostimulated with glucan or chitosan by injection or immersion. *The Progressive Fish-Culturist* **56**, 258-261.

Blazer, V.S. (1991) Piscine macrophage function and nutritional influences: a review. *Journal of Aquatic Animal Health* **3**, 77-86.

Fujiki, K., Matsuyama, H. and Yano, T. (1994) Protective effects of sodium alginates against bacterial infection in common carp, *Cyprinus carpio*, L. *Journal of Fish Diseases* **17**, 349-355.

Jeney, G. and Anderson, D.P. (1993) An *in vitro* technique for surveying immunostimulants in fish. *Aquaculture* **112**, 283-287.

Jørgensen, J.B., Sharp, G.J.E., Secombes, J. and Robertsen, B. (1993) Effect of a yeast-cell-wall glucan on the bactericidal activity of rainbow trout macrophages. *Fish and Shellfish Immunology* **3**, 267-277.

Kajita, Y., Sakai, M., Atsuta, S. and Kobayashi, M. (1990) The immunomodulatory effects of levamisole on rainbow trout, *Oncorhynchus mykiss*. *Fish Pathology* **25**, 93-98.

Robertsen, B., Rørstad, G., Engstad, R. and Raa, J. (1990) Enhancement of non-specific disease resistance in Atlantic salmon, *Salmo salar* L., by a glucan from *Saccharomyces cerevisiae* cell walls. *Journal of Fish Diseases* **13**, 391-400.

Rørstad, G., Aasjord, P.M. and Robertsen, B. (1993) Adjuvant effect of a yeast glucan in vaccines against furunculosis in Atlantic salmon (*Salmo salar* L.). *Fish and Shellfish Immunology* **3**, 179-190.

Siwicki, A.K., Anderson, D.P. and Rumsey, G.L. (1994) Dietary intake of immunostimulants by rainbow trout affects non-specific immunity and protection against furunculosis. *Veterinary Immunology and Immunopathology* **41**, 125-139.

20. VACCINES

20.1 Evaluation

20.1.1 DEFINITIONS

(a) Specific immunity

In contrast to the substances covered in the previous chapter, vaccines are medicines used to stimulate specific immunity, that is, immunity to one or a very limited number of pathogens. Specific immunity is thus limited in the range of protection given; and its development is often at a temperature-dependent rate which means that it is slow in all fish and particularly so in those inhabiting cold waters. However, it is more longlasting than non-specific immunity; and once an organism has developed a specific immunity, even though that immunity has "disappeared" it is remembered and can be redeveloped ("recalled") in a very much shorter time than the initial development.

(b) Humoral and cell-mediated immunity

Specific immunity is manifested in two ways: humoral and cell-mediated. Either type or both may be developed in response to different stimuli. Furthermore either type, both or neither may be protective.

Humoral immunity is that developed in a "humour", or body fluid, which is normally plasma but occasionally mucus. The immunity takes the form of antibodies which are high molecular weight proteins known as immuno-globulins. Immuno-globulins as a group can be measured by electophoresis; those (there is often more than one) developed in response to a particular vaccine or infection can be titrated by standard serological means.

Cell-mediated immunity is manifested as leucocytes, which may or may not be in the blood, which have developed activity against specific pathogens.

(c) Active and passive immunity

The specific immunity developed by a vaccine or by recovery from an infection is known as active. This means that the development of the immunity occurs in the individual vaccinated or recovering.

In birds and mammals there is a physiological transfer of antibodies from the mother to the embryo or neonate. These antibodies confer short-term protection on the neonate, but because they were not produced by the neonate the protection is called passive immunity. The process is routinely exploited commercially, particularly in sheep and pigs, by vaccinating the pregnant mother against diseases (including ones to which she may not, because of her age, be susceptible). It is thus possible to ensure that the neonates will be fully protected.

Passive immunization against ichthyophthiriasis is known to occur in mouthbrooding tilapia. It is thought that the larvae ingest antibodies in mucus secreted by the buccal mucosa of the parent. It is also known that fish of many species which have recovered from ichthyophthiriasis are able to facilitate the recovery of co-habiting fish from the infection. This is thought to be due to the exudation of antibody in skin mucus.

However, these instances of transfer of passive immunity in fish are rare, and they are not exploited with vaccines. In the context of fish, "immunity" normally means active immunity.

(d) Antigens and adjuvants

An antigen is a pathogen, or a substance or organism resembling a pathogen, which stimulates specific immunity to itself. By definition a vaccine always contains one or more antigens. The antigen may, particularly if it is living, be the only pharmacologically active component in the vaccine.

Adjuvants are substances which potentiate antigens and which are included in vaccines for this purpose. They are by definition non-antigenic, that is, they do not by themselves stimulate specific immunity. Nevertheless, as noted in the previous chapter, some adjuvants are immuno-stimulants and so do stimulate non-specific immunity. Being non-specific in their action most adjuvants can be used in vaccines with a range of different antigens.

(e) Types of vaccine

In some vaccines the antigen is living and able to replicate itself in the vaccinated animal. For such an antigen to be safe it must either be attenuated, that is naturally or artificially rendered innocuous, or it must be heterotypic, that is of a different species from the pathogen. The archetypal vaccine, cowpox virus, *vaccinia*, to protect against small pox, was heterotypic. There are at present no living attenuated bacterial vaccines for fish and they are rare although not unknown in mammalian medicine. However there are well established methods of attenuating viruses and attempts are being made to apply these to fish pathogens. The advantage of this type of vaccine is that the dose need be only sufficient to ensure that the antigen establishes itself in the animal; it will then replicate until destroyed by the immunity it has itself engendered.

Live vaccines are effective stimulators of cell-mediated immunity and so are particularly useful for intra-cellular pathogens such as viruses.

The commoner type of vaccine is one in which the pathogen is rendered incapable of replicating. In the case of bacteria they are said to be "killed", and in the case of viruses "inactivated". The process is achieved by heat or by chemicals, notably formalin or β-propio-lactone (BPL). Killed bacterial vaccines are often called bacterins, although strictly this term should be restricted to those containing only bacterial cell bodies. A vaccine consisting of a killed whole bacterial culture, where the antigenicity of the extra-cellular products as well as the cells is exploited, is strictly an anaculture. Inactivated or killed vaccines must be given in larger doses than living attenuated ones, and a two dose regimen often gives significantly better results than a single dose; thus they are expensive to use. However, they can often be produced with little developmental work.

A disadvantage of inactivated or killed vaccines is that the heat or chemical processing may be critical: too little leaves the vaccine pathogenic, too much renders it non-antigenic. Attempts are being made to obviate this problem by identifying the protective antigens on a pathogen, extracting these and using them alone. This makes for a safer and more reliable product but the extraction process adds to the cost.

20.1.2 EFFICACY PARAMETERS

(a) Efficacy testing problems

Vaccines are used to protect individual animals against specific natural infections, and the ultimate test of efficacy is their ability to do this. However, there are practical problems in applying this test to fish vaccines:

- In fish immunity is never absolute. Vaccine efficacy must be measured.
- There is no such thing as a consistently reproducible *natural* infection.
- Artificial infection may be reproducible but its use does not measure protection against natural infection.

In relation to the last point it should be noted that natural infection in a group of fish is normally acquired from the environment or from introduced individuals. As the infection spreads the infection pressure increases, but immunity, both non-specific and specific, develops in parallel with this. Artificial infection may be by the introduction of infected individuals ("co-habitation challenge"), but other methods tend to give the highest challenge at the beginning. Nordmo and Ramstad (1997) studied four methods of challenging Atlantic salmon vaccinated against furunculosis, *viz.* intraperitoneal inoculation, co-habitation, immersion and intramuscular inoculation. They found that the order of the degree of challenge varies with the adjuvant used in the vaccine and with the time interval between vaccination and challenge.

Table 20.1. Effect of water temperature on the onset of immunity of sockeye salmon (3-3 g) following immersion vaccination with vibrio bacterin on rainbow trout (4-5 g) vaccinated with ERM bacterin. Bacterin was diluted 1:10 with water and the fish challenged at intervals following vaccination using a bath challenge. After Johnson K.A. *et al*, 1982, with permission

	10°C			18°C			Control		
Challenge day	*n**	sl†	%	*n*	sl	%	*n*	sl	%
V. anguillarum									
5	30	13	43	30	1	3	60	34	57
10	28	2	7	30	0	0	61	31	51
15	30	0	0	30	0	0	58	30	52
20	30	1	3	30	0	0	60	36	60
120	19	2	11	40	0	0	40	30	75
Y. ruckeri									
7	44	13	30	46	6	13	47	38	81
14	59	14	24	40	7	18	58	35	60
21	50	4	8	48	3	6	47	37	79
28	59	4	7	56	4	7	58	46	79

* n = Number of fish.
† sl = Specific loss.

A further problem in evaluating the efficacy of a fish vaccine is that the rate of development of immunity varies with temperature. Johnson *et al.* (1982) studied the immersion vaccination of fry of two species of salmonids and showed earlier protection at a higher temperature (see Table 20.1). However, Lillehaug *et al.* (1993) studying the use of *Vibrio salmonicida* (Hitra disease) vaccine in Atlantic salmon found that while the rate of production of antibody was slower at low temperatures the final titre was unaffected. The authors discussed their findings in relation to the practical use of the vaccine since the immunity is needed in the early spring when the water temperature is low. For some diseases rate of onset of immunity as well as ultimate titre is a relevant parameter of vaccine potency.

(b) Factors affecting responses to vaccines

The following are among the factors affecting the responses of fish to vaccines. They should be noted in any report on vaccine efficacy and taken into account in routine use of vaccines.

- The age and size of the fish. Some salmonid species cannot develop any specific immune response until they reach 1.5 g liveweight, weight apparently being more important than age. Common carp appear to develop an immunological tolerance to antigens given orally before they are 8 weeks old.
- Pre-exising disease (see Subsection (c) below)
- For injected vaccines not only the dose but also the route of administration and any anaesthetic used;
- For immersion vaccines, the concentration, exposure time and use of any hyperosmotic infiltration procedure;
- For oral vaccines the method of incorporation in feed;
- The water temperature - which should be monitored throughout the interval between vaccination and sampling;
- For euryhaline species the salinity;
- The nutritional status of the fish (see Blazer, 1991);
- Any stressors occurring during the interval between vaccination and sampling - which will depress immune responses.

(c) Effects of pre-exising disease on responses to vaccines

The effects a pre-existing disease may have on the response to a vaccine depend on the type of disease. Some diseases may affect either the uptake of vaccine or the immunological response to it.

Fish suffering from some diseases will not feed and so will not spontaneously ingest any orally administered vaccine. For immersion vaccines gills are the route of uptake, so any gill pathology, whether due to water pollution, microbial infection or metazoan infestation, will reduce the uptake. The immune system is located mainly in the blood, spleen and head kidney and diseases affecting these tissues and organs will suppress specific immune responses to vaccines (and they will also suppress non-specific immune responses). Examples are furunculosis (which is a septicaemia), infectious haemopoietic necrosis and proliferative kidney disease.

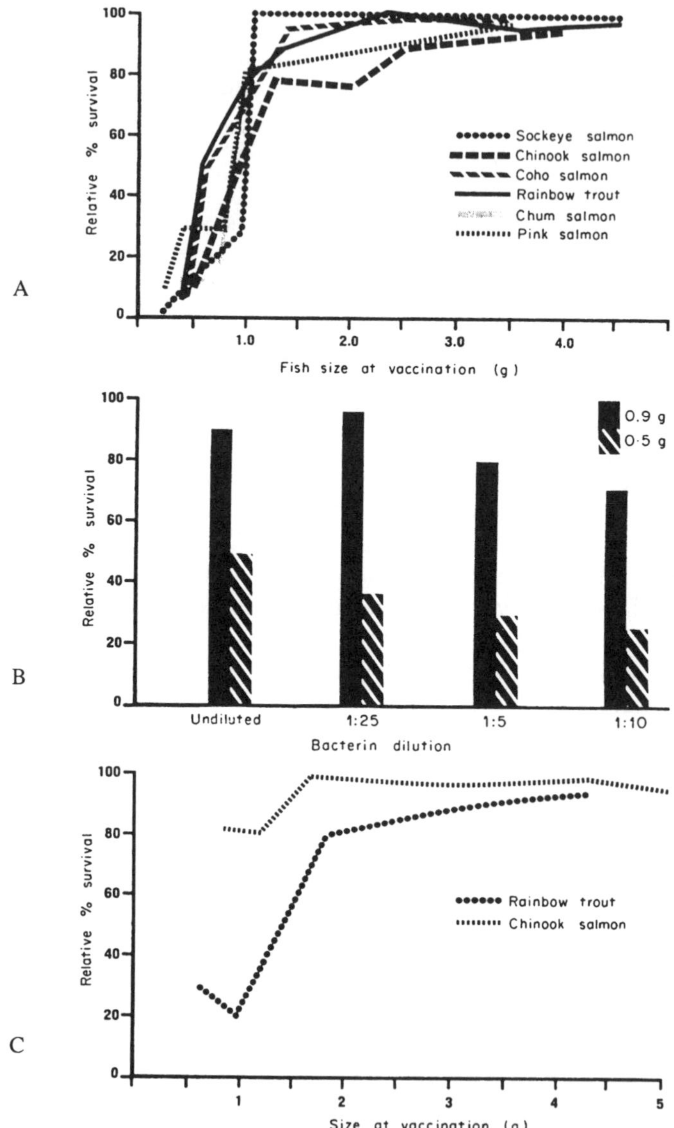

Figure 20.1. A. Onset of protective immunity in salmonods vaccinated by immersion with vibrio bacterin diluted 1:10. Fish were vaccinated at various sizes and were challenged with virulent *V. anguillarum* 30-90 days after vaccination.

B. Effect of various concentrations of vibrio bacterin on the protective immunity of chum salmon at 0.5 and 0.9 g. All fish were challenged about 30 days after vaccination. Data is expressed as survival relative to unvaccinated controls. Control mortality was 76%.

C. Onset of immunity in rainbow trout and chinook salmon vaccinated by immersion in *Yersinia ruckeri* bacterin at various sizes and challenged with virulent *Y. ruckeri* about 45 days after vaccination. After Johnson K.A. *et al*, 1982, with permission.

(d) In vitro measurement of antibody

All standard methods of titrating serum antibody are valid in fish. There can often be a wide variation in response between individual fish; so a large number should be sampled, and the geometric mean titre would be more appropriate than the arithmetic mean. It must be borne in mind that in fish serological responses rarely correlate with protection.

(e) Relative percent survival

It was noted in Section (a) above that the ultimate test of the vaccinal protection of a group of fish against a disease is to challenge them with that disease; and that a measurable degree of protection rather than absolute protection is to be expected. The measure commonly adopted is the Relative Percent Survival (RPS); it is derived from observations of mortality in groups of vaccinated and control (unvaccinated) fish given the same challenge.

$$\text{RPS} = \left(1 - \frac{(\%\ \text{mortality in the vaccinated group})}{(\%\ \text{mortality in the control group})}\right) \times 100\%$$

Lillehaug *et al.* (1993) have pointed out that this formula is inapplicable if the challenge infection pressure is so high that all the unvaccinated control fish die. For such an eventuality they have proposed an RPS_{60} which is the RPS on the day the cumulative mortality in the unvaccinated fish reaches 60%. The proponents point out that this parameter "... takes into account both the level of mortality and time to death."

(f) Other in vivo measures of immunity

For microbial pathogens which produce rapidly lethal infections, the ratio of the LD_{50} in vaccinated animals to the LD_{50} in unvaccinated ones is a valid measure of the protective immunity resulting from the vaccination. This measure, sometimes called the Protection Index, is more widely applicable in fish than in birds and mammals because, as noted above, immunity in fish is rarely absolute and so it is possible to determine an LD_{50} in vaccinates.

In addition to proposing the RPS_{60} for challenges where all the controls die, Lillehaug *et al.* (1993), proposed a mean survival time measure which would be applicable where all the vaccinates die as well. If a vaccine had any activity at all it would at least increase the time elapsing between challenge and the death of a fish. This period is called the survival time and the measure of vaccine potency is the increase of mean survival time of vaccinates over the mean survival time for controls. The authors admit that, "In experiments in which some individuals survive, this parameter is only an estimate." In their study of the effect of temperature on Hitra disease vaccination these authors found that on challenge after 2, 4 or 6 months the mean survival time was longer in Atlantic salmon vaccinated at low temperatures.

20.1.3 ECONOMIC EVALUATION

For a fish farmer the decision as to whether to use a vaccine is obviously influenced by economics - the savings to be expected as a result of use must be greater than the cost of administration.

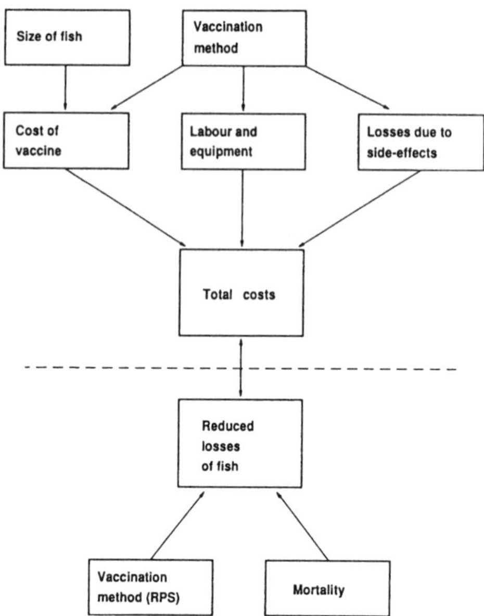

Figure 20.2. Economic model for vaccination of fish, illustrating which factors influence the costs and possible savings associated with vaccination. After Lillehaug A. *et al.* (1989), with permission.

Lillehaug (1989) enumerated the factors to be taken into account, one of which was the expected efficacy of the vaccine for which he used the RPS. Costs are the sum of:
1. Labour
2. Vaccine
3. Others, *e.g.* administration equipment, anaesthetics

The expected savings are the product of:
1. The value of a fish (*i.e.* the gross profit on it at harvest)
2. The expected reduction in fish mortality = (expected mortality in unvaccinates) x RPS

A saving which was not taken into account is in the cost of disposal of dead fish. Such fish are infected irrespective of whether they have been vaccinated; they must be disposed of in a way that does not pose a hazard to survivors.

20.2 Routes of administration

20.2.1 ORAL VACCINATION

(a) Disadvantages

For vaccines as much as for other medicaments the oral route is an easy and cheap method of administration. Furthermore, some marine species such as sea bass and sea bream are very sensitive to any form of handling, including netting, and this precludes injection or even immersion vaccination. Nevertheless the efficacy of oral vaccination in proportion to the antigenic mass used is low and it varies according to the nature of the antigen and the vaccine formulation. This is mainly due to the sensitivity of vaccines to heat, meaning that they cannot be incorporated into feed before pelleting. Surface-coating is therefore essential, and this requires the use of a large quantity of vaccine to ensure that each fish receives the minimum effective dose.

It has been found that anal intubation is immunologically more efficacious than oral administration for a wide range of antigens in many species of fish (but not all, *e.g.* tilapia). There are two plausible explanations for this, the location of the relevant immuno-competent tissues and the effect of digestive secretions on antigens. Either explanation or both may be applicable depending on the antigen. There are no Peyer's patches in fish, but there is gut-associated lymphoid tissue (GALT); this is thought to be an important part of the secretory immune system, and it is concentrated in the hindgut. *Vibrio anguillarum* bacterin given orally or by anal intubation to a variety of fish species can be detected in the hindgut mucosa for up to 3 weeks but it is not detectable in the head kidney. Orally administered antigens are subjected to the digestive secretions of which two components may be damaging: stomach acid can denature protein antigens and proteolytic enzymes will hydrolyse them. It is known that absorption of biologically active proteins is greater in species without stomachs such as cyprinids. In vaccines known to be active orally the protective antigens are probably lipo-polysaccharides (LPS).

(b) Protection of orally administered antigens

Efficacious as anal intubation may be, it is not a practical method of mass administration of vaccines to fish. It merely shows that if intact antigens can be delivered to the hindgut immunity will be stimulated.

A number of different methods of protecting orally administered antigens from digestive secretions have been reviewed by Ellis (1995). He pointed out that results from oral intubation cannot be relied upon as evidence of the effects of in-feed administration; it may allow the rapid passage of the antigen through the stomach and limit its exposure to digestive secretions. He also noted that at least one protective technique proved immunologically counter-productive, probably because it was too effective and prevented absorption of the antigen.

This latter problem was overcome by Piganelli *et al.* (1994) who tested enteric coated antigen microspheres (ECAMs) where the coating consisted of a polymer which dissolved at high pH, that is, in the intestine. Essentially the vaccine formulation consisted of dextrose beads onto which the antigen was sprayed. This was followed by spraying an aqueous latex dispersion of methacrylic-acrylic co-polymer. ECAMs containing 10 μg antigen per pellet when given in feed for 30 days to coho salmon conferred as good protection against *Vibrio anguillarum* as an intraperitoneally injected vaccine.

20.2.2 IMMERSION VACCINATION

(a) Portals of entry

The portal of entry of immersion vaccines is somewhat controversial. It was originally assumed to be the lateral line and later the gills. The intestines would be expected to be a major portal of entry for immersion vaccines in marine fish, but Robohm and Koch (1995) have suggested they should also be considered as a possible portal in freshwater species. Using botulinum toxin as a marker they showed that the goldfish drinks sufficient water for an adequate dose of soluble antigen to be ingested during immersion vaccination. (They did not show that antigen arrived at immuno-competent tissue intact, nor did they explain why immersion vaccination is normally more efficacious than oral administration). It is probable that the portals vary between particulate and soluble antigens and possibly between hyperosmotic infiltration and direct immersion. It should also be noted that the hyperosmotic infiltration technique was developed for proteins; antigens are unlikely to be proteins unless they are on the surface of particles.

Ototake *et al.* (1996) studied the disposition of radio-labelled bovine serum albumin in rainbow trout after hyperosmotic infiltration and after direct immersion for 3 minutes. Their findings suggested that for a soluble protein antigen the skin was of primary importance in both methods of administration, with T_{max} in trunk skin at about 10 minutes. The gills were the next most important portal, and the radio-activity appeared to have reached the stomach and intestines through the blood. The concentration in the lateral line skin was not significantly different from the trunk skin.

(b) Local immunity

The transference of passive immunity (see Section 20.1.1(c)) in skin mucus is evidence of the presence of immuno-competent cells in the skin, and formation of antibodies is also believed to occur in gill mucus. The stimulation of these cells is most likely to result from direct contact with antigen such as will occur in immersion vaccination, and the immunity so developed will protect by preventing the ingress of the pathogen. Thus in immersion vaccination serum antibody and activation of circulating leucocytes are not necessarily the most important criteria of immunity.

(c) Adjuvants

Surface-active adjuvants such as alum will precipitate, and hence concentrate, antigens. This is a valuable action but arguably not an adjuvant action in immersion vaccination because it is not directed at the immuno-competent tissues of the vaccinated animal. However the compounds also have true adjuvant action in that they make the antigens more hydrophobic and absorbable onto mucous surfaces.

Levamisole is variously reported as valuable and useless as an adjuvant for immersion vaccines!

(d) Factors affecting uptake

For immersion vaccines both concentration and exposure time are relevant to the dosage; but Tatner and Horne (1983) working with radio-labelled *Vibrio anguillarum* bacterin (or, more strictly, anaculture) in rainbow trout have shown that neither of these factors is directly proportional to the antigenic mass of vaccine taken up. Furthermore other factors are also relevant. In particular, for a given dosage uptake correlates with temperature in the range 5-18°C, implying that uptake is an active, pharmacokinetic process rather than a passive diffusion.

- Time — 5 seconds exposure gave a lower uptake than longer periods but extending the time beyond 10 seconds made no significant difference.
- Anatomy — There is significantly greater uptake at the head end of the fish, which includes the gills, than in the rest, which includes the lateral line.

- Fish size — There is a positive correlation between fish size and uptake.
- Particle size — The soluble fraction of the vaccine is taken up to a significantly greater extent than the bacterial cells.
- Adjuvant — The inclusion of potash alum significantly increases uptake.

(e) Spray vaccination

There is an upper limit to the size of fish which can in practice be vaccinated by immersion. Gould *et al.* (1978) described a technique of mixing bacterin into an inert carrier and including a bentonite adjuvant, and spraying this onto large fish as a slurry. High pressure was not needed. Concentrations of bacterin down to 1 mg/g gave satisfactory immunity which was longer lasting than following oral vaccination.

20.2.3 INJECTION VACCINATION

(a) Advantages and disadvantages

With a majority of antigens injection vaccination stimulates a better immune response than oral or immersion administration; and it is possible to produce efficacious multivalent vaccines. For fish weighing over 10 g injection reduces the vaccine requirement because the dose is normally independent of the size of the fish, whereas with oral and immersion vaccines it is proportional to body weight. However the technique has disadvantages, the main one being its cost. The vaccine must be sterile (or uncontaminated in the case of a live vaccine); and the administration costs are high, either in terms of labour or capital expenditure on equipment. Without automatic injection equipment the fish must be anaesthetized; and operator safety is an important consideration.

The technique cannot be applied to fish weighing less than about 15 g. Where the intraperitoneal route is used the adjuvant may cause peritoneal adhesions.

(b) Injection technique

Monovalent injectable vaccines are normally formulated to be active in a 0.1 ml dose; multivalent vaccines usually have a 0.2 ml dose. Where oil adjuvants are used the dose is 0.2 ml because the vaccine is viscous and using the larger volume reduces the error in dosage.

The normal route is intraperitoneal, for which the needle is inserted in the mid-ventral line just cranial to the vent and directed cranially. Recently a range of vaccines has been produced with a recommendation for injection into the dorso-median sinus (see Section 1.4.2(c)). This route is claimed to obviate the development of peritoneal adhesions and so to be of particular value for broodfish.

(c) Adjuvants

By definition an adjuvant potentiates the action of an antigen on cells of the specific immune system, but there are different methods of achieving this. Some immunostimulant compounds will also act as adjuvants and it may be assumed that they enhance the reactivity of immuno-competent cells to antigens.

A commoner type of adjuvant is an irritant which acts by attracting leucocytes to itself and hence to the antigen; the leucocytes phagocytose the adjuvant and the antigen which is adsorbed onto it. Freund's Complete and Incomplete Adjuvants (FCA and FIA) are extreme examples which are very popular with research workers; however they are so irritant that on grounds of target species safety a vaccine containing either of them would not be granted a market authorization. Demonstration of immunity with the use of one of these adjuvants is therefore of little practical significance. Aluminium hydroxide, which may be added as such or be precipitated from alum by the vaccine, is the commonest irritant adjuvant in commercial vaccines; the saponin, Quil A, which is a more severe irritant is also occasionally used. The use of an adjuvant of this type may actually delay the arrival of antigen at the main organs of specific immunity; but this does not necessarily mean a delay in the immune stimulus - the leucocytes conveying the vaccine to the organs are themselves immuno-competent.

When used with bacterins injected intraperitoneally into rainbow trout or brook trout, alum enhances the immunogenicity of some bacterins including *Aeromonas salmonicida* and *Vibrio salmonicida* but not *Vibrio anguillarum.* Alum has side-effects including retardation of growth, peritoneal adhesions and abnormal swim-bladder development. It should be noted that some bacterins produce peritoneal adhesions, which may be worse than those caused by alum alone. The combination of antigen and alum is always worst. Alum appears to react with bacterial toxins and overall may be said to be beneficial with toxic bacterins.

Another way in which adjuvants can act is by rendering adsorbed antigens more lipophilic and hence more easily absorbed across biological membranes. The most obvious examples of this are the mineral oils which are increasingly being used. They are irritant, but it has been shown in mammals that they are absorbed from the peritoneal cavity so rapidly that adhesions do not form. It may be assumed that a similar position obtains in fish. It is also claimed that once these oils are absorbed they form depots from which the antigen is released slowly over a long period.

20.2.4 COMPARATIVE PHARMACOKINETICS

(a) The immune system

The potency of a vaccine depends in large measure on the efficacy with which it is transported to immuno-competent cells. It has been noted that such cells exist in the mucosae of the hindgut (GALT) and gills. Fish have no bone marrow or lymph nodes and the important immunological organs are the thymus, head kidney and spleen.

(b) Oral vaccination

In fish species which have been studied in detail, antigens in the lumen of the intestine are known to be absorbed into the mucosa, especially of the hind gut where the GALT is, but there is little evidence of their being transported to other organs. An exception is the A-layer protein of *Aeromonas salmonicida* which is found in Atlantic salmon in the spleen for up to 16 weeks and in the head kidney for up to 6 weeks, but this is only in small quantities in either case.

(c) Immersion vaccination

Immediately after immersion administration vaccine can be found in the mucus under the opercula and on the gills but apparently not on the skin. Its presence in the hindgut mucosa is controversial. Antigens given by immersion do get into the bloodstream, in contrast to those given orally; but reports suggest that their further disposition varies according to their chemical nature (protein or LPS) and the fish species. Protein antigens are found in the head kidney as little as 3 hours after immersion vaccination; that of *V. anguillarum* is said to disappear by 24 hours from rainbow trout head kidney whereas the A-layer protein of *A. salmonicida* is said to persist for up to 6 weeks in Atlantic salmon. The LPS of *A. salmonicida* does not get into the spleen of Atlantic salmon whereas the A-layer protein is found there for at least 8 weeks. Neither type of antigen is found in the liver.

(d) Intraperitoneal injection

Following intraperitoneal injection of vaccines antigens are absorbed into the bloodstream very rapidly. In the rainbow trout they can be found in all tissues except the central nervous system within 1 minute. They accumulate in the kidney and spleen. LPS is also found temporarily (up to 2 weeks) in the gut and liver but the A-layer protein is not.

20.3 Available vaccines

Vaccine development is probably the most rapidly advancing aspect of fish medication, and this makes any review of available vaccines subject to obsolescence. The microbiology of antigens is outside the scope of this book but the pharmacokinetics of vaccine formulations is relevant. Recent advances have been particularly in the field of bacterial vaccines where there has been a commercial demand due to the development of bacterial resistance to authorized antibacterial drugs and the questionable economics of obtaining market authorization for new ones. The commercial availability of oil adjuvant furunculosis vaccine for Atlantic salmon has led to a dramatic fall in the use of antibacterial drugs by that sector of the aquaculture industry.

20.3.1 VIRAL VACCINES

(a) Viral haemorrhagic septicaemia (VHS)

VHS is a notifiable disease subject to legal controls in many countries. There is a strong protective immunity in survivors of the disease but vaccines are not normally used because immune individuals are often carriers of the infection.

An attenuated live vaccine has been developed, but the attenuation mechanism was not fully understood; and since the vaccine showed residual virulence it could not be accorded a market authorization. Vaccines can be produced by growing the viruses in tissue culture and inactivating with either formalin or β-propio-lactone (BPL), but they are digested if given orally and so need to be given by injection. Since the highest mortality is in fish under 6 months of age, which cannot be given injections, vaccines of this type have not so far proved economic.

(b) *Infectious pancreatic necrosis (IPN)*

IPN used to be considered a disease of first-feeding trout fry. Administration of any medication to fish of this age is difficult, and because of the time specific immunity takes to develop, vaccination of fry is impossible. However, recently the disease has become of increasing importance in Atlantic salmon post-smolts and an injectable recombinant vaccine has been developed for use in these fish. It is being included in a tetravalent injectable product together with *Aeromonas salmonicida*, *Vibrio anguillarum* and *V. salmonicida* bacterins.

(c) *Infectious haemopoietic necrosis (IHN)*

IHN is a disease of very young salmonids and, for the same reasons as for IPN, vaccination is impossible. In addition, because the virus affects the haemopoietic tissue it is immuno-suppressive.

(d) *Spring viraemia of carp (SVC)*

Survivors of SVC show a substantial but not absolute immunity to reinfection. As its name implies the disease occurs at low water temperatures and it has been shown that fish infected by cohabitation in water at a high temperature (21.5°C) had low morbidity and mortality but developed immunity. However, deliberate infection with virulent virus cannot be called vaccination and is hazardous for untreated fish in the effluent watercourse.

Experiments have been conducted with both live attenuated and inactivated vaccines but in both cases they were effective only by injection. Vaccination immediately before an expected outbreak is impossible because the low temperature of the water means that immunity would develop too slowly. Vaccines are not widely used, but where they are the recommended regimen is intraperitoneal injection in late summer or autumn with the aim of inducing an immunity which will last about 9 months.

(e) *Channel catfish virus (CCV)*

CCV affects only young catfish at high temperatures. Inactivated vaccines appear to have no action but it has been shown that an effective attenuated vaccine can be produced by passage of the virus in walking catfish (*Clarias batrachus*) cell culture. Improved results have been reported from administering this vaccine by a two step hyperosmotic infiltration procedure.

Awad *et al.* (1989) developed a sub-unit vaccine for immersion administration by separating the envelope from the rest of the virus using a surfactant. They tested this by four separate dose regimens (apart from keeping untreated control fish), *viz.*

1. Immersion of eyed eggs;
2. Immersion of eyed eggs, followed 2 weeks later when they had hatched by immersion of the fry;
3. Immersion of the fry once;
4. Immersion of the fry twice at 2 weeks interval.

The survival rates following artificial infection were:

Regimen 1	30.5%		Regimen 3	82.0%
2	80.7%		4	88.8%
Controls	Nil		Controls	44%

To vaccinate a poikilothermic animal a second time after an interval of only 2 weeks seems unjustifiable. Even in homoiotherms the recommended interval is not less than 4 weeks and preferably 6 weeks. The authors admitted that the survival rates in regimens 3 and 4 were not significantly different; they failed to note that the survival rates in regimens 2 and 3 were even closer. They described the second doses in regimens 2 and 4 as "boosters" but there was no evidence of immunity having been boosted. Nevertheless single immersions given to fry did stimulate significant immunity.

20.3.2 BACTERIAL VACCINES

(a) Furunculosis

Furunculosis was the first fish disease for which any form of artificial immunization was attempted, and it continues to attract a very substantial research effort. Midtlyng *et al.* (1992) made a comparison of nine commercially available vaccines, although three of them were multivalent products in which the furunculosis component was the same as one of the six monovalent products. All were formalin-killed whole cultures (anacultures); and all contained an adjuvant of mineral oil, aluminium compounds, glucan or levamisole. All were injectable and were given as a single intraperitoneal injection. At 6 weeks after vaccination the only products to have induced significantly better immunity than the controls were two with glucan adjuvant, one with levamisole and a trivalent vaccine with mineral oil adjuvant. At 3 and again at 6 months only the monovalent and trivalent vaccines with mineral oil adjuvant were significantly better than controls. It is in fact these last vaccines which have stood the test of time in field use.

In this trial an initial reduced appetite was noted in all vaccinated fish but this lasted for only about 3 weeks. The mineral oil adjuvant did cause temporary lesions at the injection site but peritoneal adhesions were not reported. It is now being recommended that this vaccine should be injected into the dorso- median sinus.

Inglis *et al.* (1996) studied the protection of fish known to be infected with furunculosis and stressed by transport. On the day before transport groups were given vaccine, amoxycillin or both. They were able to show that in the short term (the time for which the stress was relevant) the amoxycillin protected but the vaccine did not. In the longer term the vaccine stimulated immunity and the simultaneous use of amoxycillin did not compromise this.

(b) Carp erythrodermatitis

This disease is caused by the same bacterium, *Aeromonas salmonicida*, as furunculosis. Some of the older furunculosis vaccines formulated for immersion administration conferred a measure of protection to carp, but it is questionable whether they will remain on the market. The oil adjuvant vaccine should not be given by intraperitoneal injection to carp but it might be satisfactory if injected into the dorso- median sinus.

(c) Cold water vibriosis (Hitra disease)

Vibrio salmonicida is contributory to the aetiology of this disease if not the sole cause. An injectable vaccine with a mineral oil adjuvant gives satisfactory protection, and a combined product with furunculosis vaccine is available. Hitra disease occurs only in marine fish and since transfer to the sea is stressful vaccination should be a few weeks before transfer.

(d) Vibriosis

Vibrio anguillarum bacterins can be administered by injection, immersion or bathing. Injection is the most efficacious method but is labour-intensive and stressful to the fish. It may be the economic method if it is combined with another vaccination which has to be by injection, for example furunculosis. For immersion a 1:10 dilution of the injectable product is used and an exposure for 60 secs. is adequate. Bathing is the least efficacious method of administering vaccine but it may be the only feasible method for fish already in sea cages. A 1:1000 dilution is used for 1 hour and oxygenation of the water before, during and after treatment is essential.

Since vibriosis occurs in seawater Atlantic salmon are vaccinated before transference to the sea. A water temperature of 8°C or higher is regarded as necessary for an immune response to be made, so they are usually vaccinated in the autumn because smoltification occurs in the spring before the water has warmed sufficiently.

(e) Enteric redmouth disease (ERM)

ERM occurs in freshwater and so it is desirable to vaccinate susceptible species as early in life as possible. This precludes injection although this route gives the greater immunity. *Yersinia ruckeri* bacterins for immersion administration are available and there is now a vaccine specifically formulated for spray administration. The concentration for immersion vaccination is in the range 1:100 to 1:1000 of the injectable product. The total requirement is normally proportional to body weight so there is an economic argument for early vaccination. By contrast injected dose volumes are normally constant and irrespective of body weight so for fish not vaccinated as juveniles there may be a time when injection becomes economic.

Further reading

Anderson, D.P. (1992) Immunostimulants, adjuvants and vaccine carriers in fish: Applications to aquaculture. *Annual Review of Fish Diseases* **2**, 281-307.

Awad, M.A., Nusbaum, K.E. and Brady, Y.J. (1989) Preliminary studies of a newly developed subunit vaccine against channel catfish virus. *Journal of Aquatic Animal Health* **1**, 233-237.

Blazer, V.S. (1991) Piscine macrophage function and nutritional influences: a review. *Journal of Aquatic Animal Health* **3(2)**, 77-86.

Ellis, A.E. (1995) Recent developments in oral vaccine delivery systems. *Fish Pathology* **30**, 293-300.

Gould, R.W., O'Leary, P.J., Garrison, R.L., Rohovec, J.S. and Fryer, J.L. (1978) Spray vaccination: a method for the immunization of fish. *Fish Pathology* **13**, 63-68.

Gudding, R., Lillehaug, A., Midtlyng, P.J. and Brown, F. (eds.) (1997) Fish Vaccinology. *Developments in Biological Standardization* Vol. 90.

Inglis, V., Robertson, D., Miller, K., Thompson, K.D. and Richards, R.H. (1996) Antibiotic protection against recrudescence of latent *Aeromonas salmonicida* during furunculosis vaccination. *Journal of Fish Diseases* **19**, 341-348.

Johnson, K.A., Flynn, J.K. and Amend, D.F. (1982) Onset of immunity in salmonid fry vaccinated by direct immersion in *Vibrio anguillarum* and *Yersinia ruckeri* bacterins. *Journal of Fish Diseases* **5**, 197-205.

Lamers, C.H.J., de Haas, M.J.H. and van Muiswinkel, W.B. (1985) The reaction of the immune system of fish to vaccination: development of immunological memory in carp, *Cyprinus carpio* L., following direct immersion in *Aeromonas hydrophila* bacterin. *Journal of Fish Diseases* , 253-262.

Lillehaug, A. (1989) A cost-effectiveness study of three different methods of vaccination against vibriosis in salmonids. *Aquaculture* **83**, 227-236.

Lillehaug, A., Ramstad, A., Baekken, K. and Reiten, L.J. (1993) Protective immunity in Atlantic salmon (*Salmo salar* L.) vaccinated at different water temperatures. *Fish and Shellfish Immunology* **3**, 143-150.

Midtlyng, P.J., Reiten, L.J. and Speilberg, L. (1996) Experimental studies on the efficacy and side-effects of intraperitoneal vaccination of Atlantic salmon (*Salmo salar* L.) against furunculosis. *Fish and Shellfish Immunology* **6**, 335-350.

Nordmo, R. and Ramstad, A. (1997) Comparison of different challenge methods to evaluate the efficacy of furunculosis vaccines in Atlantic salmon. *Journal of Fish Diseases* **20**, 119-126.

Norqvist, A., Hagstrøm, A. and Wolf-Watz, H. (1989) Protection of rainbow trout against vibriosis and furunculosis by the use of attenuated strains of *Vibrio anguillarum. Applied and Environmental Microbiology* **55**, 1400-1405.

Ototake, M., Iwama, G.K. and Nakanishi, T. (1996) The uptake of bovine serum albumin by the skin of bath-immunized rainbow trout (*Oncorhynchus mykiss*). *Fish and Shellfish Immunology* **6**, 321-333.

Piganelli, J.D., Zhang, J.A., Christensen, J.M. and Kaattari, S.L. (1994) Enteric coated microspheres as an oral method for antigen delivery to salmonids. *Fish and Shellfish Immunology* **4**, 179-188.

Robohm, R.A. and Koch, R.A. (1995) Evidence for oral ingestion as the principal route of antigen entry in bath-immunized fish. *Fish and Shellfish Immunology* **5**, 137-150.

Sommer, C.V., Binkowski, F.P., Schalk, M.A. and Bartos, J.M. (1984) Some factors that can affect studies of drug metabolism in fish. *Veterinary and Human Toxicology* **28** (Supplement 1), 45-54.

Tatner, M.F. (1993) Fish Vaccines. In *Vaccines for Veterinary Applications* (ed. A.R. Peters) Butterworth-Heinemann, Oxford. pp. 199-224.

21. OSMO-REGULATORS

21.1 Physiological osmo-regulation

21.1.1 THE NEED

Like any other class of animal fish need to maintain a constant internal environment. They have a greater problem than terrestrial animals however in that they live in a liquid at a different osmotic pressure from themselves. Freshwater fish are continually absorbing water by osmosis and need to excrete it. Marine fish are continually losing water and need to replace it.

21.1.2 THE ORGANS

(a) The skin

The organs of osmo-regulation in the fish are the skin, intestine, kidneys and gills. The skin probably acts as a semi-permeable membrane allowing passive transfer of water. The scales which are embedded in the dermis are an important barrier to osmotic transfers.

(b) The intestine

The marine fish is continually drinking water to replace that lost by osmosis. A significant proportion of this water must be absorbed from the intestine and because it is against the osmotic gradient it must be an active (that is, energy-absorbing) process. Freshwater fish drink very little water although the volume is not entirely negligible (see Section 20.2.2).

(c) The kidneys

As in other vertebrates the kidneys are organs of nitrogenous excretion. They also have a role in osmo-regulation through water excretion. In seawater the urine is sparse and concentrated.

(d) The gills

The gills have roles not only in external respiration (gaseous exchange) but also in nitrogenous excretion and osmo-regulation. This latter is mediated by mucus secretion to limit water transfer and the active excretion of ions. Ionic transfer is through specialized cells in the gill epithelium called ionocytes or chloride cells. These are mitochondria-rich and contain the enzyme Na^{+}-K^{+}-ATPase.

21.1.3 SMOLTIFICATION

In anadromous and catadromous species profound physiological changes are necessary for the fish to maintain osmotic stability in migration between fresh and seawater. These physiological changes are accompanied in many species by visible anatomical changes. In salmonids a young fish migrating to the sea is known as a smolt, and the physiological and anatomical changes are called smoltification.

(a) Anatomical changes

The capacity to absorb significant quantities of water through the intestinal mucosa is established while the fish is still in freshwater but ultra-structural changes do not occur until the first 2 days in seawater.

In the kidneys there is a reduction in the number of active glomeruli during smoltification. There is a decrease in the volume of water excreted with a concomitant rise in urine concentration.

In the gills the number, size and Na^+-K^+-ATPase activity of the chloride cells rise.

(b) Endocrine control

Among anterior pituitary hormones, prolactin is important to osmo-regulation in freshwater and hypophysectomized fish do not survive in freshwater. Growth hormone (GH) secretion rises during smoltification. GH and prolactin have an antagonistic effect on adaptation to seawater. Both ovine prolactin and ovine GH (oGH) will stimulate Na^+-K^+-ATPase activity in the gills and it may be assumed that during smoltification endogenous GH has the same effect. Salt excretion uses energy and it is known that injections of GH increase oxygen consumption. oGH produces smoltification changes in the gills of both Atlantic and Pacific salmon at 5 mg/kg but rainbow trout GH (rtGH) is at least 10 times as active. In a trial of the use of oGH at different seasons of the year it was found to promote seawater adaptation in late autumn and winter but to be of no particular benefit to fish transferred to seawater in the spring or summer. The result was attributed to the high endogenous secretion of GH in the summer.

Cortisol, which, rather than cortisone, is the normal adrenocortical hormone in fish, also stimulates smoltification changes in the gills and intestinal mucosa and hence saltwater adaptation. Cortisol and GH act together in the gills since GH increases the number of cortisol receptor sites. At supra-physiological doses either hormone will increase the exposed area of chloride cells; the combination will increase both the surface area of each cell and the density of cells.

21.2 Medicinal osmo-regulation

21.2.1 INDICATIONS

(a) Life support for ailing fish

Loss of osmo-regulation is potentially life-threatening to a fish, and impairment of function of any of the four organs of osmo-regulation must be taken seriously. Life-threatening lesions of the intestine alone are probably rare but compromised intestinal function will place additional stress on the other osmo-regulatory organs.

Uncontrolled transfers of water will occur at any traumatic injury or ulceration of the skin. Thus osmo-regulatory failure may be the ultimate cause of death from bacterial skin disease such as erythrodermatitis of carp (where the failure is manifested as dropsy) or external parasitism such as sea-lice on Atlantic salmon. Similarly, kidney infections such as bacterial kidney disease and proliferative kidney disease, and gill infections such as bacterial gill disease or gill fluke infestation will all cause mortality due to failure of osmo-regulation. Any form of therapy may need to be supplemented by assisting the fish to maintain its osmo-regulation until the affected organ has healed sufficiently.

(b) Induction of smoltification

In modern salmonid farming there is economic pressure to maintain a steady supply of harvested fish at all seasons, and there is also an economic gain from restocking netpens with young fish soon after harvesting. This means a requirement for smolts at all seasons of the year. As natural smoltification is seasonal there is a need for artificial induction of this process or support of fish transferred to seawater when the process is incomplete.

In addition to these points there is an economic advantage in inducing precocious smoltification because fish in seawater have higher growth rates and better food conversion ratios than those in freshwater.

21.2.2 COMMON SALT

(a) Indications

Sodium chloride is one of the most widely used substances in the medication of freshwater fish. It has been recommended for all of the following indications:

- Reduction of osmotic stress - a "nursing" use in fish with disease in any osmo-regulatory organ;
- Prevention of loss of chloride ions as a result of stress during handling and transport;
- Mitigation of ammonia or nitrite toxicity in tanks, and of acid rain toxicity in ponds;
- Treatment of bacterial gill disease - this is a "nursing" treatment, reducing the mucus secretion on the gills and hence aiding gill function; it is a supportive therapy, not a replacement for antibacterial treatment;
- Prevention of columnaris disease;
- Control of saprolegniosis in both stressed fish and eggs at hatcheries;
- Control of protozoan ectoparasites, *e.g. Costia (Ichthyobodo), Trichodina, Ichthyophthirius, Scyphidia, Epistylis, Vorticella, Trichophyra, Tetrahymena, Chilodonella* and *Oodinium*;
- Control of monogenean fluke infestations;
- Control of crustacean ectoparasites (benzylureas (see Section 14.0.0) are more active but not always available in formulations suitable for use in fish tanks);
- Promotion of the absorption of other drugs in hyperosmotic infiltration (see Section 1.1.3)
- Potentiation of the anti-parasitic action of formalin and mitigation of its side effects.

(b) Mode of action

Paradoxically for such a widely used drug the value of sodium chloride lies in its lack of any pharmacological action on fish. It is used only in freshwater where its sole action is to raise the osmotic pressure.

The nursing or life-support action of salt results from its reducing the osmotic gradient between the water and the fish and hence reducing osmotic flux of water into the fish. The presence of other (non-toxic) ions in the water will reduce the absorption of soluble toxins including ammonia, nitrite and formalin. The effect on mucus is also one of water extraction - an astringent action. The action on ectoparasites depends on their having a lower tolerance to high osmotic pressure than fish have.

(c) Composition for medicinal use

Common salt is commercially available either crystallized from evaporation of seawater or mined (from crystallized deposits of seawater in past geological eras). In either case it is impure, but the impurities are non-toxic in the concentrations usually found and the salt could safely be used for fish medication.

Salt is hygroscopic and when damp will cake in lumps. For culinary use it is normally treated with one of a number of anti-caking agents; most of these are harmless but yellow prussate of soda (YPS) is used in some countries and this is best avoided for fish. Another common additive is sodium iodide, and while this may be valuable for human nutrition it is toxic to fish in the concentrations normally added. Iodized salt should never be used for fish. Commercial forms of salt suitable for use in fish medication include:

- Any intended for human consumption except iodized salt,
- Any intended for animal consumption which does not contain added minerals,
- Salt intended for recharging water softeners,
- Artificial sea salts.

(d) Recommended concentrations

The concentration of solutes in fish plasma is approximately the same as in mammalian plasma, 0.9% w/v. However for routine supportive therapy 0.3% w/v is normally recommended. This is equivalent to 3 g/l, 3 kg/m^3 or 3 lb/100 (imperial) gallons. This concentration may safely be maintained for as long as the supportive therapy is needed. Up to 0.5% w/v is sometimes used but where this is done the salinity of the water should be raised gradually, over 2-3 days, to avoid shock to the fish (and to the biological filter if any). In an osmo-regulatory emergency a fish can be put directly into 0.5% salt but it must be expected to show distress for a few hours.

For the control of ectoparasites short term dip treatments at higher concentrations are preferable. Alevins and fry can be given up to 1% for 10 minutes; larger fish can be given progressively higher concentrations up to 3% for fish weighing not less than 100 g. Fish should always be monitored during exposure and the water should be diluted if signs of distress are seen. *Ichthyophthirius* is more resistant than other protozoan parasites and it is usually necessary to dip infected fish in 2.5% salt for 3-15 minutes on three or four occasions at weekly intervals.

21.2.3 BETAINE

Betaine is a quaternary amine derivative of glycine. When given in feed it concentrates in the cells of the fish body but not in the plasma; in the cells it exerts an osmotic pressure ensuring cell turgor without increasing the potassium ion concentration. Betaine elevates Na^+-K^+-ATPase activity in the gills and so increases the active excretion of ions in seawater.

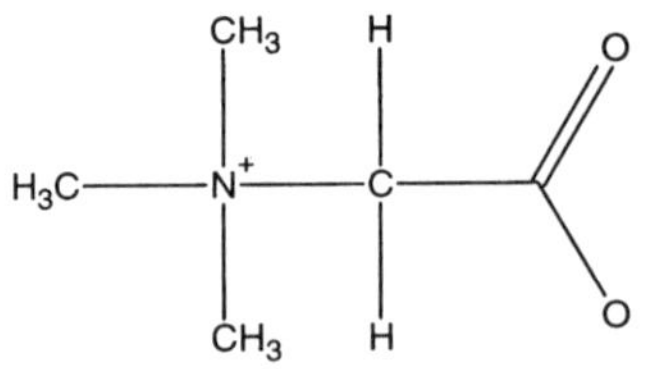

$H_3C-N^+(CH_3)_2-CH_2-COO^-$

Betaine

Figure 21.1.

It also affects excretion of water and Mg^{++} ions, indicating that it has actions on the gut and kidneys.

Betaine is commercially available as a mixture with various amino-acids. The product is used at 1.5% in the diet to enhance the adaptation of smolts to seawater. Virtanen *et al.* (1989) showed that in Atlantic salmon it made no significant difference in the first 2 weeks at sea, but that thereafter there was a significant reduction in mortality in treated fish.

21.2.4 GROWTH HORMONE

GH is the hormone controlling the osmo-regulation in smolts and is presumed to be the hormone which stimulates the smoltification process. A salmon GH produced by a genetically modified organism has been developed and given the non-proprietary name, somatosalm; it has been tested for activity as a smoltification inducer. Further commercial development is in abeyance at present due to regulatory requirements in the EU but it is conceivable that it, or another similar product, may be commercialized elsewhere in the future.

The proposed method of use of somatosalm was to give it to presmolts and move them to sea 5 days later. The dose was 10 μg in 0.2 ml of 0.9% saline by intraperitoneal injection.

Further reading

Boeuf, G. (1993) Salmonid smolting: a pre-adaptation to the oceanic environment. In: *Fish Ecophysiology* (eds. J. C. Rankin and F.B. Jensen) Chapman and Hall, London, pp. 105-135.

Edgell, P., Lawseth, D., McLean, W.E. and Britton, E.W. (1993) The use of salt solutions to control fungus (*Saprolegnia*) infestations on salmon eggs. *The Progressive Fish-Culturist* **55**, 48-52.

Virtanen, E., Junnila, M. and Soivio, A. (1989) Effects of food containing betaine/aminoacid additive on the osmotic adaptation of young Atlantic salmon, *Salmo salar* L. *Aquaculture* **83**, 109-122.

Sakamoto, T., McCormick, S.D. and Hirano, T. (1993) Osmo-regulatory actions of growth hormone and its mode of action in salmonids: a review. *Fish Physiology and Biochemistry* **11**, 155-164.

22. DISINFECTANTS

22.1 Uses of disinfectants

22.1.1 DEFINITIONS

"Disinfection means the freeing of an article from some or all of its burden of live pathogenic microorganisms which might cause infection during its use." (Cruickshank *et al.*, 1973). It may be noted that the aim of disinfection is not to remove or inactivate all microorganisms but only pathogenic ones, which may vary in size from virus particles to helminth eggs.

Disinfection must be distinguished from cleansing which is the removal of organic matter, including faeces and uneaten food, from surfaces. Such matter will contain a very large microbial flora only a small proportion of which will be pathogenic; but it is virtually impossible to sterilize dirt, so cleansing is an essential preliminary to disinfection. Iodophors are examples of a loose chemical combination of a cleansing agent with a disinfectant.

Methods of disinfection include not only chemical treatments but also physical ones such as heat and gamma-irradiation. There are in practice no procedures efficacious against the whole range of pathogens which are not damaging to the material to be disinfected, or hazardous to the operator and/or the fish subsequently managed with the disinfected equipment. The choice of method must therefore take into account the material to be disinfected and the nature of the pathogens likely to be present.

Because disinfectants are not applied to fish they are not medicinal substances under some legal systems. However disinfection of the surfaces of eggs in hatcheries is a normal procedure in aquaculture and whether this use is legally medicinal depends on the legal status of fish eggs.

22.1.2 MATERIALS TO BE DISINFECTED

(a) Tanks and utensils

Vessels holding fish (including netpens) and the utensils placed in the water are usually disinfected by chemical treatment of the relevant surfaces. Cleansing is an important aspect of the total procedure.

(b) Earth ponds

True disinfection of earth ponds is impossible. It is nevertheless essential to minimize the population of pathogens such as bacteria, parasite eggs and vectors in the organic detritus which inevitably accumulates on the bottom. One way in which this can be achieved is to use butyl linings to ponds as these can be cleaned and disinfected between crops of fish.

(c) Water

Disinfection of water must be distinguished from its chemical purification which is the management of pH and such toxic materials as ammonia and heavy metal ions. Disinfection is the inactivation of infectious pathogens and may be necessary in the following circumstances:

- In tank culture where the water is recirculated;
- In freshwater farms where the inflowing water may be contaminated, for example from other farms upstream;
- In all hatcheries except those using well water;
- In laboratories and fish processing factories where the effluent water may be contaminated.

22.2 Chemical disinfectants

22.2.1 IODOPHORS

(a) Chemistry

Iodophor is a general name for any compound in which a surfactant acts as a carrier and solubilizing agent for elementary iodine. The combinations retain the cleansing properties of the detergents; and because the iodine is slowly released from them they reduce its vapour pressure, odour and staining effect. They may be corrosive to some metals due to the iodine release and pH, but are non-toxic and non-irritant when applied in dilute solutions to skin and mucosal surfaces. The disinfectant action ultimately depends on the iodine released from them and they are usually measured in terms of available iodine.

The following chemical reactions are thought to occur when iodine is dissolved in water:

$$
\begin{aligned}
I_2 + H_2O &= (H_2OI)^+ + I^- \\
(H_2OI)^+ &= HOI + H^+ \\
HOI &= OI^- + H^+ \\
3HOI &= IO_3^- + 2I^- + 3H^+ \\
I_2 + I^- &= I_3^-
\end{aligned}
$$

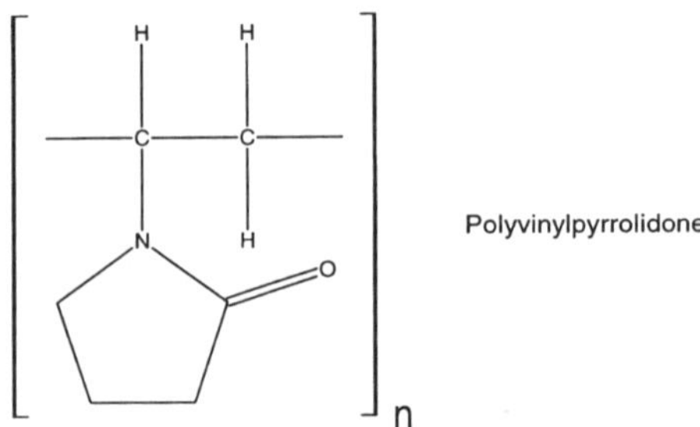

Figure 22.1.

Of the seven chemical species formed the molecules, I_2 and HOI, and the cation, $(H_2OI)^+$, are potent disinfectants. They have good bactericidal action over a wide range of pH, and are also active against viruses and fungi. There are several different iodophors available on the market, some with undisclosed chemical structures. Typically the commercial concentrates contain 1% or 1.6% w/v available iodine. One of the most widely used is povidone-iodine, based on the surfactant polyvinylpyrrolidone which is a polymer of vinylpyrrolidone. It has a virtually neutral pH, which makes it suitable for use on fish eggs, although possibly less effective than high or low pH iodophors as a cleansing agent for utensils.

(b) Mode of action

The disinfectant action of iodophors is thought to depend on the following chemical reactions:

- The formation of N-iodo radicals on the side chains of basic amino-acids. This prevents hydrogen bonding at these points and so reduces the secondary and tertiary folding of proteins;
- The oxidation of -SH groups in cysteine, preventing the formation of -S-S- links in proteins;
- Reaction with the phenolic groups in tyrosine;
- The oxidation of unsaturated fatty acids in membranes altering their permeability.

All except the last of these reactions are with proteins and all involve consumption of available iodine. It is a general characteristic of iodophors that they are inactivated by proteins, and so to obtain satisfactory disinfection of surfaces all organic material and especially proteinaceous material must be removed first. The activity of iodophors decreases in use and care must be taken to ensure adequate residual activity in disinfection procedures. Many iodophors change colour when their bactericidal potency is exhausted.

(c) Disinfection of tanks and utensils

The spectrum of activity of iodophors makes them useful disinfectants for tanks and utensils, and in particular any containers used to bring fish or eggs onto a premises. Only firm, non-absorbent surfaces such as fibreglass, timber and cast materials can be disinfected, and care must be taken to clean them thoroughly before the iodophor is used.

Iodophors are toxic to fish at concentrations as low as 50 ppm available iodine; care must be taken to ensure that used disinfectant and rinsings do not contaminate rivers or water supplies to fish farms. They should be discharged into sewers where they will eventually be inactivated by proteins.

Use recommendations vary widely between different iodophors. A well-known brand of povidone-iodine is recommended for use at 50 ppm available iodine with no exposure time specified; another acidic iodophor is used at 250 ppm available iodine with the recommendation that absorbent materials such as ropes and nets should be steeped in the solution for not less than 15 minutes. For viral disinfection 100 ppm available iodine is usually recommended. In following recommendations it is important to distinguish between concentrations of iodophor and concentrations of available iodine. If an iodophor contains 1% available iodine, 50 ppm available iodine will be contained in 0.5% iodophor.

(d) Disinfection of eggs

Disinfection of eggs is necessary not only for prevention of fungal (*Saprolegnia* spp.) infection but also to prevent the importation of infections into hatcheries on eyed eggs. It is for the latter purpose that iodophors with their high activity against viruses are particularly valuable.

Fish eggs are relatively insensitive to iodophors at neutral pH. Embryos and fry are sensitive, and eggs are sensitive to low pH. In consequence it is safe to use iodophors for most of the incubation period but they must either be buffered to near neutrality or be naturally neutral, such as povidone-iodine. If iodophors are used on newly fertilized eggs before they have finished absorbing water and hardening, some iodine will be absorbed and be lethal to the embryo. For freshwater species the concentrated iodophor must be diluted in 0.9% saline instead of water, and saline must be used to rinse the eggs after disinfection. If iodophors are used at the very end of the incubation period it is possible that some of the iodine will wash off the surface of some of the eggs and harm the first sac-fry to hatch.

While eggs are insensitive to iodophors to the extent that the compounds can be used safely and effectively as disinfectants for most of the incubation period, the sensitivity is affected by:

- pH - eggs are not only sensitive to acids *per se* but acids increase their sensitivity to iodophors;
- The species of fish - 100 ppm available iodine for 10 minutes is normally recommended for salmonid eggs, but this would cause high mortality in walking catfish eggs for which 20 ppm for 10 minutes is regarded as ideal;
- The individual female fish spawning the eggs - in at least some species there are significant differences in the sensitivity of eggs spawned by different females (Alderman, 1984).

22.2.2 LIME

(a) Chemistry

The word, “lime”, comprehends three distinct chemical species, *viz.*

1. Limestone - this may be calcite ($CaCO_3$), dolomite ($CaMg(CO_3)_2$) or a mixture of the two.
2. Burnt lime or Quicklime - calcium oxide (CaO)
3. Hydrated lime or Slaked lime - calcium hydroxide ($Ca(OH)_2$)

All three can be used to raise the pH of acid water although burnt lime cannot be used where there are fish because of the heat produced in its reaction with water. Limes used for this purpose are never pure and their efficacy is measured in terms of “neutralizing value” (NV) compared to a standard of pure calcium carbonate being 100%.

Theoretical NVs are	pure calcium oxide	179%
	pure calcium hydroxide	135%

In practice NVs are:

agricultural limestone	80-110% (the higher NV samples containing dolomite)
burnt lime	150-165%
slaked lime	115-135%

In addition ground limestones are assigned an "efficiency rating" (ER) which depends on the particle size distribution. An ER of 100% means that all particles are below 0.24 mm. Larger particles of limestone dissolve only slowly and are of little practical value.

(b) Use as disinfectants

Limes are used for the treatment of earth ponds. While disinfection is strictly impossible, a pond which has contained fish will have a considerable organic sediment to be eliminated, and there are two possible and not entirely compatible approaches. If there is reason to believe there is a significant population of pathogens in the sediment these can be killed by a drastic raising of the pH using burnt lime or hydrated lime. This will sterilize the entire bacterial flora and the sediment will then have to be dredged out mechanically. If pathogens are not thought to be a particular problem the bacterial decomposition of the sediment can be accelerated by buffering it, and here a high ER limestone would be ideal.

22.2.3 QUATERNARY AMINES

(a) As disinfectants

Quaternary amines are cationic surfactants available only as salts. Benzalkonium chloride, which is actually a mixture, is a typical example. As surfactants they have a powerful cleansing action which is an important preliminary to disinfection. As disinfectants they are active, particularly at slightly alkaline pH, against most bacteria and some fungi but not against viruses. The mode of action appears to be to alter the permeability of cell membranes.

Quaternary amines are used as cleansing and disinfecting agents on items with hard surfaces such as tanks and utensils, but their activity is reduced on fibrous or porous surfaces so they are of little value for ropes or nets. They are inactivated by soaps, and by other anionic substances such as proteins and fatty acids so they cannot be used in the disinfection of processing plants. The normal use concentration is 0.1% w/v.

(b) Medicinal use

The combined cleansing and antibacterial action of quaternary amines has been used in the treatment of bacterial gill disease in fish. In this infection the bacteria multiply in the mucus film covering the gills, and the irritant action of the bacteria on the gill epithelium increases the secretion of mucus. Benzalkonium chloride used as a dip has a dual action: it lifts the mucus off the gills thus improving gaseous exchange and it kills the bacteria. 10 ppm for 5-10 minutes is usually satisfactory; but where a dip regimen is not practical, for example in a pond, 0.5 ppm may be used as a long term treatment.

22.2.4 OZONE

(a) Uses

Ozone, in the form of ozonized oxygen, is used for the disinfection of water as it has rapid activity against both bacteria and viruses. When bubbled into water it equilibrates at 0.2 ppm ozone in freshwater and 0.15 ppm in brackish or sea water. In freshwater and more particularly seawater it reacts to form other oxidizing antibacterial chemical species. In seawater some of these species are toxic; in consequence ozone should not be used on influent water to marine installations but it remains very useful for effluent water especially where, as in laboratories, the effluent may be expected to be infected.

At the equilibrium concentrations ozone effects 10^{-4} reductions in bacterial counts in freshwater in 60 secs and in seawater in 180 secs; it has this degree of action on infectious pancreatic necrosis (IPN) virion counts in both types of water. However this is in water which has not been used for aquaculture. Summerfelt and Hochheimer (1997) point out that the half-life of ozone in water is, at best, 165 minutes at 20°C but very much less in the presence of organic compounds and/or nitrites; in water used in recirculating aquacultural systems the half-life may be less than 15 seconds. They comment, "Achieving large microbial reductions in recirculation systems requires much more ozone than would be needed to disinfect the influent of typical single-pass aquaculture systems."

Summerfelt and Hochheimer (1997) discuss the methods and costs of generating and applying ozone and conclude that the optimum cost:benefit ratio is achieved at 0.013 mg ozone per kg "feed" (*i.e.* influent freshwater).

(b) Safety aspects

Ozone is harmful to fish, its main effect being destruction of the gill epithelium, see Table 22.1.

Table 22.1. Toxicity of dissolved ozone to fish; LC50 = the concentration lethal to 50% of sample fish

Species	Ozone concentration (mg/L)	Effect	Reference
Rainbow trout *Oncorhynchus mykiss*	0.0093	96-h LC50	Wedemeyer *et al.* (1979)
Rainbow trout	0.01-0.06	Lethal	Roseland (1975)
Bluegill *Lepomis machrochirus*	0.01	60% mortality	Palter and Heidenger (1979)
Fathead minnow *Pimephales promelas*	0.2-0.3	Lethal	Arthur and Mount (1975)
White perch *Morone americana*	0.38	24-h LC50	Richardson *et al.* (1983)
Bluegill	0.06	24-h LC50	Paller and Heidenger (1979)
Striped bass (larvae) *Morone saxatilis*	0.08	96-h LC50	Hall *et al.* (1981)

Before treated water is re-used for fish the residue of ozone should be specifically removed. This may be done with reducing agents but since this leads to chemical pollution physical processes such as ultra-violet radiation or activated carbon are more commonly used. The physical processes do not decompose the ozone and since it is harmful to man, causing pulmonary oedema and haemorrhage, care has to be taken to ensure that it does not escape, either from the elimination plant or between the ozonizer and elimination plant.

Further reading

Alderman, D.J. (1984) The toxicity of iodophors to salmonid eggs. *Aquaculture* **40**, 7-16.

Cruickshank, R., Duguid, J.P., Marmion, B.P. and Swain, R.H.A. (1973) In: *Medical Microbiology. A Guide to the Diagnosis and Control of Infection.* Churchill Livingstone, Edinburgh and London. pg 59.

Liltved, H., Hektoen, H. and Efraimsen, H. (1995) Inactivation of bacterial and viral fish pathogens by ozonation or UV irradiation in water of different salinity. *Aquacultural Engineering* **14**, 107-122.

Summerfelt, S.T. and Hochheimer, J.N. (1997) Review of ozone processes and applications as an oxidizing agent in aquaculture. *The Progressive Fish-Culturist* **59**, 94-105.

APPENDIX I

TRADE NAMES OF PRODUCTS SPECIFICALLY FORMULATED FOR USE IN FISH

Trade Name	Manufacturer	Active Ingredient(s)
Alpha-Jec	Alpharma	vaccines
Apoject	Alpharma	vaccines
Apoquin	Alpharma	flumequine
Apoxolon	Alpharma	oxolinic acid
Aquacil	Grampian	amoxycillin
Aquaflor	Schering-Plough	florfenicol
Aquagard	Novartis	dichlorvos
Aqualets	Alpharma	Pellet formulation of drugs which minimizes leaching
Aqualinic powder	Vetrepharm	oxolinic acid
Aquaswab	Technical Aquatic Products	ethanol and iodophor
Aquatet	Vetrepharm	oxytetracycline
Aquavac	A.V.L.	emulsion vaccines for oral administration
Aquinox	Grampian	oxolinic acid
Argentyne	Argent	povidone-iodine
Ark-klens	Vetark	benzalkonium chloride
Biojec Biomed Biovax	Alpharma	vaccines
Buffodine	Evans Vanodine International	buffered iodophor
Calicide	Nutreco	teflubenzuron
Chitosan	Polish Sea Fisheries Institute	deacylated chitin
Cyprinopur	Sera	1,3 dihydroxybenzol, ethanol, phenol
Ectobann	Skretting	teflubenzuron
Excis	Grampian	cypermethrin
FAM 30	Evans Vanodine International	iodophor
Finquel	Argent	MS-222
FinnStim	Finnsugar Bioproducts	betaine and amino-acids

Flumisol	Intervet	flumequine
Fumaqua	Sanofi	fumagillin
Furanace	Dainippon	nifurpirinol
MacroGard	Mackzymal	glucan
Marinil	Wildlife Laboratories	metomidate
Norvax Protect	Intervet Norbio	Combined furunculosis, vibriosis and cold water vibriosis vaccine
Ovadine	Syndel	povidone-iodine
Ovaprim	Syndel	GnRH-A and domperidone
Paramove	Solvay-Interox	hydrogen peroxide
Paracide-F	Argent	formalin
Parasite-S	Syndel	formalin
Piyersivac	Impfstoffwerk Dessau-Tornau	ERM spray vaccine
Protoban	Vetark	formalin and malachite green
Reproboost	Aquapharm Technologies	GnRH-A in sustained release implants
Salartect	Brenntag	hydrogen peroxide
Salmosan	Novartis	azamethiphos
Sarafin	Vetrepharm	sarafloxacin
Sulfatrim	Grampian	trimethoprim and sulphadiazine
Synahorin	Teikoku Hormone Manufacturing	Human chorionic gonadotrophin and mammalian pituitary extract
Tamodine	Vetark	iodophor
Tamodine-E	Vetark	buffered iodophor
Tetraplex	Grampian	oxytetracycline
TMS	Syndel	MS-222
Vetremox	Vetrepharm	amoxycillin
Vibriffa-Bain	Rhône-Merieux	*Vibrio anguillarum* vaccine
Wescodyne	Novartis	povidone-iodine

Trade Mark Proprietors

Alpharma Aquatic Animal Health Division
Bellevue
WA., USA

Alpharma Laboratorium A.S.
Skøyen,
Oslo, Norway

Argent Chemical Laboratories
Redmond, WA, USA

A.V.L. (=Aquaculture Vaccines Ltd.)
Saffron Walden,
Essex, UK

Aquapharm Technologies Corp.
Columbia,
MD, USA

Brenntag (UK) Ltd.,
Hampton,
Middx., UK

Finnsugar Bioproducts
Helsinki
Finland

Grampian Pharmaceuticals Ltd.,
Leyland,
Lancs., UK

Intervet Norbio AS
Bergen
Norway

Mackzymal
Tromsø
Norway

Novartis Animal Health UK Ltd.,
Whittlesford
Cambs., UK

Nutreco
Forus
Norway

Schering-Plough Animal Health
Kenilworth,
N.J., USA

Sanofi Sant, Nutrition Animale
Libourne,
France

Sea Fisheries Institute
Gdynia
Poland

T. Skretting AS
Stavanger
Norway

Solvay-Interox Ltd.,
Warrington,
Lancs., UK

Syndel Laboratories
Vancouver,
BC, Canada

Technical Aquatic Products Ltd.
Bristol, UK

Vetark Professional
Winchester,
Hants., UK

Vetrepharm Ltd.,
Fordingbridge
Hants., UK

APPENDIX II

COMMONLY CULTURED SPECIES OF FISH

Acipensoidei

ACIPENSIDAE - STURGEONS

Acipenser brevirostrum	- shortnosed sturgeon
Acipenser oxyrhynchus	- Atlantic sturgeon
Acipenser ruthenus	- sterlet

Angilloidei

ANGUILLIDAE - EELS

Anguilla anguilla	- European eel
Anguilla japonica	- Japanese eel
Anguilla rostrata	- American eel

Salmonoidei

SALMONIDAE

Salmo salar	- Atlantic salmon
Salmo trutta	- brown trout
Oncorhynchus clarki	- cutthroat trout
Oncorhynchus gorbuscha	- pink salmon
Oncorhynchus keta	- chum salmon
Oncorhynchus kisutch	- coho salmon, silver salmon
Oncorhynchus masou (Syn. *O. rhodurus*)	- (lacustrine) masu salmon, cherry salmon, amago salmon, Sakura salmon, (anadromous) yamame salmon
Oncorhynchus mykiss	- (lacustrine) rainbow trout (anadromous) steelhead trout
Oncorhynchus nerka	- (lacustrine) kokanee salmon (anadromous) sockeye salmon
Oncorhynchus tschawytscha	- chinook salmon
Salvelinus alpinus	- arctic charr
Salvelinus confluentus	- bull trout
Salvelinus fontinalis	- brook trout, brook charr
Salvelinus leucomaenis	- Japanese charr, white spotted charr
Salvelinus malma	- Miyabe charr

Salvelinus namaycush	-	lake trout
Thymallus thymallus	-	grayling

PLECOGLOSSIDAE - AYU

Plecoglossus altivelis	-	ayu

Clupeoidei

CHANIDAE - MILKFISH

Chanos chanos	-	milkfish

Cyprinoidei

COBITIDAE - LOACHES

Paramisgurnus dabryanus	-	Chinese loach

CYPRINIDAE - CYPRINIDS

Aristichthys nobilis	-	bighead carp
Carassius auratus	-	goldfish
Carassius carassius	-	crucian carp
Catla catla	-	catla
Cirrhinus mrigala	-	mrigal
Cirrhinus molitorella	-	mud carp
Ctenopharyngodon idella	-	grass carp
Cyprinus carpio	-	common carp (mirror carp and koi are breeds within this species)
Hypothalamichthys molitrix	-	silver carp
Labeo calbasu	-	kalbasu
Labeo rohita	-	rohu
Leuciscus idus	-	golden orfe
Mylopharyngodon piceus	-	black carp
Parabramis pekinensis	-	bream
Puntius goniotus	-	Thai carp
Tinca tinca	-	tench

Characoidei

SERRASALMIDAE

Colossoma macropomum	-	tambaqui
Pieractus mesopotamicus	-	pacu

CURIMATIDAE - CURIMATAS

Prochilodus affinis	-	curimata pioa
Prochilodus margravii	-	curimata pacu

Siluroidei

ICTALURIDAE - CATFISH

Clarias batrachus	-	Asian catfish, walking catfish
Clarias gariepinus (Syn. *C. lazera*)	-	African catfish sharptooth catfish
Clarias macrocephalus	-	hito
Ictalurus catus	-	white catfish
Ictalurus furcatus	-	blue catfish
Ictalurus melas	-	black catfish
Ictalurus natalis	-	yellow bullhead
Ictalurus nebulosus	-	brown bullhead
Ictalurus punctatus	-	channel catfish
Pylodictis olivaris	-	flathead catfish
Silurus glanis	-	wels, sheatfish

Percoidei

CARANGIDAE - JACKS

Seriola dumerili	-	amberjack
Seriola quinqueradiata	-	yellowtail
Pseudocaranx dentex	-	striped jack

CENTRARCHIDAE - SUNFISHES

Lepomis macrochirus	-	bluegill
Micropterus salmoides	-	largemouth bass

CENTROPOMIDAE

Lates calcarifer	-	barramundi

ESOCIDAE - PIKES

Esox lucius	-	northern pike

PERCIDAE - TEMPERATE PERCHES

Stizostedion lucioperca	-	pike-perch
Stizostedion vitreum	-	walleye

SCAENIDAE

Scaeniops ocellatus	-	red drum

SERRANIDAE - SEA BASSES

Dicentrarchus labrax	-	sea bass
Epinephelus fario	-	blue-spotted grouper
Epinephelus fuscoguttatus	-	Red Sea grouper
Epinephelus salmoides	-	estuary grouper
Epinephelus suillus	-	Pacific grouper

Epinephelus tauvina	-	common grouper
Morone chrysops	-	white bass
Morone saxatilis	-	striped bass
M. chrysops female x *M. saxatilis* male	-	sunshine bass
M. saxatilis female x *M. chrysops* male	-	palmetto bass

SIGANIDAE - RABBITFISHES

Siganus guttatus	-	rabbitfish

SPARIDAE - BREAMS

Pagrus major	-	red sea-bream
Sparus aurata	-	gilthead sea-bream

CICHLIDAE - CICHLIDS

Sarotherodon (Syn. *Oreochromis*) *aurea*	-	golden tilapia
Sarotherodon galilaeus	-	Galilean tilapia
Sarotherodon mossambicus	-	Mozambique tilapia
Sarotherodon niloticus	-	Nile tilapia
Sarotherodon spilurus	-	saltwater tilapia
Cichlasoma urophthalmus	-	Mexican mojarra

Mugiloidei

MUGILIDAE - MULLETS

Mugil cephalus	-	grey mullet

Pleuronectoidei - Flatfish

Paralichthys olivaceus	-	Japanese flounder
Hippoglossus hippoglossus	-	halibut
Scophthalmus maximus (Syn. *Psetta maxima*)	-	turbot

APPENDIX III

GLOSSARY

Alevin	- Sac-fry (*q.v.*) of a salmonid species
Anaculture	- Inactivated whole culture of bacteria (for use as the antigen in a vaccine), containing inactivated bacterial exudates as well as cells (*c/f* bacterin).
Anadromous	- (Of fish species) migrating from seawater up rivers to spawn (*e.g.* Atlantic salmon).
Bacterin	- Killed bacterial cells for use as the antigen in a vaccine; vaccine in which the antigen consists only of killed bacterial cells.
Catadromous	- (Of fish species) migrating from freshwater to seawater to spawn (*e.g.* eels).
Euryhaline	- (Of fish species) able to live in waters of a wide range of salinities.
Excipient	- Component of a formulated medicinal product other than an active ingredient (*e.g.* carrier, solvent, solubilizer, buffer).
Fry	- Juvenile fish which have just started feeding.
Grilse	- Salmonid developing sexual maturity precociously, usually after only one year in seawater.
Lacustrine	- (Of fish species) living in lakes, *i.e.* in static freshwater.
Multivalent	- (Of vaccines) stimulating immunity against two or more infections.
Parr	- Young salmonid of an anadromous (*q.v.*) species which is still living in freshwater.
Sac-fry	- Newly hatched fish which has a yolk-sac from which it takes its nourishment.
Smolt	- Young anadromous salmonid which has undergone the physiological changes necessary for life in seawater.
Smoltification	- The process of changing from a parr (*q.v.*) to a smolt (*q.v.*)
Stenohaline	- (Of fish species) able to survive only in waters within a narrow range of salinities.
Viscus	- An abdominal organ (plural viscera).

INDEX

32893809